T0254479

Pathway Analysis and Optimization in Metabolic Engineering

Rapid advances in functional genomics and proteomics have created a platform from which pressing problems in biotechnology can be addressed. Facility in the targeted manipulation of the genetic and metabolic composition of organisms, combined with unprecedented computational power, is forging a niche for a new subspecialty of biotechnology called metabolic engineering. This emerging field has metabolic pathways and gene networks as its targets, and optimization as its ultimate goal.

Pathway Analysis and Optimization in Metabolic Engineering introduces researchers and advanced students in biology and engineering to methods of optimizing biochemical systems of biotechnological relevance. It examines the development of strategies for manipulating metabolic pathways, demonstrates the need for effective systems models, and discusses their design and analysis, while placing special emphasis on optimization. The authors propose power-law models and methods of Biochemical Systems Theory toward these ends. All concepts are derived from first principles, and the text is richly illustrated with numerous graphs and examples throughout. Special features include:

- both nontechnical and technical introductions to models of biochemical systems
- a review of basic methods of model design and analysis
- concepts of optimization, both generic and specific to biotechnology
- detailed case studies

Applications of metabolic engineering range from the industrial production of organic acids and enzymes to antibiotics and agricultural products. It is a truly interdisciplinary field, and while biotechnologists will find this introductory book a highly valuable reference, it will also be of utility to other scientists and engineers interested in biological systems.

Néstor V. Torres is Associate Professor of Biochemistry and Molecular Biology at the Universidad de La Laguna, Canary Islands, Spain. His research group has pioneered research in metabolic control analysis and biochemical systems theory, and in the application of mathematical and computational modeling studies to the optimization of biochemical systems.

Eberhard O. Voit is Professor in the Department of Biometry and Epidemiology, and in the Department of Biochemistry and Molecular Biology, at the Medical University of South Carolina. He is the editor of *Canonical Modeling* (Van Nostrand/Reinholt, 1991) and the author of *Computational Analysis of Biochemical Systems* (Cambridge University Press, 2000).

Pathway Analysis and Optimization in Metabolic Engineering

NÉSTOR V. TORRES

Universidad de La Laguna, Canary Islands, Spain

EBERHARD O. VOIT

Medical University of South Carolina

CAMBRIDGE
UNIVERSITY PRESS

CAMBRIDGE UNIVERSITY PRESS
Cambridge, New York, Melbourne, Madrid, Cape Town,
Singapore, São Paulo, Delhi, Tokyo, Mexico City

Cambridge University Press
The Edinburgh Building, Cambridge CB2 8RU, UK

Published in the United States of America by Cambridge University Press, New York

www.cambridge.org
Information on this title: www.cambridge.org/9780521177481

First published 2002
First paperback edition 2011

A catalogue record for this publication is available from the British Library

Library of Congress Cataloguing in Publication data
Torres, Néstor V., 1958–
    Pathway analysis and optimization in metabolic engineering / Néstor V. Torres,
Eberhard O. Voit.
        p.   cm.
    Includes bibliographical references and index.
    ISBN 0-521-80038-2
        1. Biochemical engineering – Mathematical models.   2. Mathematical optimization.
    3. System theory.   I. Voit, Eberhard O.   II. Title.
    TP248.3 .T67   2002
    660.6′3 – dc21   2002071511

ISBN 978-0-521-80038-9 Hardback
ISBN 978-0-521-17748-1 Paperback

Additional resources for this publication at www.cambridge.org/9780521177481

To those who taught us and those who learn from us

Contents

Preface

If the 1990s were the decade of molecular biology, culminating in the completion of the Human Genome Project, the first decade of the twenty-first century can be envisaged as the time of applying that new knowledge to pressing problems in medicine, pharmaceutical sciences, agriculture, and, generally, the biotech industry. Advances in functional genomics and the targeted manipulation of the genetic composition of organisms, combined with unprecedented computational power and access, are in the process of forging a new subspecialty of biotechnology, called *Metabolic Engineering*. Metabolic engineering offers a different perspective, a fresh approach with a focus not only on the development of better tools for the biological manipulation of organisms, but also on a deeper understanding of metabolic pathways in a systemic, integrative way.

Metabolic engineering is an interdisciplinary science that has grown from numerous roots in molecular biology and biochemistry, genetics, chemical engineering, biotechnology, mathematical modeling, and systems analysis. Its targets are metabolic pathways and gene networks, often in microorganisms, and its goal is the optimization of the yield of some desired metabolite. Microorganisms are manipulated and grown at an industrial scale to produce energy (for instance, in the form of ethanol), organic solvents, gums, and pigments. The food industry relies heavily on preservatives such as citric acid, which is currently produced by microbes at a rate of several hundred thousand tons per year. The pharmaceutical industry uses microorganisms for the production of vitamins, amino acids, lipids, enzymes, and precursors for antibiotics.

Metabolic engineering has two major components. One is the development of strategies for manipulating pathways in organisms, and the other is the actual biotechnological implementation of such strategies. This book exclusively treats the strategic component and does not discuss the astonishing advances in molecular biology and biotechnology that make it worthwhile even developing strategies for pathway manipulation. The numerous novel options for specifically altering the genetic or genomic make-up of organisms, the seemingly unlimited possibilities of modern biology, the methods of fermentation technology, and chemical process control reach far beyond the scope of what we want to present in this text.

Much of this book will directly or indirectly discuss optimization applied to metabolic networks. Of course, neither the development of optimization methods nor the study of metabolic networks is new, but the specific and customized application of one to the other was not much considered until about a decade ago. In the olden days of fermentation technology, yield in microorganisms was improved successively through chemically induced mutagenesis and subsequent selection of the best-suited strains. Combined with simultaneous innovations in experimental conditions and growth media, the strategy of mutation and selection was very fruitful, often increasing the yield of the original parent strain more than a hundredfold. In recent years, however, the performance of well-studied systems like ethanol or citric acid production has begun to stagnate, and the mutation-selection approach in general has apparently reached an impasse in many cases. The former strategies of removing, one by one, the most restrictive bottlenecks in a metabolic pathway have become insufficient.

Faced with this situation, some forward-looking metabolic engineers have shifted their attention toward a deeper mechanistic understanding of the biochemical processes that are involved in the synthesis of a desired product. Initially targeting specific biochemical or transport steps that seemed to limit the microbial production process, they soon realized that a more global view of metabolism is required. This view must consider integrated pathways or, going one step further, networks of pathways, because alterations in one pathway very often have direct or indirect consequences that affect other pathways. If a microbe uses larger than normal amounts of NAD^+ to generate ethanol, the demand for NAD^+ is increased and will most likely be satisfied by other pathways. The metabolic engineering question thus becomes how one should use available kinetic information to redirect material flux toward a desired end product, without compromising other pathways and, thereby, the viability of the organism.

The realization of the intricate connectedness of enzyme-catalyzed reactions, individual pathways, and integrated networks of pathways makes it clear that mathematical and computational methods must be employed. Mathematics facilitates the bookkeeping issues arising from systems with many constituents and allows the quantitative specification of processes with a power that is simply unachievable by the unaided human mind. Efficacious computational tools must accompany the theoretical advances in mathematical representation and analysis. Only a few decades ago, the mathematical analysis of a realistically sized metabolic system would have been infeasible or even impossible, because the underlying equations were unsolvable. And although computer power has doubled about every eighteen months, we are still not at a point where every mathematical problem can be solved numerically in a reasonable period of time. This implies that the choice of "good" mathematical representations and effective methods of analysis is an important issue. As an example, nonlinear regression and the search for maxima in nonlinear systems are theoretically well understood in principle. However, even large computers and the best available software do not always solve such problems in an expeditious and satisfactory manner.

These computational challenges are especially relevant for the analysis and optimization of metabolic pathways, as they contain many substrates, intermediate metabolites, products, enzymes, cofactors, and modulators. Additionally, enzyme-catalyzed steps have nonlinear characteristics that preclude simple scaling. For instance, most proceed quickly for low substrate concentrations and tend to saturate for high concentrations. Activators and inhibitors affect reactions in a nonlinear fashion and may lead to damped or increasing oscillations, stable limit cycles, or chaotic time courses. All these phenomena are found in living systems, so a valid mathematical description of a metabolic system must be able to account for them. Thus, if our goal is the optimization of some flux or metabolite under practically relevant conditions, it is clear that we are faced with a nonlinear search problem.

The optimization of nonlinear systems is notoriously difficult, but metabolic engineers have developed strategies to minimize or even avoid nonlinearities. After the brief discussion in the previous paragraph, one may think that such attempts would be futile or head in the wrong direction. However, this conclusion is faulty. By focusing on specific properties of metabolic networks and by using prudent approximations, different groups of researchers have proposed two types of linear approaches, each with a number of extensions and generalizations. One approach is based on the stoichiometry of the metabolic network. It studies fluxes characterizing the production and degradation of individual metabolites. At steady state, where the concentrations of all metabolites are constant and all incoming fluxes exactly balance all outgoing fluxes, the stoichiometry leads to a linear system that is readily analyzed with methods of matrix algebra. Extensions of this stoichiometric strategy include the consideration of physiological, physicochemical, and thermodynamic constraints that the organism must satisfy in addition to the purely stoichiometric balance. Other extensions study features of hypothetical paths through the metabolic network and the feasibility of optimally rerouting and redistributing flux. Very importantly, the linearity of the stoichiometric system allows this approach to be scaled up to essentially any size.

The stoichiometric approach more or less ignores the nonlinearities that govern enzymatic steps. On one hand, one may be willing to pay this price for the advantages of linearity and unlimited size. On the other hand, one must wonder to what degree it is valid to ignore the regulatory features of pathways, which obviously exist and often lead to other systems responses than in an unregulated system. One approach that fully acknowledges the nonlinearities, yet ultimately leads to linear optimization tasks, is based on a modeling framework called *Biochemical Systems Theory* (BST). The hallmark of this theory is approximation of rate laws and other processes with products of power-law functions. Mathematically, this type of representation is very convenient and solidly grounded in Taylor's theory of approximating functions with polynomials. Further theoretical support is found in Bode analysis of control engineering and in the laws of allometry, which pervade biology from biochemical and physiological levels to phenomena in organismic growth and population biology.

Power laws also appear in elemental chemical kinetics and have time and again been shown to provide valid representations of cellular processes in intact organisms.

In this book, we propose power-law approximations and methods of BST as tools for pathway analysis and optimization in metabolic engineering. Pathway analysis with methods of BST began in the late 1960s and has been successfully applied to a wide variety of phenomena within and outside biology and biotechnology. The optimization aspect is more recent, with the first article on the subject published in 1992. As of yet, most optimization work has been of a theoretical or predictive nature, but experimental confirmations of predictions based on optimization with methods of BST have lately begun to appear in the literature.

As a highly interdisciplinary venture, metabolic engineering requires some basic knowledge in several areas of biology, chemistry, mathematics, computing, and engineering. A text on a subject this widespread creates the challenge of suitable presentation. Indeed, we expect that there will be few readers who will not find some parts of this book too easy or too difficult. Our personal involvement with biologists influenced our decision to write the book with a quantitatively interested biologist in mind who has had some basic calculus or at least is not afraid of unfamiliar symbols and terminologies. Of course, we also hope to attract other scientists and engineers interested in metabolic pathways. Quantitatively trained readers may find some material on optimization methods rather trivial, but then again, they may enjoy a quick refresher or the option of simply skipping the material.

We have organized the book in the following fashion. Chapter 1 discusses the search for useful models. What are we looking for when we want to analyze and optimize pathways? What are our options? How can we distinguish between a useful model and a purely academic exercise in applied mathematics? The chapter describes some established and some more recent modeling strategies for biochemical and metabolic networks and explores to what degree these models are sufficient for the tasks of metabolic engineering.

Chapter 2 reviews key properties and methods of BST. It outlines the process of model design and discusses which components should be included and to what degree simplifying assumptions are valid and necessary. The first step of model design leads to a symbolic model, which is subsequently parameterized with information from the lab or the literature. Once a numerical model is established, it is subjected to an array of analyses. Some address the features of the model at steady state, some assess the robustness and stability of the model, and some diagnose dynamic responses to perturbations or structural changes.

Chapter 3 applies the methods of model design and analysis to a specific pathway of relevance, namely citric acid production in the mold *Aspergillus niger*. The chapter begins with the construction of a model that ultimately turns out to be unsatisfactory. We consider the inclusion of this "failed attempt" beneficial, because it reflects the reality of modeling, yet is seldom documented in the literature. Furthermore, even though the initial model is not suited for practical application, it contains a number of good and bad results that provide clues for modifying the model. The good results indicate which information and model structures should be retained. The bad results are just as important. They pinpoint problem areas in the initial model, the diagnosis of which provides guidance for the revision process toward an improved model. In practice, the modeling process cycles through many rounds of design, analysis, interpretation, and the inevitable realization that some aspects of the investigated phenomenon are still not adequately addressed. Chapter 3 shows only two models, but these should suffice to illustrate the procedure of amending and fine-tuning models.

Chapter 4 prepares us for the optimization of metabolic networks that are discussed in later chapters. This chapter begins with the basics of optimization, discussing the determination of maxima and minima in simple univariate functions. Most readers will remember this topic from a college calculus class or even from high school. It is included here because the determination of maxima and minima in multivariate functions is quite similar in principle. Thus, the optimization of big systems should be an extension of straightforward calculus. However, the theoretical development reveals how quickly the barriers to an actual optimization of a large network become overwhelming. As an alternative to the theoretical approach, Chapter 4 introduces the concept of numerical optimization of large nonlinear systems. The algorithms that implement these concepts should hypothetically solve problems of any size, but practice again demonstrates the challenges of nonlinear systems.

Happily, Chapter 5 provides relief. After all the complications uncovered in Chapter 4, this chapter explains how metabolic systems can be optimized with linear methods. In the end, the stoichiometric and the BST approach both lead to a constrained optimization task known in the field as a *linear program*. Linear programs can be solved with numerous software packages, and it is not the intent of the chapter to discuss computational details. Instead, the philosophy is that the reader should develop a feel for the concepts of linear programs and either consult the rich literature for mathematical and computational details or trust the available methods and software packages. The chapter presents the components of linear programs with special emphasis on objective functions and constraints that are relevant in metabolic engineering.

Equipped with knowledge about biochemical systems theory from Chapter 2 and at least superficial expertise in linear programming from Chapter 5, Chapter 6 describes a full-blown optimization of the citric acid pathway discussed in Chapter 3. It derives suitable objective functions, discusses in detail constraints on metabolites and fluxes, as well as issues of metabolic burden, and analyzes a variety of results that are obtained under different conditions. It also touches on issues of experimental implementation.

Chapter 7 presents a second detailed case study, the glycolytic pathway, and poses different tasks for optimization. First, ethanol yield is the target of optimization, but subsequently, the same system is used for the optimization of glycerol and carbohydrate production. The chapter furthermore discusses methods for simultaneously satisfying multiple and often contradictory tasks. Initially, two tasks are reformulated in a manner that still allows optimization of a single objective that combines the two tasks. This strategy and its results are then compared with a true multiobjective optimization.

The concluding chapter is an attempt to predict the immediate future of pathway analysis and optimization in metabolic engineering. It seems rather evident that modern techniques of genome analysis will strongly influence metabolic engineering. The chapter indicates where we see the role of these techniques in terms of pathway analysis and optimization.

Many individuals have helped us create this book, and we would be remiss if we neglected to mention at least those who directly and significantly assisted us. The following have given us particularly valuable feedback and comments, and we would like to express to them our sincere gratitude: Fernando Alvarez-Vasquez, Julio Avila, Julio Banga, Agnieszka Bronowska, Carlos M. Gónzález-Alcón, W. Chip Hood, Alberto Marín-Sanguino, Kellie Sims, and Julio Vera.

Uncounted others have influenced our thinking, sparked our interest, and supported us throughout our careers, allowing us to develop and share the ideas proposed in this book. Thank you all!

Néstor V. Torres   Eberhard O. Voit
La Laguna, Tenerife   *Charleston, SC*

## CHAPTER ONE

Target: A Useful Model

CRITERIA FOR MODEL SELECTION

Before we get lost in the technical details of manipulating functions, approximating complicated phenomena, or designing and analyzing models, it is useful to establish more clearly what exactly our target is. At a superficial level, this is easily stated. Our target is a mathematical model of the biotechnological phenomenon of interest, and this model should be valid, yet convenient for analysis, manipulation, and optimization. Once we have such a model, we can screen hypotheses and perform test runs on the computer, which is much faster and cheaper than implementing and executing the actual experiments in the lab.

While the target is obvious, the difficulty is that no unique, optimal model entirely satisfies all items on our wish list. Why is that? The complications begin with the question of validity. What is a valid representation of a particular phenomenon? Although initially surprising, the question of validity is not something absolute. Instead, validity depends heavily on the purpose of the model analysis. A model for studying the aerodynamics of a butterfly will normally not account for the color patterns of its wings, and that is probably a valid omission. By contrast, the coloration may be crucial for ecological questions of camouflaging and predation by birds.

As a familiar, yet illustrative example, consider the growth of a bacterial population (Thornton 1922), as discussed by Lotka (1924, pp. 70–1; see Table 1.1 and Figure 1.1).

The symbols in the figure show the observed size of the bacterial colony over time, and the line is the graph of the logistic function

$$S(t) = \frac{0.2524}{\exp(-2.128t) + 0.005125}. \tag{1.1}$$

Inspection of Figure 1.1 suggests that the function $S(t)$ fits very well. Nonetheless, as the table indicates, there are discrepancies between observed and computed values. Can the logistic function be considered a valid representation, even though it underestimates the true colony size at day 1 by almost 30%? There is no definite answer.

Table 1.1. Observed and Calculated Growth of a Bacterial Population (Adapted from Lotka 1924)

Age of Colony (Days)	Area in cm^2	
	Observed	Calculated
0	0.24	0.2511
1	2.78	2.0324
2	13.53	13.0761
3	36.3	37.0479
4	47.5	47.3930
5	49.4	49.0231

Obviously, the function captures the saturating trend in colony growth and returns sizes reasonably close to those observed. Furthermore, some error no doubt exists in the data, which might account for the inaccuracies. After all, even in the ideal case of perfectly circular colonies, the calculated and observed colonies at day 1 differ by merely one-twentieth of an inch in radius. This simple analysis suggests that good data fit alone is not a reliable criterion for the validity of a model.

Another aspect of the same example is whether the logistic function "explains" anything. On one hand, the function allows us to make relatively accurate predictions about the size of the colony between observations and beyond the observation period. For instance, the function would have predicted the colony size at day 5 quite accurately, even if it had not been measured. This power of prediction implies that the model provides a certain degree of explanation. On the other hand, the parameter values of the function (0.2524, 0.005125, and −2.128) are not meaningful, or even measurable quantities that could be obtained from the bacterial colony itself. If we had a new colony, these parameter values would most probably not be optimal. Also, the simple logistic "model" does not capture any of the biological phenomena that underlie the growth of the bacteria, their biochemistry, or their physiology. If we wanted to predict the growth of the same type of colony under the influence of a growth inhibitor, this model alone would not be too helpful.

In conclusion of the present discussion of validity, we must distinguish what is important to capture in a given model and what can be ignored. In the biotechnological

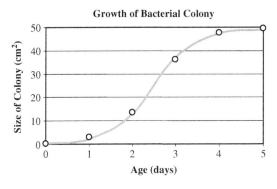

Figure 1.1. Growth of a colony of bacteria over a time period of five days. Circles indicate observations, the line is the graph of a logistic function (see text for details).

setting of a microbial culture or chemostat, typical variables of interest are metabolites, enzyme activities, fluxes, and controls such as pH, temperature, oxygen content, cofactors, and of course substrates. The physical shape of the involved molecules is clearly of importance for the enzyme-catalyzed reactions to proceed, but for typical batch processes it is often of secondary interest. Thus, returning to the issue of identifying our target, we envision a model that allows us to ask questions about changes in fluxes, metabolite concentrations, microbial population sizes, the relative importance of fluxes that funnel material through the system of metabolic pathways, and the effects of substrates and modulators. The model should enable us to map relevant observations into a mathematical realm, which would then allow us to execute "virtual experiments" and explore hypothetical scenarios.

Validity is certainly important, but it is not the only criterion in the selection of a model. A second requirement for a good model is its mathematical tractability. Only a model that permits effective evaluation, preferably both algebraically and computationally, has the potential of becoming a general tool in an applied science. We shall see in the next sections that some of the traditional models of enzyme kinetics are very useful for studying individual processes, but that they can become mathematically unwieldy in a network of just a few pathways.

If we had no history of modeling biochemical phenomena, our first stab at a useful model would probably be some linear system. The reason for this choice would be that no other branch of mathematics offers as rich a repertoire of theorems, methods, and tools as linear mathematics, and validly representing our phenomena of interest with linear methods would be half the battle. However, it has been said that focusing on linear functions within the huge realm of nonlinearities is like dividing the animal kingdom into elephants and non-elephants. Indeed, if one goes by the number of all possible mathematical structures, linear models are negligible. Even so, it is still often well worth considering linear systems, because they have very many unique and desirable features. This will become apparent throughout the book.

The drawback with linear functions in a biotechnological context is that they are often simply not appropriate descriptions of natural phenomena. For instance, essentially all processes in living organisms saturate if the dependent variable becomes very large. The growth of a population may be linear or exponential (i.e., linear in the logarithm) for small population sizes, but eventually the growth cannot continue unabated and the growth function ultimately flattens. Another limitation of linear models is their inability to represent stable oscillations that return to their original dynamics after a small perturbation. There are many oscillatory phenomena in biology, and many of them are stable in this sense. The inability of linear models to capture these oscillations is therefore a drawback. The same applies for chaotic responses, which also require nonlinear descriptions.

If linear functions are not justified, one has three options for model selection. First, one could use linear functions anyway and accept the consequent inaccuracies. This may sound like too drastic a simplification, but much of engineering is based on linear systems, and the enormous accomplishments of engineering attest to the validity of this approach. However, a crucial difference between engineering and

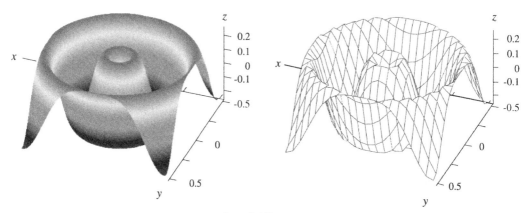

Figure 1.2. Function $z = 0.25 \cdot \sin[5\pi(x^2 + y^2)^{1/2}]$ of two independent variables, x and y, and its "wiremesh" representation, which corresponds to a two-dimensional piecewise linear approximation, if each quadrilateral is dissected into two triangles.

biology is that engineers are often in a position to design systems according to their own specifications, and these may include linear response functions. Biologists, by contrast, must live with what nature presents, and that is usually nonlinear.

The second option is using a piecewise linear representation. To represent a function in this fashion, one replaces it with sufficiently many small linear pieces, and the analysis shifts from piece to piece, depending on the value of the dependent variable. The analogous procedure may be applied to higher-dimensional functions (Figure 1.2).

Obviously, the smaller the individual pieces, the closer the agreement between the piecewise linear model and the modeled nonlinear reality. But, of course, there is again a drawback: smaller individual pieces require more breakpoints or "breaklines" between adjacent pieces and thus complicate the computational implementation and analysis.

The third option is to accept and confront the nonlinearities as they appear. The challenge here is that the true structure of the nonlinearity is seldom known. Even smooth and nearly error-free data that show a nonlinear trend do not uniquely identify the mathematical form of the underlying function. As an example, consider fabricated "data" with modest experimental error from a simple saturated process (Figure 1.3). Without further information, these data could be adequately modeled by a variety of functions, such as a shifted Hill function $f_1(x)$, an arctangent $f_2(x)$ or logistic function $f_3(x)$, or even a statistical distribution function $f_4(x)$. The function $f_4(x)$ is the cumulative of the normal distribution $N(1.7, 1.2)$, multiplied with 1.85. Similarly, other cumulative frequencies, like Student's t-distribution, could be used to model the data. Even the sine function $f_5(x) = 0.8 \cdot \sin(0.82 \cdot x + 4.8) + 1$ fits rather well.

The point of this comparison is that the criterion of a close data fit is rather weak and unreliable. In particular, a good data fit alone does not provide strong guidance for model selection. Sorribas, March, and Voit (2000) reached a similar conclusion in the context of identifying the best-fitting statistical distribution for a given data

$f(x)$

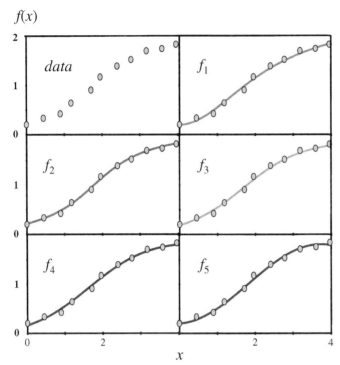

Figure 1.3. Numerous functions provide pleasing fits to the fabricated data in the top left panel. Shown here are a shifted Hill function ($f_1(x) = \frac{2.2x^2}{2.3^2+x^2} + 0.2$), an arctangent function ($f_2(x) = 0.77 \cdot \text{arctg}(0.9x - 1.6) + 1$), a logistic function ($f_3(x) = \frac{1.9}{1+\exp(2.2-1.3x)}$), a stretched normal cumulative ($f_4(x) = \frac{1.85}{\sqrt{2.4\pi}} \int_{-\infty}^{x} \exp(-\frac{1}{2.88}(u - 1.7)^2)\,du$), and a sine function ($f_5(x) = 0.8 \cdot \sin(0.82 \cdot x + 4.8) + 1$). The data fit alone is not sufficient to prefer any of these functions to an alternative.

set. They drew random numbers from a given distribution and showed that in a high percentage of cases these data were better fitted by a different distribution than the original that had been used to generate the random numbers.

With no reliable guidance from a graph of data, one has two choices of model selection. One may use a black-box model that fits the data as the logistic function fits the measured bacterial growth data and the functions f_1 through f_5 fit the manufactured data in Figure 1.3. A function found this way is often simple and robust, but its explanatory value is limited, as discussed previously. The alternative approach is to search for valid descriptions of the underlying processes. In most biological phenomena, these processes are not isolated but highly interconnected, and if we are to capture the essential features of these phenomena, we must find effective mathematical ways of representing systems.

The following sections describe some approaches to developing explanatory candidate models. We begin with a very brief review of the derivation of the best-known biochemical model, the Michaelis–Menten rate law. This rate law has proven extremely useful for the analysis of individual reactions in vitro, and thousands of articles deal with the characteristic parameters of this rate law, the Michaelis constant

K_M, and the maximal velocity of the reaction, V_{max}. Although widely used, the Michaelis–Menten rate law has serious disadvantages. Again, they fall in two categories: validity and tractability. Some generalizations of the original rate law overcome problems with the validity of the underlying assumptions. However, these generalizations exacerbate the challenges of tractability. Ultimately, rate laws of this type lead to so many mathematical and biological challenges that we have to search for other solutions.

The proposed solution in this chapter, and indeed throughout this book, is to use power-law approximations of processes. The advantages of these approximations include a relatively wide range of validity, mathematical justification, a good fit to observations, and the feature of scale-invariance, which has also been called the *telescopic property* (Savageau 1979a, 1985). Whether the system is small or large, whether a process involves two or 200 variables, whether the model addresses a phenomenon at a low or a high level of biological organization, the mathematical structure of these representations remains the same. This scale-invariance has tremendous implications, from both a conceptual and a practical point of view.

MODELS OF BIOCHEMICAL PROCESSES

Fortunately, our search for the best mathematical representation of a biochemical system does not have to start from scratch. In fact, the history of quantitatively studying chemical and biochemical processes is almost 200 years old. Three major roots of today's understanding of biochemical processes and networks are thermodynamics, kinetics, and stoichiometry. We summarize some key findings of these disciplines, and proceed by identifying approaches that offer a good balance between validity, justifiability, interpretability, and mathematical effectiveness and efficiency.

Thermodynamics

Biological systems are based on physical principles. They must satisfy the laws of physics just like any other entity in the physical world. Among the laws and principles of physics, the results of thermodynamics are of particular interest for chemical and biochemical processes, because they govern the relationships between mass, energy, work, and heat. They describe which processes are possible and which are energetically infeasible. As Callen (1960) put it, "Thermodynamics is the study of the macroscopic consequences of myriads of atomic coordinates, which, by virtue of statistical averaging, do not appear explicitly in a macroscopic description of a system." Many theoretical results of thermodynamics were first observed in experiments addressing questions that related energy to pressure, temperature, and volume in ideal gases. Only later were the results formulated as laws and corollaries. "The basic problem of [classical] thermodynamics is the determination of the equilibrium state that eventually results after the removal of internal constraints in a closed composite system" (Callen 1960, p. 7). A simple example is the observation that coffee, if left to its own devices, after a while assumes room temperature.

A significant finding of thermodynamics is that systems tend toward states of minimal energy. In terms of biological systems, this is quite counterintuitive, because organisms apparently violate these laws all the time. Plants generate high-energy compounds like complex sugars out of lower-energy substrates like CO_2 and water, which contrasts simple thermodynamic arguments that would postulate the opposite, namely the degradation of high-energy compounds into lower-energy compounds. Even though organisms have tasks that require increases in energy, they cannot simply ignore or violate the laws of thermodynamics but instead must find ways to complete their tasks in spite of them.

Many books have been written on thermodynamics, some by celebrated scientists such as Max Planck (1945) and Enrico Fermi (1956). These treatises address the topic from a purely physical viewpoint and often use mathematical concepts of considerable sophistication. A reasonably intuitive introduction to the area from a biological perspective is Jou and Llebot (1990); the discussion below more or less follows this book. Katchalsky and Curran (1967), Westerhoff and van Dam (1987), and Heijnen (2001) also treated the topic in the context of biological processes. Two more general texts are Callen (1960) and Kestin (1966).

The first law of thermodynamics asserts that the change in energy within a closed system, which does not gain or lose matter, is equivalent to the sum of heat and of energy-increasing work that the system receives from the outside. Hermann von Helmholtz (1821–94) formulated this law as the widely acknowledged impossibility of constructing a *perpetuum mobile*, a machine of perpetual motion (of the first kind) that would indefinitely produce more energy than it received (Gerthsen and Kneser 1971).

The second law further limits possible transitions within a system. It asserts that heat does not pass spontaneously from a cold to a hot body, unless other processes are in effect. Such a transition would be possible according to the first law, according to which the overall state of energy remains constant, but is not possible according to the second law. The second law is not only valid for isolated systems but also can be formulated to include more realistic, nonisolated systems. Such systems are closed with respect to matter, but exchange heat and work with the environment. The vast majority of biological systems are nonisolated and, furthermore, open with respect to the flux of matter. Their openness presents a significant challenge for thermodynamic considerations.

One reformulation of the second law of thermodynamics is based on Gibbs' (1839–1903) concept of *free energy*, which is defined as the sum of the internal energy of a state of the system and the product of pressure and volume, from which the product of temperature and entropy is subtracted. The entropy characterizes the current state of the system and, in some sense, is a measure of the disorder or statistical homogeneity of its molecules (see Callen 1960, Appendix B for intuitive explanations of the concept of entropy). Highly structured systems have low entropy, whereas unstructured systems have high entropy. In this terminology, the second law of thermodynamics states that the change in Gibbs free energy must be less than zero for any spontaneous process in a closed system that operates under constant pressure and constant temperature. Expressed differently, the entropy in such a system cannot decrease and the order cannot increase.

Instead of supposing that pressure and temperature are constant, other variables, such as the internal energy of the system and the volume could be considered constant. Each set of variables that are assumed to be constant yields a different constraint in the other variables, and these constraints are called *thermodynamic potentials*. In closed systems, the different thermodynamic potentials are in a sense equivalent, and each contains all the thermodynamic information about the systems, such as specifications of the equilibrium state to which the system moves, including stability of this state, characteristics of phase changes, and relationships between the various thermodynamic quantities.

If expressing similar phenomena in different ways seems confusing, be consoled by Callen (1960, p. 85), "The peculiar multiplicity of formulation and reformulation of the basic thermodynamic formalism is responsible for the apparent complexity of a subject which in its naked form is quite simple." This simplicity stems from the fact that essentially all thermodynamic constraints are computed as partial derivatives of one and the same function, which describes the energy status of the system.

If the system is open, the thermodynamic potentials are not only functions of temperature, volume pressure, and free energy, but also of the numbers of moles of each chemical species present. The change in energy that accompanies a change in the number of moles is called the *chemical potential*. The chemical potentials of all species (types of molecules) relate to the internal energy of the system in a fashion analogous to temperature. For instance, two connected systems exchange energy until both attain the same temperature, and two connected compartments containing different chemical species move toward molecular homogeneity. In both cases, the systems move toward the state of maximal entropy, and one can show that this state corresponds to one of minimal energy. This has significant implications for the characterization of such phenomena as osmotic pressure, diffusion, freezing point depression, and boiling point elevation. The tendency of systems toward a state of lower energy can be observed macroscopically. As early as 1872, Ludwig Boltzmann (1844–1906) furthermore showed that this tendency could be explained in terms of *microstates*, which correspond to locations and properties of individual molecules. The use of probabilistic arguments for predictions of transitions between macrostates has led to the subfield of statistical thermodynamics.

Alberty (1994) reviewed thermodynamic concepts for systems consisting of chemical and biochemical reactions. The equilibrium in such a system is affected by the ionic milieu and typically changes if the pH is altered. Ion effects on changes in free energy are particularly important for reactions involving nucleic acids, proteins, and other polyelectrolytes. To account for these effects, Alberty (1994, 2000) proposed to include the pH as an additional independent variable in thermodynamic computations, just like temperature, volume, pressure, and chemical potentials, and to redefine fundamental quantities like the Gibbs free energy and entropy correspondingly.

Although classical thermodynamics has been called "one of the outstanding achievements of the scientific mind" (Katchalsky and Curran 1967), it is limited in scope, because it provides a theory for systems that are either in equilibrium or are undergoing reversible processes. Idealized physical systems may satisfy these requirements, but biological systems rarely do so. Organisms are nonisolated and open, regularly exchanging not only heat, but also matter. Equilibrium states and transitions between them, which constitute the cornerstone of classical thermodynamics, are not as relevant for an organism, because equilibrium would mean death. Instead, biological systems operate at *nonequilibrium stationary states*. These states are characterized by influxes and effluxes of matter and energy that are in balance and keep the concentrations of all chemical species (more or less) constant over time. Such states can only be maintained by an external supply of energy.

Nonequilibrium thermodynamics thus deals with questions of energy that are linked to transport and metabolism in open, *dissipative* systems, which require the influx of energy. These systems usually contain *irreversible* processes, which Callen (1960) defines as requiring an increase in entropy. All systems in the real world are of this type. Dissipative systems must overcome the constraints given by the laws of thermodynamics through the coupling of two or more processes. One process leads to a structure with higher energy and lower entropy, thereby running in the direction opposite to the one predicted by its thermodynamic affinity. The energetic gain in this process, which apparently violates the second law of thermodynamics, is "paid for" by energy released in a concomitant reaction that supplies energy and moves in the direction predicted by its thermodynamic affinity. A typical example is the coupling of phosphorylation with oxidation. Phosphorylation is crucial for the storage of chemical energy; probably the most prominent case is the conversion of ADP into ATP. The increase in energy during this process is coupled with energy transfer from another reaction, such as the oxidation of NADH to NAD^+. In animals, the external energy supply is chemical, whereas plants, of course, may also use sunlight for some of the reactions that lead to higher-energy compounds.

Nonequilibrium thermodynamics allows the estimation of energy requirements in dissipative systems, the extent and stoichiometry of reactions or pathways, and the degree of coupling among them. It also characterizes the efficiency of coupled reactions, which is defined as the ratio of free energy consumed in one direction over the energy liberated in the opposite direction. For example, it permits the estimation of the number of moles of oxygen needed for the phosphorylation of one mole of ADP and the computation of the efficiency of photosynthesis. Deductions from the principles of nonequilibrium thermodynamics have also led to the insight that, in the vicinity of the thermodynamic equilibrium, stationary nonequilibrium states are characterized by minimum entropy production (Prigogine 1947/1955). At such a state, the system loses minimal amounts of free energy and is energetically most economical, which might be a rationale for the uncounted control mechanisms with which organisms tend to preserve this state (Katchalsky and Curran 1967). Overall, nonequilibrium thermodynamics provides a collection of constraints that observed or hypothetical reactions in vitro and in vivo must satisfy. Ricard (1999) provides good explanations of the relationships between nonequilibrium thermodynamics and chemical and biochemical rate equations.

An intrinsic feature of the thermodynamic approach is its strict focus on energy, which almost completely excludes time. Thus, a typical result characterizes the energetic possibility or likelihood of a reaction, but it gives no indication whether this reaction occurs on a time scale of seconds or years. Sometimes, temporal considerations are not necessary, but in other cases timing is crucial. If that is the case, thermodynamics is usually not the optimal approach, and one will instead focus on *kinetic* representations of reactions.

Kinetics

Thermodynamics and kinetics are not entirely unrelated. In fact, kinetics may be considered an "empirically based form of generalized thermodynamics" (Westerhoff and van Dam 1987). Kinetics does not focus on energy levels as much as thermodynamics. Instead, it addresses the *temporal* aspects of a reaction. How fast does the reaction proceed? What is the half-life of a metabolite? What affects the speed of the reaction? Certainly, answers to these questions involve thermodynamics at a deeper level, but kinetic studies minimize aspects of energy and instead center directly on metabolite concentrations and fluxes, as well as their fluctuations over time.

In the overwhelming majority of studies, kinetic analyses ignore spatial features. Instead, it is implicitly assumed that all participants of a reaction are available in a homogeneous mix. It is also typically assumed that the substrate concentration is much higher than the enzyme concentration, so that the availability of enzyme drives the process. Some newer studies have questioned some of these assumptions and propose alternative descriptions. We will discuss a few of them throughout the chapter.

The typical chemical or biochemical rate function relates the temporal change in a chemical compound or metabolite concentration to the concentration itself. In the simplest case of a first-order degradation process that does not involve an enzyme, the rate is directly proportional to the concentration. In straightforward notation, this elemental chemical reaction reads

$$v(X) = -kX. \tag{1.2}$$

The rate constant k is positive by definition, because it represents the turnover per time unit, which cannot be negative. The negative sign indicates that material X is actually lost from the existing pool. The mathematical form of the equation results from considerations of statistical thermodynamics that go back more than a hundred years to Svante Arrhenius (1859–1927). Details can be found in textbooks on thermodynamics and kinetics.

Because the rate $v(X)$ represents the change in concentration over time, we can write it as a derivative with respect to time, namely

$$\frac{dX}{dt} = \dot{X} = -kX. \tag{1.3}$$

In kinetic studies, the differentiation variable is almost always time, t, and it is becoming customary to denote the derivative with the dot notation, as indicated in Eq. (1.3). This notation is somewhat simpler, but does not explicitly identify time as the independent variable.

It might be illustrative to relate the kinetic equation to concepts of statistical thermodynamics. Following the exposition of Westerhoff and van Dam (1987), we consider a simple irreversible reaction

$$A \rightarrow B + C$$

and denote the number of molecules of substrate A at time t with x. From a mechanistic point of view, it is almost impossible to predict when a given substrate molecule will be converted into molecules of types B and C. However, if there are very many molecules of type A, statistical thermodynamics allows us to consider the conversion of any given molecule as a probabilistic process. Because no mechanism replenishes A, the number x can only decrease over time. Specifically, one can formulate the probability of a drop from x to $x - 1$ during a small time interval Δt as

$$P(x, t \rightarrow x - 1, t + \Delta t) = k \cdot x \cdot \Delta t, \tag{1.4}$$

where k is some constant. If Δt is chosen to be small enough, there will be no drop from x to $x - 2$, $x - 3$, etc. within Δt. If we let $P_x(t)$ denote the probability that there are x substrate molecules at time t, we can formulate the probability for x molecules at time $t + \Delta t$ as

$$P_x(t + \Delta t) = k \cdot (x + 1) \cdot \Delta t \cdot P_{x+1}(t) + (1 - k \cdot x \cdot \Delta t) \cdot P_x(t). \tag{1.5}$$

This so-called *master equation* expresses in the first term on the right-hand side the probability of a transition from $x + 1$ to x molecules and in the second term the probability that the number does not fall further from x to $x - 1$. If Δt is decreased toward zero, Eq. (1.5) becomes a set of differential equations, each of which describes the dynamics of a given number x of molecules in the system. These equations can be solved, and one can compute the expected (average) number $\bar{x}$ of substrate molecules at any given time. The result is

$$\bar{x} = x_0 \cdot \exp(-k \cdot t), \tag{1.6}$$

where x_0 is the initial number of substrate molecules. This equation is exactly the solution to the differential equation (1.3), which constitutes the macroscopic, kinetic description of the reaction. The derivation based on master equations demonstrates that the introduction of stochastic elements compensates for our lack of precise knowledge of the molecular state of the system. The details of this derivation and further extensions are found, for instance, in Westerhoff and van Dam (1987).

As an embellishment to the simple degradation of a substrate A, suppose X_1 is converted into X_2 in an elemental chemical reaction. Because no material is lost in this ideal situation, the production of X_2 equals the degradation of X_1, but with opposite sign, because one increases and the other decreases. Thus, we can write the process as a small system with two variables:

$$\dot{X}_1 = -kX_1$$
$$\dot{X}_2 = kX_1. \tag{1.7}$$

(Note that *time* does not count as a variable in these types of discussions.) The solution of the linear differential equation (1.3) is an exponential function of time. Thus, the time dependence of X is nonlinear, but the function describing the process corresponds to a differential equation whose right-hand side is linear.

The linear differential equation (1.3) is not only used for elemental chemical reactions. Of relevance to our discussion are simple reactor and chemostat designs (Bailey and Ollis 1977). It is also often assumed as the default for outflow from well-stirred reactors, for generic decay and growth processes, and for standard compartment models (e.g., Edelstein-Keshet 1988; Jacquez 1996).

If two metabolites are involved in a *bimolecular* reaction, their concentrations enter the right-hand side of the (differential equation) rate law as a product. In the generic case where X_1 and X_2 are substrates of such a reaction that generates product X_3, the increase in the concentration of X_3 is given as

$$\dot{X}_3 = k_3 X_1 X_2. \tag{1.8}$$

It is noteworthy that this type of process leads to a *product* in its substrate concentrations and not a sum, even though one might speak of "adding a second substrate" or formulate the reaction as $X_1 + X_2 \rightarrow X_3$. The reason for the product form of the rate law lies partly in thermodynamics and partly in the fact that the two molecules have to come into physical contact. In a homogeneous mixture, the latter is a matter of probability, which suggests as the simplest model a formulation as product.

In the case of Eq. (1.8), the concentration of X_3 increases, and therefore the right-hand side does not carry a minus sign. Notice that X_3 does not appear on the right-hand side. It does not affect its synthesis and depends entirely on the availability of both X_1 and X_2. For every molecule of X_3 that is produced, one molecule of X_1 and one molecule of X_2 disappear. Therefore, the loss in either one substrate is

$$\dot{X}_1 = \dot{X}_2 = -\dot{X}_3 = -k_3 X_1 X_2. \tag{1.9}$$

One may discuss whether there are reactions that truly involve more than two substrates or whether such reactions are in fact sequences of bimolecular reactions. Regardless, theory suggests that such reactions would again be described with differential equations whose right-hand sides consist of products of substrates. For instance, suppose one molecule of X_1 and two molecules of X_2 would be converted into product X_3. The describing rate equations would be

$$\dot{X}_3 = k_3 X_1 X_2^2,$$
$$\dot{X}_1 = -k_3 X_1 X_2^2, \tag{1.10}$$
$$\dot{X}_2 = -2k_3 X_1 X_2^2.$$

The exponent 2 associated with X_2 indicates that two molecules are used. It is consistent with the product of X_1 and X_2 in the bimolecular reaction. One says that the process is of (kinetic) order 1 with respect to X_1 and of (kinetic) order 2 with respect to X_2. Note that the rate constant is doubled in the last equation to indicate that X_2 disappears at twice the rate of X_1 and at twice the rate of the synthesis of X_3.

Westerhoff and van Dam (1987) pointed out that kinetic details of elemental processes could always be incorporated in thermodynamic descriptions. As an example, they considered a simple reaction with one substrate S and one product P. The typical kinetic description of the process is some rate function of the two independent variables S and P. However, one could also express the independent variables as $\log([S]/[P])$ and $[S] + [P]$. The first of these terms is directly related to the thermodynamic concept of the affinity of the reaction, which demonstrates that concentrations can readily be transformed into thermodynamic forces.

The Michaelis–Menten Rate Law

In the early 1900s, Michaelis and Menten (1913) proposed a reaction scheme for enzyme-catalyzed reactions, based on earlier ideas of Henri (1903). They postulated that a substrate S and its catalyzing enzyme E form an intermediate complex (ES) in a reversible reaction. Once formed, this complex would break apart and either return substrate and enzyme or yield product P while simultaneously releasing the enzyme molecule unchanged. The typical diagram including rate constants for the process is shown in Figure 1.4. In this simple form, synthesis of product from the intermediate complex is assumed to be irreversible; in other words, there is no reaction with rate k_{-2}.

Equipped with rate equations for bimolecular reactions (between substrate and enzyme), the formulation of equations is straightforward. One obtains

$$\dot{S} = -k_2(ES), \tag{1.11a}$$

$$(\dot{E}S) = k_1 S \cdot E - (k_{-1} + k_2) \cdot (ES), \tag{1.11b}$$

$$\dot{P} = k_2(ES). \tag{1.11c}$$

The easiest way to understand the equations is to go backward. The equation for the synthesis of product P is a simple elemental reaction with rate k_2, whose substrate is the intermediate complex (ES). The change in the concentration of the intermediate complex is governed by three processes: the bimolecular reaction involving substrate S and free enzyme E; the reverse reaction with rate k_{-1}, in which the intermediate complex (ES) breaks down into enzyme and substrate; and the forward reaction with rate k_2 that leads to the generation of product. Stoichiometry between substrate and product and the quasi-steady-state assumption discussed next require the change in substrate to be the negative of the synthesis of product.

Figure 1.4. Diagram of an enzyme-catalyzed reaction mechanism according to Henri, Michaelis, and Menten.

$$S + E \underset{k_{-1}}{\overset{k_1}{\rightleftharpoons}} (ES) \xrightarrow{k_2} P + E$$

The system of equations describes the conversion of substrate into product over time. Its main assumption is that the laws of elemental chemical kinetics apply, which require that the reaction proceeds in a homogeneous mixture and that enough substrate and enzyme are available to justify the probabilistic arguments underlying the theory of chemical kinetics. The differential equations can be solved with a numerical integrator, such as the Runge–Kutta or Gear algorithms, which are available in standard software packages for mathematical analysis.

Further assumptions are necessary to obtain the familiar form of the (algebraic) Michaelis–Menten rate law. Like many authors, Schulz (1994, Chapter 1, p. 8) lists the following four

Assumption 1: $E_T = E + (ES)$

Assumption 2: $S_T \gg E_T$

Assumption 3: $\dot{E}_T = (\dot{E}S) = 0$

Assumption 4: $P_0 = 0$

The first assumption is an *enzyme conservation expression* indicating that the total enzyme concentration, E_T, can be divided into free enzyme and enzyme bound in the intermediate complex. The second assumption states that the total substrate concentration S_T is much larger than the total enzyme concentration E_T. The third statement is called the *quasi-steady-state assumption*. Its first part, $\dot{E}_T = 0$, asserts that no enzyme is formed or lost during the process; it is usually accepted as true, at least in vitro. The second part, $(\dot{E}S) = 0$, is more critical. It presumes that the concentration of the intermediate complex is constant. This assumption is clearly violated at the beginning of the experiment. Furthermore, it implies that the reactions forming and destroying the intermediate complex are much faster than the overall conversion of substrate into product. The fourth assumption is made for convenience and could be dropped. It just says that no product is present at the beginning of the experiment.

The algebraic form of the Michaelis–Menten rate law derives from Eqs. (1.11b) and (1.11c), the quasi-steady-state assumption, and the formulation of $\dot{P}$ as the rate v_p. The result is

$$0 = k_1 S \cdot E - (k_{-1} + k_2) \cdot (ES) \tag{1.12a}$$

$$v_p = k_2(ES). \tag{1.12b}$$

Assumption 1 allows us to replace (ES) with the difference between the constant quantity E_T and E. Simple algebra then leads to the so-called Briggs–Haldane (Briggs and Haldane 1925) formulation of the rate as

$$v_p = \frac{k_1 k_2 E_T S}{k_{-1} + k_2 + k_1 S}. \tag{1.13}$$

It is customary to rearrange the parameters and define the *Michaelis constant*

$$K_M = (k_{-1} + k_2)/k_1 \tag{1.14}$$

and the *maximum velocity*

$$V_{\max} = k_2 E_T. \tag{1.15}$$

With these new parameters, the Michaelis–Menten rate law has the familiar form

$$v_p = \frac{V_{\max} S}{K_M + S}. \tag{1.16}$$

This form of the rate law, along with many generalizations, has been immensely successful in the characterization of enzymes and the analysis of simple pathways in vitro.

As modelers, we must ask whether this formulation is optimal. Following the arguments earlier in the chapter, we should judge the rate law against two criteria: validity and mathematical convenience.

Validity

The Michaelis–Menten rate law was conceived almost one hundred years ago. For several decades afterward, it was more or less accepted as probably true, and its applicability and validity were not much questioned. However, beginning in the 1970s, the critical assumptions leading to the rate law were more thoroughly scrutinized. Doubts arose as to whether the mechanisms that seemed to work well in vitro would also be operational in vivo. Specific questions targeted the alleged homogeneity within the cell, the validity of the steady-state assumption, the ample availability of substrate, and the dependence of the rate law on conditions like the pH (Roberts 1977). The culmination of these doubts might have been the title of an article by Hill, Waight, and Bardsley (1977), "Does any enzyme follow the Michaelis–Menten equation?" Detailed discussions of these issues can be found in Savageau (1992a, 1995a,b) and Schulz (1994).

Schulz (1994, p. 22ff) analyzed Assumptions 2 and 3 in some detail. Assumption 2 requires the substrate to be available in such excess that its concentration is more or less constant. In particular, is supposes that the fraction of substrate bound to enzyme is insignificant throughout the assay. Most in vitro systems probably satisfy this condition. However, if the substrate is a large molecule and the reaction occurs in vivo, Assumption 2 may easily be violated. In this case, it seems to be more reasonable to include in the set of assumptions something like a substrate conservation expression, analogous to Assumption 1. Such an expression complicates the mathematical formulation and results in a rate law of the form

$$v_p = \frac{k_2}{2}[(K_m + A_t + E_t) - \sqrt{(K_m + A_t + E_t)^2 - 4E_t A_t}] \tag{1.17}$$

(Goldstein 1944; Cha and Cha 1965; Reiner 1969, pp. 82–90), which is "not very convenient" and "rather unwieldy" (Schulz, 1994, pp. 24–5).

If the substrate concentration is not available in essentially unlimited amounts, the rate law becomes a function of time. Schulz (1994, p. 25) proposed to simplify the differential equation model with a model of product synthesis that has the form

$$P(t) = \frac{k_1 k_2 E_T S_T}{k_{-1} + k_2 + k_1 A_T} + \frac{k_1 k_2 E_T S_T}{(k_{-1} + k_2 + k_1 S_T)^2} \left(e^{-(k_{-1} + k_2 + k_1 S_T)t} - 1\right). \qquad (1.18)$$

This model is approximately quadratic in the millisecond domain and becomes essentially linear in the range of seconds and minutes. Even though noticeably more complex than the original rate law, Schulz warns that this formulation cannot be entirely valid either. The product concentration cannot grow indefinitely, as the approximate differential equations and this solution would predict, but should show sigmoidal, saturated dynamics.

Schnell and Mendoza (1997) pursued a different approach. They developed a closed-form solution of the Michaelis–Menten reaction scheme for the entire time course of substrate. Their result is

$$[S'](t) = \omega([S_0'] \exp(-kt + [S_0']), \qquad (1.19)$$

where ω is the *Omega function* (Wright 1959; Corless et al. 1996), which satisfies the transcendental equation $\omega(x) \exp(\omega(x)) = x$, $[S'] = [S]/K_M$ is the "reduced concentration," and $k = V_{\max}/K_M$ is the first-order rate constant. The subscript zero refers to the initial concentration.

Heinrich and Schuster (1996) and Schnell and Maini (2000) provided good reviews of work treating the quasi-steady-state assumption (QSSA). Earlier analyses by Segel (1988) and Segel and Slemrod (1989) resulted in a succinct condition for the validity of the QSSA in vitro:

$$\frac{[E_0]}{K_m + [S_0]} \ll 1. \qquad (1.20)$$

In vivo, this condition tends to break down (Sols and Marco 1970), motivating Segel and Slemrod (1989) to propose a *reverse QSSA*, in which the substrate, rather than the enzyme is in a quasi-steady state with respect to the overall reaction.

Schnell and Maini (2000) asked what might happen if S is *not* much higher than E. They started with elemental chemical kinetics and assumed that the reverse QSSA is valid. The solution for the entire time course then is

$$[S](t) = [S_0] \exp(-k_1 [E_0] t) + \frac{K_S [S_0]}{[E_0]} (\exp(-k_2 t) - \exp(-k_1 [E_0] t)), \qquad (1.21)$$

$$[(ES)](t) = [S_0](\exp(-k_2 t) - \exp(-k_1 [E_0] t)), \qquad (1.22)$$

where $K_S = k_{-1}/k_1$ is the equilibrium dissociation constant of S from the intermediate substrate–enzyme complex. This solution closely approximates numerical solutions of the original differential equation system describing the mechanistic Michaelis–Menten scheme. Schnell and Maini showed that the reverse QSSA is satisfied if K_M and $[S_0]$ are both much smaller than $[E_0]$.

In summary, the algebraic Michaelis–Menten rate law is very useful and valid if the substrate is not limiting and the QSSA is satisfied. If this is not the case, substrate and product concentrations are functions of time, and the dynamic time courses

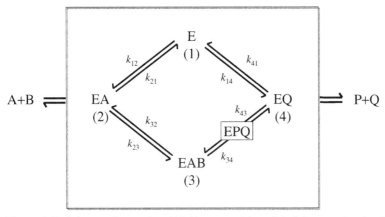

Figure 1.5. Scheme of an ordered bi–bi sequential model of the reaction $A + B \rightleftharpoons P + Q$.

of $P(t)$ and $S(t)$ become complicated. If the reverse QSSA is satisfied, $S(t)$ can be approximated, but the solution is still much more complicated than the original rate law.

Mathematical Convenience

In its original form, the Michaelis–Menten rate law is compact and easy to use, analyze, and interpret. Even some generalizations, for instance accounting for competitive inhibition of the reaction, are easily implemented, and the result retains its simple algebraic character. However, if several substrates or reactions are involved, and if several modulators affect the pathway, the rate law quickly becomes unwieldy.

As an example, consider a reversible *ordered bi–bi sequential reaction* scheme as shown in Figure 1.5. The reaction is called *ordered* because the binding between enzyme and substrates and the dissociation of products from the enzyme occur in a sequential fashion rather than at random. Specifically, the first step binds substrate A and the enzyme, followed by association of the complex EA with substrate B. The three-component complex EAB is converted into EPQ, a process that involves no interaction of the enzyme with a reactant and therefore does not appear in the rate law. Product P must dissociate from the complex before Q. Examples of this mechanism include some pyridine nucleotide dehydrogenases (Schulz 1994, p. 60).

The reaction involves only two substrates and two products, yet the kinetic representation in the tradition of Henri, Michaelis, Menten, Briggs, and Haldane is already quite complicated (Figure 1.6).

If the reaction involves inhibitors and other modulators, the complexity of the rate law becomes overwhelming. Savageau (1976, p. 75) illustrates this with the enzyme glutamine synthetase, which is affected by at least eight reactants and modifiers (Woolfolk and Stadtman 1967). Even under the restrictive assumption that none of the reactants or modifiers enters the rate law with a power higher than one, this rate law would consist of about 500 terms and would take at the order of one hundred million experimental assays to establish. It is hard to imagine that mathematical

$$v = \cfrac{\left(\dfrac{num.1}{coef.\,AB}\right)(A)(B) - \left(\dfrac{num.1}{coef.\,AB} \times \dfrac{num.2}{num.1}\right)(P)(Q)}{\begin{array}{l} \left(\dfrac{constant}{coef.\,A} \times \dfrac{coef.\,A}{coef.\,AB}\right) + \left(\dfrac{coef.\,A}{coef.\,AB}\right)(A) + \left(\dfrac{coef.\,B}{coef.\,AB}\right)(B) \\[2.5ex] + \left(\dfrac{coef.\,AB}{coef.\,AB}\right)(A)(B) + \left(\dfrac{coef.P}{coef.\,AP} \times \dfrac{coef.\,AP}{coef.\,A} \times \dfrac{coef.\,A}{coef.\,AB}\right)(P) \\[2.5ex] + \left(\dfrac{coef.Q}{constant} \times \dfrac{constant}{coef.\,A} \times \dfrac{coef.\,A}{coef.\,AB}\right)(Q) \\[2.5ex] + \left(\dfrac{coef.AP}{coef.\,A} \times \dfrac{coef.\,A}{coef.\,AB}\right)(A)(P) + \left(\dfrac{coef.BQ}{coef.\,B} \times \dfrac{coef.\,B}{coef.\,AB}\right)(B)(Q) \\[2.5ex] + \left(\dfrac{coef.PQ}{coef.Q} \times \dfrac{coef.Q}{constant} \times \dfrac{constant}{coef.\,A} \times \dfrac{coef.\,A}{coef.\,AB}\right)(P)(Q) \\[2.5ex] + \left(\dfrac{coef.ABP}{coef.\,AB}\right)(A)(B)(P) \\[2.5ex] + \left(\dfrac{coef.BPQ}{coef.\,BQ} \times \dfrac{coef.BQ}{coef.\,B} \times \dfrac{coef.\,B}{coef.\,AB}\right)(B)(P)(Q) \end{array}}$$

Figure 1.6. Rate equation of the ordered bi–bi sequential reactions scheme from Figure 1.5 in kinetic form (Schulz 1994). The expressions *num.*, *constant*, and *coef.* refer to various combinations of sums and products of rate constants and the total enzyme concentration.

models of complex pathways could be successfully described and numerically identified in terms of these detailed enzyme mechanisms.

As Schulz (1994) summarized it:

Useful as these [kinetic] studies [under experimentally controlled conditions] have been, the concept of an enzyme catalyzing a reaction in isolation runs counter to the purpose for which enzymes have been provided in nature. The purpose of an enzyme in nature is to catalyze a reaction in concert with the other enzymes in the metabolic pathway, and the purpose of a metabolic pathway is to catalyze a series of reactions in concert with the many other pathways with which it interacts.... If coordination of the action of... enzymes becomes flawed, it is certain that the living organism is going to encounter serious difficulty.

One could argue that the rate laws might become messier for more complex systems, but that increased computing power would more than compensate. Indeed, as early as thirty years ago, Garfinkel (*e.g.*, 1968, 1980, 1985) and others constructed computer models in which the enzyme-catalyzed reactions were represented by rate laws in the tradition of Michaelis and Menten. These early studies and the enormous improvements in computation since Garfinkel's work suggest that computer-aided model design itself is not the bottleneck. Rather, the main problem is the analysis of these models and the interpretation of results. Heinrich and Rapoport (1974) commented on this topic with this remark:

Firstly, from the computer output it appears difficult to differentiate between important and unimportant effects, enzymes, metabolites, etc. Secondly, it is difficult to see how some effects are brought about. Thirdly, such a computation is often impracticable for experimentalists. Fourthly, many ad hoc assumptions are even now necessary for the mechanisms of several of the constituent enzymes of a chain. The strong dependence of the model of an enzymatic chain on the detailed mechanisms of single enzymes is unfavorable. Analytical approaches, on the other hand, circumvent some of these difficulties, as they are usually connected with a reduction to the essential parameters. They may show clearly the reasons for regulatory effects and they are more practicable for experimentalists.

The problems with computer models are not just academic. A typical model is so complex that hundreds or thousands of simulations are necessary to explore just its main features. Time constraints and the sheer size of the simulation task may lead to a failure to detect relevant scenarios and sometimes prompt incorrect conclusions. For example, Wright, Butler, and Albe (1992) collected an enormous amount of information for the construction of a very detailed kinetic model of the tricarboxylic acid cycle in the slime mold *Dictyostelium discoideum*. Formulated in the tradition of Michaelis and Menten, the reactions were expressed as rational functions, which precluded the authors from executing algebraic analyses. The model represented many aspects of this pathway well, yet some conclusions of the numerical model analysis were later shown to be faulty, because an unexpectedly large range in time scales among the governing processes masked some of the model's slow dynamics (Shiraishi and Savageau 1992a–d, 1993). The main cause of the problem was not the size of the system. After all, linear systems with thousands of variables can be completely analyzed and optimized with very high efficiency. Rather, the problems arose because of the inconvenient nonlinear nature of the rational functions involved.

Stoichiometric Systems

The discussion in the previous section suggests that Michaelis–Menten models even of moderately large biochemical systems become overwhelmingly complicated and require a considerable amount of information, which often is not available in sufficient detail to estimate all involved parameters, such as K_Ms, K_Is, and V_{max}s. As an illustration, Curto et al. (1997, 1998a,b) attempted unsuccessfully to construct and parameterize a model of purine metabolism purely with traditional rate laws. It turned out that many reactions were ill characterized, which necessitated complementation of the rational function rate laws with power-law functions.

One fruitful alternative to establishing a full kinetic model is the exploration of the *stoichiometry* of the system. According to the Merriam-Webster Dictionary, the word *stoichiometry* derives from the etymological roots for *element* and *measurement* and "deals with the application of the laws of definite proportions and of the conservation of mass and energy to chemical activity." Interpreted in a slightly different way, this approach establishes constraints that reactions and pathways have to satisfy. These constraints occur at several levels. First, there is again thermodynamics, which governs aspects of energy. At the second level, there are clear conservation laws among all atoms involved. For instance, the incoming and outgoing carbon and nitrogen atoms have to balance for every chain of reactions. The third level addresses the level of chemical and biochemical reactions. All constraints at these three levels limit the possibilities of conversions between substrates and products. In terms of mathematics, stoichiometric constraints reduce the degrees of freedom in the system, and because the conservation relationships are typically sums and differences, this loss in degrees corresponds to the reduction of the rank of a matrix that describes the topology of the reaction network.

A crucial idea in the development of methods for stoichiometric analyses at the pathway level was the summation of single enzymatic reactions in the butanediol fermentation pathway, which led to an overall balance between substrate input and product output. This balance was formulated as a *stoichiometric fermentation equation*, which was suggested, for instance, as a "gateway sensor" for online monitoring of fermentations (Papoutsakis 1984; Papoutsakis and Meyer 1985). Tsai and Lee (1988) showed that this type of summation and simplification applies analogously to large reaction networks, where it is executed with methods of simple matrix algebra.

The stoichiometric "wiring diagram" of a biochemical system may be represented as a directed graph, which in turn corresponds mathematically to a balance equation that may be formulated as a set of linear ordinary differential equations. This set expresses the dynamics of metabolite concentrations, collected in a vector $\mathbf{S}$, in terms of a stoichiometry matrix $\mathbf{N}$, and a vector $\mathbf{v}$ of reaction rates (Gavalas 1968; Glansdorff and Prigogine 1971; Horn and Jackson 1972; Heinrich, Rapoport, and Rapoport 1977; Stephanopoulos, Aristidou, and Nielsen 1998). In typical notation, such a balance equation reads

$$\frac{d\mathbf{S}}{dt} = \mathbf{N} \cdot \mathbf{v}. \tag{1.23}$$

The stoichiometry matrix $\mathbf{N}$ consists of one row for each metabolite and one column for each reaction. If a reaction generates a metabolite, the corresponding element is $+1$, and if a reaction uses a metabolite as substrate, the matrix element is -1. If a metabolite and a reaction are unrelated, the corresponding matrix element is zero. The matrix elements also indicate stoichiometric relationships that are not one-to-one. For instance, if two substrate molecules are used to form one product molecule, the loss in substrate is coded as -2. Generally, the element S_{ij} represents the stoichiometric coefficient of the ith chemical species in the jth reaction.

An example, adapted from Schauer and Heinrich (1983), is shown in Figure 1.7; bigger, more realistic stoichiometry matrices are found, for example, in Savinell (1991), Heinrich and Schuster (1996), and Alberty (1996, 2000).

The great appeal of the stoichiometric approach is its underlying mathematical representation in terms of matrix equations. These are very advantageous, because a multitude of theorems and methods supports them. For instance, the rank of the stoichiometry matrix, which is easily computed with standard methods of linear algebra, is a reflection of conservation relationships between metabolites at steady state (Gavalas 1968; Heinrich et al. 1977; Waller and Mäkllä 1981; Reder 1988; Heinrich and Schuster 1996, 1998). In other words, the rank signifies how many reactions are actually linearly independent.

Fell and Small (1986) used a stoichiometric approach for the analysis of lipogenesis. Savinell and Palsson (1992a–d) computed optimal flux patterns in *Hybridoma*

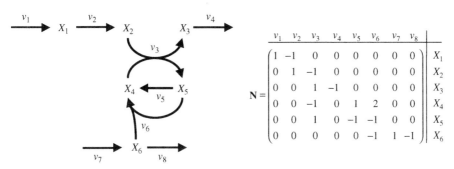

Figure 1.7. Partial system of reactions involved in erythrocyte glycolysis and corresponding stoichiometry matrix **N**. Some of the reactions are reversible, but show a preferred direction as indicated by arrows. X_1: FP$_2$, X_2: 1, 3-P$_2$G, X_3: 3-PG, X_4: ADP, X_5: ATP, X_6: AMP (adapted from Schauer and Heinrich 1983).

cells from a large stoichiometric model. Vallino and Stephanopoulos (1993) and Varma and Palsson (1995) used the stoichiometric formulation to predict reaction rates within a pathway when only input and output fluxes are known. Alberty (1996, 2000) related the stoichiometry of pathways to thermodynamic principles (see also Heijnen 2001). The linear structure even facilitated the development of computer algorithms that design possible metabolic pathways leading from a given substrate to a desired product (Seressiotis and Bailey 1988; Mavrovouniotis, Stephanopoulos and Stephanopoulos 1990). Efforts of optimizing stoichiometric systems are discussed in Chapter 5.

The consideration of stoichiometric mass balances, together with other physiological constraints, has led to a novel approach to analyzing metabolic networks that is called *Flux Balance Analysis* (FBA). The greatest advantage of FBA is perhaps that it can be applied to networks of essentially any size. To illustrate the point, Varma, Boesch, and Palsson (1993), Pramanik and Keasling (1997), and Edwards and Palsson (2000) developed stoichiometric balance models for extensive pathways in *E. coli*. Similarly, Jorgensen, Nielsen, and Villadsen (1995), Pons et al. (1996), Schilling and Palsson (2000), and Covert et al. (2000) constructed flux balance models for large portions of the metabolism of *Penicillium chrysogenum*, *Corynebacterium melassecola*, *Haemophilus influenzae*, and *Helicobacter pylori*, respectively. The stoichiometric system itself usually has no unique solution but allows for an entire solution ("null") space. To establish biologically feasible solutions, this null space is further restricted by thermodynamic and other physicochemical constraints (Varma, Boesch, and Palsson 1994; Schilling and Palsson 1998; Schilling et al. 1999; Schilling, Letscher, and Palsson 2000). The idea behind this strategy is that a sufficient set of constraints would ultimately lead to a complete set of biologically feasible network configurations that could be used for metabolic engineering tasks.

The stoichiometric approaches at the level of biochemical systems focus almost exclusively on the topology of the systems and record which metabolites may be converted into which other metabolites. They ignore processes like diffusion and convection, as well as most kinetic details (Heinrich and Schuster 1996, Chapter 2). If such details are indeed known, they cannot be used and important information is lost. This triggers the question of whether one could retain some of the advantages of the stoichiometric approach, most importantly its underlying linearity, while still making full use of all available kinetic information. As we discuss in the following, there is a compromise between the inherent nonlinearity of biochemical systems and the power of linear mathematics. To arrive at this compromise, we need to start over again and begin with first principles.

Simplified Representations

In our search for suitable representations of biochemical systems, linear models outside stoichiometric matrix formulations had to be considered with caution, because more often than not rate laws and biochemical processes simply are not linear. Process formulation in the tradition of Michaelis and Menten proved successful for individual reaction steps in vitro, but the previous discussion made it clear that this type of model becomes too cumbersome for reactions that are even slightly more complex than the simple, unmodulated one-substrate one-product reaction.

Taken together, these observations suggest that we need a different approach that leads to a simpler mathematical form and can readily be scaled up to large systems. It is not surprising that such an approach must be based on approximation, because it is simply impossible to account for every process in mechanistic detail. Because linear functions are not feasible for our purposes, we must look for nonlinear approximations. Experience of the past three decades has singled out a most useful candidate, the *power-law approximation*, which is equivalent to linear approximation in a logarithmic coordinate system.

Several observations support this type of approximation. First, rate laws in power-law form are direct generalizations of the rate laws for elemental chemical reactions, in which the original integer kinetic orders are replaced by real numbers. Even if chemical reactions follow different rate laws, it has been suggested to use power laws as "effective" representations for the analysis of chemical networks (Clarke 1980, p. 13). Second, some measurements in vivo are apparently consistent with this form of rate law (Kohen et al. 1973, 1974). Third, the phenomenon of allometry, which almost universally seems to govern the relationships between the growth of different parts of an organism and between different metabolic and physiological functions, is a direct manifestation of power-law functions (Savageau 1979b).

In addition to support from biological observations, the power-law approximation receives mathematical justification from two sources. The first is a method developed for the analysis of rational functions in control engineering, called Bode analysis. The second is the fact that a power-law approximation is equivalent to a first-order Taylor approximation in logarithmic space. It is useful to discuss these two roots in more detail. However, the section on Bode analysis may be skipped without loss of understanding subsequent sections and chapters.

Bode Analysis

Control systems play a vital role in engineering. They are used in essentially all settings that require close-to-constant temperature, pressure, humidity, viscosity, or flow. They are indispensable in robotics, space exploration, and aircraft and missile guidance, and have become a crucial component of industrial and manufacturing processes (Ogata 1990). A particular topic of relevance in control engineering is the analysis and industrial design of stable linear systems that are subjected to sinusoidal input (such as vibrations) over some frequency range of interest. The steady-state response of such a single-input, single-output system is called *frequency response*. Because sine wave generators and precise response measurement instruments are readily available, frequency response tests are relatively easy to execute and have become the core of classical control theory. The quantitative assessment of frequency responses has enabled engineers to design closed-loop control systems that satisfy almost arbitrary, prespecified performance requirements (Ogata 1990).

Analytical methods for characterizing the frequency response are directly or indirectly based on the fact that the frequency-response characteristics of a linear system are closely related to the input function. In particular, if the input is sinusoidal, then the steady-state output is also sinusoidal with the same frequency, though not necessarily the same magnitude and phase angle. The quantitative analysis of the input–output relationships uses a sinusoidal transfer function, in which the typical variable $s = \sigma + j\omega$ of the Laplace transformed system is replaced by the purely imaginary component $j\omega$, where ω is the frequency. If the original transfer function is $G(s)$, then the Laplace-transformed output $Y(s)$ is obtained from the Laplace transformed input $X(s)$ as

$$Y(s) = G(s)X(s), \tag{1.24}$$

and one can show that for sinusoidal input this relationship becomes

$$G(j\omega) = Y(j\omega)/X(j\omega), \tag{1.25}$$

which is a complex quantity that may be represented by the magnitude and phase angle with ω as a parameter (Ogata 1990).

Bode (1945) suggested plotting magnitude and phase angle against frequency in logarithmic coordinates. This representation presents three significant advantages. First, it shows responses to both low and high frequencies in one plot. Second, multiplication of magnitudes is converted into sums (of log magnitudes). Third, the plots in log space are well approximated by straight lines, which are readily added to yield a piecewise linear approximation of (log) magnitudes over a wide range of (log) frequencies. If only rough information about the frequency response is needed, this piecewise approximation is sufficient. If more detail is desired, the piecewise linear approximation presents asymptotes and breakpoints (sometimes called *corner frequencies*), from which the response function can be deduced with a fair degree of accuracy. The discrepancies between the true response and its approximation are greatest for the corner frequencies. Correction factors have been established for improving the accuracy at these points. These features of Bode representation in log space facilitate the construction of composite plots for arbitrary transfer functions of the type $G(j\omega)H(j\omega)$ (Ogata 1990).

The question now is what Bode plots have to do with enzyme kinetics. The connection is that both the sinusoidal transfer functions in control engineering and the rate laws of enzyme-catalyzed reactions have the mathematical form of rational functions (i.e., of ratios of polynomials). Savageau (1969a, 1976) therefore suggested using the approximation results obtained from Bode's method for the analysis of enzymatic rate laws. In particular, he concluded that it should be beneficial to represent enzymatic rate laws in logarithmic coordinates, that is, in the form

log(reaction rate)
= function of {log(substrates), log(inhibitors), log(modulators), ...}

In this logarithmic space, the formerly nonlinear rate law should approximately follow a piecewise linear function, just as sinusoidal transfer functions in the Bode plot. It is well known that many substrates and modulators of enzyme-catalyzed reactions do not vary all that much about their *nominal* value. Therefore, it should in many cases be possible to represent enzymatic reactions by only one piece of the piecewise linear approximation, which corresponds to a tremendous mathematical simplification.

By applying Bode's concepts to enzymatic rate laws, Savageau (1969a) concluded that a linear approximation in logarithmic coordinates should provide a useful representation of biochemical processes. Supporting this conclusion, it was later documented in numerous ways that the range of this simple linear approximation in logarithmic space is much wider than one might expect. For instance, Savageau (1976) demonstrated that the steady-state induction characteristics of some operons are linear over one, two, or more orders of magnitude.

Power-Law Approximation

In the search of a valid and effective mathematical representation, several observations and considerations point toward a power-law formulation, which is equivalent to linearization in logarithmic coordinates. The appropriateness of this mathematical form is no coincidence. It is based mathematically on the theorem of Brook Taylor (1685–1731), which guarantees that differentiable functions can be validly linearized about an operating point. The Taylor linearization may be executed in log space, and diverse experience suggests that this often widens the range of valid approximation for biological phenomena.

Formalizing this idea, the strategy of developing an effective model representation is as follows. First, represent all variables in the form of their logarithm; that is, define for each variable X_i a new variable Y_i as the logarithm of X_i. The base one selects for the logarithmic function does not really matter, but we will use the natural logarithm (ln) throughout this text and thus define $Y_i = \ln(X_i)$. This definition applies to all variables in the system. The process v is also represented in terms of its natural logarithm as $w = \ln(v)$.

The second step is the development of the Taylor approximation for w as a function of all Y_i. This approximation is computed at one point of choice, the *operating point*. In many cases, this point represents the nominal or typical state of the system. Taylor's theorem guarantees that every sufficiently differentiable function can be approximated by a polynomial. Typically, the accuracy of this polynomial increases as the degree of the polynomial increases. But even a very low degree is sufficient if the approximation is used relatively close to the operating point. How far one may move away from the operating point depends on the function itself and on the magnitude of acceptable deviation between the function and its approximation.

Retaining exclusively the constant and linear terms of the Taylor approximation has proven most useful in the present context, even though a second-order model has been considered as a more accurate, yet significantly more cumbersome alternative (Cascante et al. 1991). For our purposes, all higher-order terms are ignored, and this corresponds exactly to the straight-line representation in Bode's plots, or, in the multivariate case, to the analogous linear representation in higher dimensions. The final step is to return the result of this Taylor linearization in logarithmic space to Cartesian coordinates, where the result is nonlinear and in the form of a product of power-law functions.

Let us execute the procedure step by step, beginning with a function v of only one dependent variable, x, which has at least n continuous derivatives. According to Taylor's theorem, $v(x)$ may be approximated at the operating point p by a polynomial of up to the nth degree

$$v(x) \approx v(p) + v'(p)(x - p) + \frac{1}{2!}v''(p)(x - p)^2 + \frac{1}{3!}v'''(p)(x - p)^3$$
$$+ \cdots + \frac{1}{n!}v^{(n)}(p)(x - p)^n. \tag{1.26}$$

If all derivatives of v exist, n could grow to infinity, and the Taylor polynomial would be an exact representation of v. We are interested in the other extreme, namely in $n = 1$, which corresponds to the linear approximation

$$v(x) \approx v(p) + v'(p)(x - p). \tag{1.27}$$

Because we intend to execute the linearization in logarithmic coordinates, which may cause additional confusion, it might be helpful to demonstrate the process with an example. It is noted, however, that ultimately we do not have to follow these steps, because a shortcut allows us a quicker and easier computation leading to the same result. Nonetheless, to show the linearization once in detail, consider a simple Michaelis–Menten rate law of the form

$$v = v(X) = \frac{V_{\max} X}{K_M + X}. \tag{1.28}$$

Using logarithmic variables w and Y leads to

$$w = w(Y) = \ln \frac{V_{\max} \exp(Y)}{K_M + \exp(Y)}. \tag{1.29}$$

The derivative of w with respect to Y yields the slope of the rate law in logarithmic space and thus the slope of the desired approximation. The derivative is

$$\frac{dw}{dy} = \frac{K_M + \exp(Y)}{V_{\max} \exp(Y)} \cdot \frac{V_{\max} \exp(Y) \cdot (K_M + \exp(Y)) - V_{\max} \cdot \exp(Y) \cdot \exp(Y)}{(K_M + \exp(Y))^2}$$

$$= \frac{K_M}{K_M + \exp(Y)}. \tag{1.30}$$

Thus, the linearization of the Michaelis–Menten rate law in logarithmic coordinates at the operating point Y_0 is

$$w \approx w(Y_0) + \frac{K_M}{K_M + \exp(Y_0)} \cdot (Y - Y_0)$$

$$= [w(Y_0) - \frac{K_M}{K_M + \exp(Y_0)} \cdot Y_0] + \frac{K_M}{K_M + \exp(Y_0)} \cdot Y$$

$$= \ln(\alpha) + g \cdot Y. \tag{1.31}$$

The terms in brackets depend on the operating point but not on the variable Y. They are constant and collected in the term $\ln(\alpha)$. Similarly, g is constant, but depends on the operating point. To return to Cartesian coordinates, we simple exponentiate both sides and obtain

$$v \approx \alpha \cdot X^g. \tag{1.32}$$

Thus, the original rate law is approximated by a power-law function with a real-valued exponent and a positive multiplier. Reminiscent of elemental chemical kinetics, the exponent is called *kinetic order* and the multiplier is called *rate constant*.

A shortcut simplifies the computation considerably. The derivation above shows that the kinetic order is the derivative of the logarithm of the function v with respect to the log of the variable X. One can thus directly compute

$$g = \frac{dw}{dY} = \frac{d \ln v}{d \ln X} = \frac{dv}{dX} \cdot \frac{X}{v}, \tag{1.33}$$

which is computed at the operating point of choice.

Once the kinetic order is computed, the rate constant is obtained as

$$\alpha = v(X_0) \cdot (X_0)^{-g}, \tag{1.34}$$

which holds, because the function and its approximation must be equal at the operating point X_0.

By design, the numerical characteristics of the approximation depend on the operating point. One easily checks, for instance, that the kinetic order g for the approximation of the Michaelis–Menten rate law decreases from 1 to 0 if the operating point

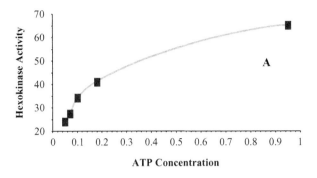

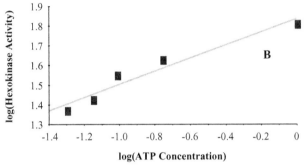

Figure 1.8. Glucose 6-phosphate production is shown as a function of ATP concentration in Cartesian and logarithmic coordinates (data from Danenberg and Cleland 1975; redrawn from Galazzo and Bailey 1990, 1991). The Cartesian representation (Panel A) suggests saturation in hexokinase activity. The linearity of the logarithmic representation (Panel B) facilitates the determination of the corresponding kinetic order. Note that linearity holds for more than an order of magnitude in ATP concentration.

is chosen at increasing substrate concentrations. Notably, for $X_0 = K_M$, the kinetic order is 0.5.

Although the numerical values of kinetic order and rate constant change with the choice of the operating point, the *structure* of the approximation is always the same, namely that of a power-law function of the type αX^g. This implies that one may set up the approximation symbolically without even knowing the exact structure of the underlying process. Furthermore, the kinetic order can be obtained from experimental data, which are plotted as rate against substrate concentration on logarithmic axes (Figure 1.8). In such a plot, the kinetic order equals the slope of the experimental relationship at the operating point.

A crucially significant benefit of this approach (including the shortcut) is that it translates directly into higher dimensions. Specifically, to compute the power-law approximation of the multivariate rate law $V(X_1, X_2, X_3, X_4, \ldots, X_n)$, one again chooses an operating point P, which now has n coordinates $p_1, p_2, \ldots, p_n$. The analogous steps of linearization in $n + 1$-dimensional logarithmic space lead to the following result: a product of power-law functions approximates V with the form

$$V = \alpha \cdot X_1^{g_1} X_2^{g_2} X_3^{g_3} X_4^{g_4} \cdots X_n^{g_n}. \tag{1.35}$$

Each kinetic order g_i can be directly calculated through *partial* differentiation of V with respect to X_i:

$$g_i = \frac{\partial V}{\partial X_i} \cdot \frac{X_i}{V}, \tag{1.36}$$

which is evaluated at the operating point P. Also in strict analogy to the case of one substrate, the rate constant is given as

$$\alpha = V(p_1, p_2, \ldots, p_n) p_1^{-g_1} p_2^{-g_2} \cdots p_n^{-g_n}. \tag{1.37}$$

Only those variables that *directly* affect the process under consideration enter the power-law representation. If a variable X_k does not affect V, the partial derivative of V with respect to X_k is zero. Consequently, g_k is zero and the expression $X_k^{g_k}$ equals 1, regardless of the value of X_k. The product structure of the representation thus makes all such factors $X_k^{g_k} = 1$ ineffectual and superfluous.

As an example for the multivariate power-law representation, consider a batch culture of yeast cells that are fermenting glucose under anaerobic conditions. The growth characteristics of the population obviously depend on the limiting substrate glucose, but are also affected by the alcohol that is being produced (Bailey and Ollis 1977, p. 348). Aiba, Shoda, and Nagatani (1968; Aiba and Shoda 1969) suggested modeling growth and product inhibition with the generalized Monod model

$$\mu = \mu_{\max} \frac{c_1}{K_1 + c_1} \cdot \frac{K_2}{K_2 + c_2}, \tag{1.38}$$

where c_1 and c_2 are the concentrations of the substrate glucose and the product ethanol, respectively [see also Nielsen (2001) for background and applications of Monod's model]. K_1 is the value of the glucose concentration for which the specific growth rate is half maximal, and K_2 characterizes product inhibition. The specific-growth function μ thus depends on two variables, and we know immediately that its power-law approximation takes the symbolic form

$$\mu \approx \alpha \cdot c_1^{g_1} \cdot c_2^{g_2}. \tag{1.39}$$

Using the shortcut of partial differentiation, the kinetic orders and the rate constant are calculated as

$$g_1 = \frac{\partial \mu}{\partial c_1} \frac{c_1}{\mu} = \mu_{\max} \cdot \frac{K_2}{K_2 + c_2} \cdot \frac{K_1}{(K_1 + c_1)^2} \cdot \frac{c_1}{\mu} = \frac{K_1}{K_1 + c_1}, \tag{1.40}$$

$$g_2 = \frac{\partial \mu}{\partial c_2} \frac{c_2}{\mu} = \frac{-c_2}{K_2 + c_2}, \tag{1.41}$$

$$\alpha = \mu \cdot c_1^{-g_1} \cdot c_2^{-g_2}. \tag{1.42}$$

These quantities are to be evaluated at some operating point $P = (c_{10}, c_{20})$ of choice. Note that g_1 is positive, which indicates that higher substrate concentrations lead to a higher specific growth function, whereas g_2 is negative, reflecting the inhibiting effect of ethanol. The rate constant is always nonnegative.

The example reiterates comments made earlier in this section. Once we are satisfied with the validity of the derivation of the linear approximation in log space, we do not need to execute any of the details involved in the theory and implementation of Bode's or Taylor's methods. Instead, with minimal information about the physical and regulatory connectivity of the system, we can directly write down a symbolical power-law model, even if the original, approximated function is not known in detail. This fact allows us to construct mathematical models for very complex biochemical systems. When the original function is unknown, we may not know the values of the parameters, but we do know the mathematical structure of the approximate representation.

S-Systems

We now use the power-law representation to develop a general system description, which is the foundation of a modeling framework called *Biochemical Systems Theory* (BST; Savageau 1969a,b, 1970). Typical of systems analysis, we formulate the dynamics of the system with a set of differential equations. Each represents the change in one dependent variable X_i, which is assumed to have a positive value. The system may also contain independent variables, which do not change during any given experiment, but may differ from one experiment to the next. Differences between the two types are discussed in greater detail in the next chapter (see also Voit 2000a).

In very general terms, the dynamics of a system can thus be formulated as

$$\dot{X}_i = V_i^+(X_1, X_2, \ldots, X_n, X_{n+1}, \ldots, X_{n+m})$$
$$- V_i^-(X_1, X_2, \ldots, X_n, X_{n+1}, \ldots, X_{n+m}) \qquad i = 1, 2, \ldots, n, \qquad (1.43)$$

where the first n variables $X_1, X_2, \ldots, X_n$ are dependent and the following m variables $X_{n+1}, X_{n+2}, \ldots, X_{n+m}$, are independent. The positive-valued function V_i^+ represents all effects that increase or augment the production of X_i, and the positive-valued function V_i^- represents all effects that increase or augment the degradation of X_i. As discussed in the previous sections, it is straightforward to compute the multivariate power-law representations of V_i^+ and V_i^-. Entering them in Eq. (1.43), we obtain directly the S-system form

$$\dot{X}_i = \alpha_i \prod_{j=1}^{n+m} X_j^{g_{ij}} - \beta_i \prod_{j=1}^{n+m} X_j^{h_{ij}} \qquad i = 1, 2, \ldots, n. \qquad (1.44)$$

Each term contains only those variables, with their associate kinetic orders, that have a direct effect on this term. For instance, if only X_1, X_3, and X_6 affect V_2^-, then its representation is

$$V_2^- = \beta_2 X_1^{h_{21}} X_3^{h_{23}} X_6^{h_{26}}. \qquad (1.45)$$

Detailed examples are given throughout the remainder of the book. Recent reviews on S-systems and their applications are found in Voit (2000a,b).

Generalized Mass Action (GMA) Systems

In the derivation of the S-system in Eq. (1.44), all processes entering a variable or affecting its production were aggregated into one process V_i^+ and all processes leaving the pool or affecting its degradation were similarly grouped into another process V_i^-. These overall processes were subsequently approximated with power-law terms, leading to right-hand sides of the S-system equations that contained exactly one difference of products of power-law functions.

An alternative representation within BST, which is also based on power-law functions, is obtained when the aggregation step is skipped. In this *Generalized Mass Action* (GMA) representation, each process entering or leaving a variable or pool is replaced individually with a product of power-law functions, leading to the formulation

$$\dot{X}_i = \gamma_{i1} \prod_{j=1}^{n+m} X_j^{f_{ij1}} \pm \gamma_{i2} \prod_{j=1}^{n+m} X_j^{f_{ij2}} \pm \cdots \pm \gamma_{ik} \prod_{j=1}^{n+m} X_j^{f_{ijk}} \qquad i = 1, 2, \ldots, n \quad (1.46)$$

in which each equation may have any number of terms. As in the S-system form, the rate constants γ_{il} are positive or zero and the kinetic orders f_{ijk} may have any real values.

GMA systems present a very attractive alternative to S-systems. They are often closer to biochemical intuition, because each process is explicitly represented and easily identified. However, this form also has drawbacks. Most important is that the GMA form does not permit the algebraic calculation of steady states, as does the S-system form. This drawback is of crucial importance for questions of optimization (see Chapters 4 and 5).

GMA- and S-systems differ only at branch points, where two processes yield the same product or where the same substrate is used for two reactions. Linear sections of pathways or cascaded systems, irrespective of whether they are modulated by other constituents of the system, are modeled in exactly the same fashion. At branch points, the two representations differ. Evaluated at the operating point, the two actually produce exactly the same numerical value, but as soon as the system deviates from the operating point, the two system descriptions differ. In many applications where the two representations were compared, the differences were typically negligible in comparison to the precision of experimental data. However, for larger deviations from the operating point, the differences may become significant. Intuition suggests that the GMA form would be more precise, because it retains the exact stoichiometry of influxes and effluxes at branch points. Although it is true that the S-system form leads to discrepancies in the stoichiometry at branch points, it was shown in careful studies that the overall accuracy of S-systems in terms of metabolite levels is usually better than that of GMA systems (Voit and Savageau 1987). The reason for this surprising finding is that the power-law approximation tends to overestimate hyperbolic functions that are typical for biochemical processes, with the Michaelis–Menten rate law being a good example. This overestimation is the same for GMA- and S-system models. The aggregation of several fluxes into a difference of two terms, which is the hallmark of the S-system form, leads to a slight underestimation of these fluxes at branch points, and this underestimation compensates for some of the inaccuracy introduced by the power-law approximation.

Thus, both forms have advantages and disadvantages, which need to be weighed in a given situation. The GMA form may be closer to biochemical intuition because it explicitly shows every flux in the pathway. It also preserves stoichiometry of fluxes at branch points without error. The S-system form requires an additional step, namely the aggregation of fluxes at branch points. This step is mathematically straightforward but requires a little bit of effort not needed in the GMA form. Because of compensation of errors, the S-system form approximates branches of traditional rate laws more accurately than the GMA form, but introduces slight discrepancies in flux stoichiometry. Of special importance for questions of optimization, the S-system form allows for a direct computation of steady states. For unbranched systems with any degree of modulation, the GMA and S-system forms are equivalent. For branched systems, the numerical differences between the two forms are usually small. As a precaution, it may be advisable to formulate both variants of power-law models and to compare their accuracy with simulations of relevant situations. If the differences are insignificant, either form may be used. If steady-state analyses are of importance, the S-system form may be more advantageous.

One could argue that the two terms of each S-system equation could be aggregated and approximated by a single power-law term. Mathematically, this is possible without any conceptual or technical difficulty and leads to a simpler structure. However, this *Riccati* (Peschel and Mende 1986) or *Half-system* (Savageau and Voit 1987; Voit 1991) form has many disadvantages that make it awkward for modeling purposes. For instance, these systems can only have steady states if one or more variables have values of zero.

As with linear systems, S-, GMA-, and Half-systems can be formulated in a piecewise fashion. Whenever a variable or set of variables leaves some range, a new operating point may be chosen, the power-law approximation is recomputed, and the analysis switches to the new parameterization. At this point, not many analyses have made use of this option. A notable exception is Savageau (2001).

Metabolic Control Analysis (MCA)

Until the 1970s, metabolic pathways were thought to be controlled by just one or by a small number of "rate-limiting" enzymatic steps. Often the alleged controlling enzyme was positioned at the beginning of a chain of reactions and perhaps inhibited through feedback exerted by the final product of the sequence. Parallel to BST, *Metabolic Control Analysis* (MCA; Kacser and Burns 1973; Heinrich and Rapoport 1974) has made numerous contributions to our current understanding of enzymatic networks by demonstrating that control of a pathway is shared among the contributing enzymes and modulators. A succinct review of MCA was recently provided by Ricard (1999); more detailed descriptions are found in the books by Westerhoff and van Dam (1987), Cornish-Bowden and Cárdenas (1990), Heinrich and Schuster (1996), Fell (1997), and Stephanopoulos et al. (1998), and in an extensive body of scientific articles. As Ricard summarized MCA, "it is not really a theory, but rather a

$$A \xrightarrow[v_1]{E_1} S_1 \xrightarrow[v_2]{E_2} S_2 \xrightarrow{\quad} \cdots\cdots \xrightarrow[v_n]{E_n} S_n \xrightarrow[v_{n+1}]{E_{n+1}} B$$

Figure 1.9. Linear pathway with constant source and sink.

formal model that allows the expression of parameters that describe, in quantitative terms, the global behaviour of an enzyme system."

The starting point for understanding the rationale of MCA is a linear pathway as shown in Figure 1.9. The source and sink, A and B, are considered constant, $S_1, \ldots, S_n$ are intermediate metabolites, and $E_1, \ldots, E_{n+1}$ are enzymes that catalyze reactions with rates $v_1, \ldots, v_{n+1}$. At least in the original formulation of MCA, all features and theorems addressed systems at steady state. An important consequence in this situation is that the flux through all metabolites must be the same, namely $J = v_i$ $(i = 1, \ldots, n+1)$. Almost all original results of MCA were based on three concepts, the *flux control coefficient*, the *concentration control coefficient*, and the *elasticity*. The flux control coefficient quantifies the effect of an enzyme E_i on the steady-state flux through the pathway. It is defined in terms of relative changes as

$$C_i^J = \frac{E_i}{J} \cdot \frac{\partial J}{\partial E_i} = \frac{\partial \ln J}{\partial \ln E_i}. \tag{1.47}$$

Similarly, the concentration control coefficient quantifies the effect of a change in enzyme E_i on the steady-state concentration of metabolite S_j:

$$C_i^{S_j} = \frac{E_i}{S_j} \cdot \frac{\partial S_j}{\partial E_i} = \frac{\partial \ln S_j}{\partial \ln E_i}. \tag{1.48}$$

These two control coefficients characterize systemic properties, because they relate a global system response to a slight local perturbation in an enzyme. The elasticity, by contrast, is a local property that relates a perturbation in a metabolite to the subsequent change in one particular reaction rate. It is also defined as a logarithmic derivative:

$$\varepsilon_{S_j}^{v_i} = \frac{S_j}{v_i} \cdot \frac{\partial v_i}{\partial S_j} = \frac{\partial \ln v_i}{\partial \ln S_j}. \tag{1.49}$$

The fundamental insights of MCA derive from relationships between these quantities. For instance, consider the pathway flux J as a function of the $n+1$ enzymes. The entire change in the flux is mathematically given as the sum of responses to changes in enzymes. Expressed in the form of the total differential of the flux, one obtains

$$dJ = \sum_{i=1}^{n+1} \frac{\partial J}{\partial E_i} dE_i. \tag{1.50}$$

The partial derivatives may be expressed in terms of the flux control coefficients, according to Eq. (1.47) as

$$\frac{\partial J}{\partial E_i} = \frac{J}{E_i} \cdot C_i^J. \tag{1.51}$$

The original version of MCA assumed that the flux is a homogeneous function of degree one in all enzyme concentrations. This implies that scaling all independent variables (enzymes) by an equal amount leads to scaling of the dependent variable (flux) by the same amount. Leonhard Euler (1707–83) showed that a homogeneous function of degree one may be represented as the sum of partial derivatives with respect to all independent variables, multiplied with the variables themselves. In terms of enzymes and overall flux, Euler's theorem asserts

$$J = \sum_{i=1}^{n+1} \frac{\partial J}{\partial E_i} E_i. \tag{1.52}$$

Substitution of Eq. (1.51) in Eq. (1.52) and division by J leads to the *summation theorem*

$$\sum_{i=1}^{n+1} C_i^J = 1. \tag{1.53}$$

The assumption of homogeneity of degree one is probably valid for many real systems, in which all enzymes operate independently of each other and affect the system in direct proportionality to their activities. If the system contains enzyme–enzyme interactions, if there are association–dissociation processes among enzymes, if the system is cascaded, or if the system response to a change in enzyme is not linear, the assumption is no longer valid [see Savageau (1991, 1992b) and Ricard (1999) for further discussion of these cases].

A similar summation theorem holds for concentration control coefficients. Analogous to the response in flux, the change in a substrate concentration S_j is given as the total differential

$$dS_j = \sum_{i=1}^{n+1} \frac{\partial S_j}{\partial E_i} dE_i. \tag{1.54}$$

Substitution of Eq. (1.48) in Eq. (1.54) and division by S_j yields

$$\frac{dS_j}{S_j} = \sum_{i=1}^{n+1} C_i^{S_j} \cdot \frac{dE_i}{E_i}. \tag{1.55}$$

It is not too difficult to see that the steady-state value of S_j is unaffected if all enzyme activities are changed by the same amount. This implies that the left-hand side equals zero and results in the simple summation theorem for concentration control coefficients:

$$\sum_{i=1}^{n+1} C_i^{S_j} = 0. \tag{1.56}$$

The systemic responses to local perturbations may be expressed as *connectivity relationships* between control coefficients and elasticities. They are consequences of the total differential that relates a change in a rate v_i to changes in the corresponding enzyme and in the metabolite concentrations:

$$dv_i = \frac{\partial v_i}{\partial E_i} \cdot dE_i + \sum_{j=1}^{n} \frac{\partial v_i}{\partial S_j} \cdot dS_j. \tag{1.57}$$

Without giving the derivation, the connectivity relationships with respect to overall flux and to metabolite concentrations are given as

$$\sum_{i=1}^{n+1} C_i^J \varepsilon_{S_j}^{v_i} = 0, \tag{1.58}$$

and

$$\sum_{i=1}^{n+1} C_i^{S_k} \varepsilon_{S_j}^{v_i} = -\delta_{jk}, \tag{1.59}$$

where δ_{jk} is the Kronecker symbol that equals 1 for $j = k$ and 0 otherwise.

The summation and connectivity relationships have appeal among experimentalists because they have been interpreted as conservation relationships. If the control coefficients and elasticities are determined experimentally, their respective sums can be tested against the theoretically expected results, as given in Eqs. (1.53), (1.56), and (1.59). In cases of discrepancies, some aspects of the experimental system have been overlooked or need to be revisited. Caution is necessary, however, with this type of interpretation. Because the control coefficients and elasticities in some systems are not necessarily positive, the summation and connectivity relationships may be satisfied, even though two or more effects, which happen to cancel each other in the summation, may have been overlooked (Savageau 1992b).

Extensions to the original theorems of MCA have relaxed the requirements of flux homogeneity with respect to enzymes. They were reviewed in Ricard (1999) and extensively discussed in the literature (Fell and Sauro 1985; Sauro, Small, and Fell 1987; Giersch 1988; Reder 1988; Kacser, Sauro, and Acerenza 1990; Fell 1997). It is noted that the summation and connectivity theorems, under valid assumptions, can be derived directly and completely from the structure of BST (Savageau, Voit, and Irvine, 1987a,b; Savageau and Sorribas 1989; Savageau 1991, 1992b).

Richness of Structure

Linear systems are very useful but limited in the repertoire of phenomena they can represent. For instance, linear functions do not saturate but continue to grow or decrease at a constant rate. Also, as mentioned earlier, they cannot describe stable oscillations or chaos. This is a relevant concern, because it is becoming increasingly clear that many chemical and biochemical systems in vivo and even in vitro can exhibit complex temporal behaviors, such as stable periodic oscillations and even chaos. Several reviews and books have addressed this topic in detail (Tyson 1976; Richter, Procaccia, and Ross 1980; Segel 1991; Goldbeter 1996; Hess 1997; Ricard 1999).

In light of these limitations, it is legitimate to ask about the boundaries of power-law systems. Surprisingly, methods of mathematical equivalence transformations have shown that systems of the form (1.44) and (1.46) are rich enough to model virtually any phenomenon that can be represented with any type or system of ordinary differential equations (Savageau and Voit 1987). In other words, if one is given a system of arbitrary ordinary differential equations, one can introduce new variables and formulate an S-system or GMA system using the new and some of the old variables. This system is not an approximation but exactly equivalent mathematically with the initial system.

The methods of *recasting*, which are used to execute these transformations, have applications in function classification and numerical analysis. Recasting, however, is no panacea for modeling. Although the result of recasting is equivalent with the original system, the number of variables often increases drastically, and analyses that are straightforward in genuine S-systems, such as computations of steady states, have to be executed for recast systems in higher-dimensional spaces under the consideration of nonlinear recasting constraints. Thus, for purposes of modeling, the method of recasting is of limited use, but it does demonstrate that the power-law structure is rich enough to model essentially any differentiable nonlinearity one may encounter in practical applications.

SUMMARY

The choice and design of a mathematical model depend on the purpose of the analysis. The purpose determines the complexity of the model, the types and degrees of simplifications, the inclusion or exclusion of variables, factors, and modulators, the interactions between them, and even the nature of the time scale. Indeed, model design is the crucial component of the entire modeling process. As Albert Einstein put it:

> The mere formulation of a problem is far more often essential than its solution, which may be merely a matter of mathematical or experimental skill. To raise new questions, new possibilities, to regard old problems from a new angle requires creative imagination and makes real advances in science. (cited in Duncan and Reimer 1998)

The selection of the "best" model is not a trivial matter, because the number of intrinsically different model structures is infinite. This is comforting, because we are not likely to run out of options. However, the overwhelming variety poses the challenge of identifying models that are particularly well suited for our purposes of representing, analyzing, and optimizing biochemical systems. This chapter shows that the quality of data fit is not a reliable criterion for model selection, because many functions may model the data set of interest with similar accuracy. Furthermore, a good data fit alone does not provide insight or explanations concerning the phenomenon under investigation. Linear models have been used successfully in engineering, but with the exception of stoichiometric models, they are rarely valid for biological processes.

Traditionally, biochemical processes have been analyzed with methods of thermodynamics or with kinetic rate laws that are generalizations of the mechanism proposed by Michaelis and Menten. Thermodynamics focuses on questions of energy transfer and essentially ignores time. For our purposes it is therefore limited in scope. Traditional rate laws often work well in vitro, but lead to mathematical difficulties if the analyzed pathways are of moderate to large size. The solution proposed here is a power-law approximation of all processes characterizing the system. This approximation is anchored in solid mathematics and has favorable features; its accuracy and validity are supported by experimental observations and by different types of applications from engineering and biology. Among alternative power-law representations for integrated systems, the S-system form has the unique feature of combining almost unlimited richness in structure with linearity at the steady state. Later chapters will show that this linearity is crucial for model diagnostics and for streamlined optimization.

REFERENCES

Aiba, S., and M. Shoda: Reassessment of the product inhibition in alcohol fermentation. *J. Ferment. Technol. Jpn.* 47, 790–805, 1969.

Aiba, S., M. Shoda, and M. Nagatani: Kinetics of product inhibition in alcohol fermentation. *Biotechnol. Bioeng.* 10, 845–64, 1968.

Alberty, R.A.: Biochemical thermodynamics. *Biochim. Biophys. Acta* 1207, 1–11, 1994.

Alberty, R.A.: Calculation of biochemical net reactions and pathways by using matrix operations. *Biophys. J.* 71, 507–15, 1996.

Alberty, R.A.: Calculation of equilibrium compositions of large systems of biochemical reactions. *J. Phys. Chem.* 104(19), 4808–14, 2000.

Bailey, J.E., and D.F. Ollis: *Biochemical Engineering Fundamentals.* McGraw-Hill, New York, 1977.

Bode, H.W.: *Network Analysis and Feedback Amplifier Design.* Van Nostrand, Princeton, NJ, 1945.

Briggs, G.E., and J.B.S. Haldane: A note on the kinetics of enzyme action. *Biochem. J.* 19, 338–9, 1925.

Callen, H.B.: *Thermodynamics*, John Wiley & Sons, New York, 1960.

Cascante, M., A. Sorribas, R. Franco, and E.I. Canela: Biochemical systems theory: Increasing predictive power by using second-order derivatives measurements. *J. Theor. Biol.* 149, 521–35, 1991.

Cha, S., and C.-J.M. Cha: Kinetics of cyclic enzyme systems. *Mol. Pharmacol.* 1, 178–89, 1965.

Clarke, B.L.: Stability of complex reaction networks. *Adv. Chem. Phys.* 43, 1–215, 1980.

Corless, R.M., G.H. Gonnett, D.E.G. Hare, D.J. Jeffrey, and D.E. Knuth: On the Lambert W function. *Adv. Comp. Math.* 5, 329–59, 1996.

Cornish-Bowden, A., and M.L. Cárdenas (Eds.): *Control of Metabolic Processes*, NATO ASI Series A (Vol. 190). Plenum Press, New York, 1990.

Covert, M.W., C.H. Schilling, J.S. Edwards, I. Famili, and B.Ø. Palsson: Constraints-based metabolic analysis of *Helicobacter pylori.* Session 69 of the AIChE Annual Meeting, Los Angeles, CA, 2000.

Curto, R., E.O. Voit, A. Sorribas, and M. Cascante: Validation and steady-state analysis of a power-law model of purine metabolism. *Biochem. J.* 324, 761–75, 1997.

Curto, R., E.O. Voit, A. Sorribas, and M. Cascante: Mathematical models of purine metabolism in man. *Math. Biosci.* 151, 1–49, 1998a.

Curto, R., E.O. Voit, A. Sorribas, and M. Cascante: Analysis of abnormalities in purine metabolism leading to gout and to neurological dysfunctions in man. *Biochem. J.* 329, 477–87, 1998b.

Danenberg, K.D., and W.W. Cleland: Use of chromium-adenosine triphosphate and lyxose to elucidate the kinetic mechanism and coordination state of the nucleotide substrate for yeast hexokinase. *Biochemistry* 14(1), 28–39, 1975.

Duncan, T.M., and J.A. Reimer: *Chemical Engineering Design and Analysis: An Introduction.* Cambridge University Press, Cambridge, U.K., 1998.

Edelstein-Keshet, L.: *Mathematical Models in Biology*, Birkhäuser Mathematics Series. McGraw-Hill, New York, 1988.

Edwards, J.S., and B.Ø. Palsson: The *Escherichia coli* MG1655 in silico metabolic genotype: Its definition, characteristics, and capabilities. *Proc. Natl. Acad. Sci. USA* 97, 5528–33, 2000.

Fell, D.A.: *Understanding the Control of Metabolism.* Portland Press, London, 1997.

Fell, D.A., and H.M. Sauro: Metabolic control and its analysis. Additional relationships between elasticities and control coefficients. *Eur. J. Biochem.* 148, 555–61, 1985.

Fell, D.A., and J.R. Small: Fat synthesis in adipose tissue. An examination of stoichiometric constraints. *Biochem. J.* 238, 781–6, 1986.

Fermi, E.: *Thermodynamics.* Dover, New York, 1956 (originally published in 1937).

Galazzo, J.L., and J.E. Bailey: Fermentation pathway kinetics and metabolic flux control in suspended and immobilized *S. cerevisiae. Enzyme Microbiol. Technol.* 12, 162–72, 1990.

Galazzo, J.L., and J.E. Bailey: Errata. *Enzyme Microbiol. Technol.* 13, 363–71, 1991.

Garfinkel, D.: The role of computer simulation in biochemistry. *Comp. Biomed. Res.* 2(1), 31–44, 1968.

Garfinkel, D.: Computer modeling, complex biological systems, and their simplifications. *Am. J. Phys.* 239(1), R1–6, 1980.

Garfinkel, D.: Computer-based modeling of biological systems which are inherently complex: Problems, strategies, and methods. *Biomed. Biochim. Acta* 44(6), 823–9, 1985.

Gavalas, G.R.: *Nonlinear Differential Equations of Chemically Reacting Systems.* Springer-Verlag, Berlin, 1968.

Gerthsen, C., and H.O. Kneser: *Physik* (11th ed.). Springer-Verlag, Berlin, 1971.

Giersch, C.: Control analysis of metabolic networks. 2. Total differentials and general formulation of the connectivity relations. *Eur. J. Biochem.* 174, 515–19, 1988.

Glansdorff, P., and I. Prigogine: *Thermodynamics of Structure, Stability, and Fluctuations.* Wiley, New York, 1971.

Goldbeter, A.: *Biochemical Oscillations and Biological Rhythms.* Cambridge University Press, Cambridge, U.K., 1996.

Goldstein, A.: The mechanism of enzyme-inhibitor-substrate reactions. *J. Gen. Physiol.* 27, 529–80, 1944.

Heijnen, J.J.: Stoichiometry and kinetics of microbial growth from a thermodynamic perspective. In: C. Ratledge and B. Kristiansen (Eds.), *Basic Biotechnology* (Chapter 3). Cambridge University Press, Cambridge, U.K., 2001.

Heinrich, R., and T.A. Rapoport: A linear steady-state treatment of enzymatic chains: General properties, control and effector strength. *Eur. J. Biochem.* 42, 89–95, 1974.

Heinrich, R., and S. Schuster: *The Regulation of Cellular Systems.* Chapman and Hall, New York, 1996.

Heinrich, R., and S. Schuster: The modeling of metabolic systems. Structure, control, and optimality. *BioSystems* 47, 61–77, 1998.

Heinrich, R., S.M. Rapoport, and T.A. Rapoport: Metabolic regulation and mathematical models. *Prog. Biophys. Mol. Biol.* 32, 1–82, 1977.

Henri, M.V.: *Lois générales de l'action des diastases*. Hermann, Paris, 1903.

Hess, B.: Periodical patterns in biochemical reactions. *Quart. Rev. Biophys.* 30, 121–76, 1997.

Hill, C.M., R.D. Waight, and W.G. Bardsley: Does any enzyme follow the Michaelis-Menten equation? *Mol. Cell. Biochem.* 15, 173–8, 1977.

Horn, F., and R. Jackson: General mass action kinetics. *Arch. Rational Mech. Anal.* 47, 81–116, 1972.

Jacquez, J.A.: *Compartmental Analysis in Biology and Medicine* (3rd ed.). Thomson-Shore, Dexter, MI, 1996.

Jorgensen, H., J. Nielsen, and J. Villadsen: Metabolic flux distribution in Penicillium chrysogenum during fed-batch cultivations. *Biotechnol. Bioeng.* 46, 117–31, 1995.

Jou, D., and J.E., Llebot: *Introduction to the Thermodynamics of Biological Processes*. Prentice Hall, Englewood Cliffs, NJ, 1990.

Kacser, H., and J.A. Burns: The control of flux. *Symp. Soc. Exp. Biol.* 27, 65–104, 1973.

Kacser, H., H.M. Sauro, and L. Acerenza: Enzyme-enzyme interactions and control analysis. *Eur. J. Biochem.* 187, 481–91, 1990.

Katchalsky, A., and P.F. Curran: *Nonequilibrium Thermodynamics in Biophysics*. Harvard University Press, Cambridge, MA, 1967.

Kestin, J.: *A Course in Thermodynamics*. Blaisdell Publishing Company, Waltham, MA, 1966.

Kohen, E., C. Kohen, B. Thorell, and G. Wagner: Quantitative aspects of rapid microfluorometry for the study of enzyme reactions and transport mechanisms in single living cells. In: A.A. Thaer and M. Sernetz (Eds.), *Fluorescence Techniques in Cell Biology*. Springer-Verlag, New York, pp. 207–18, 1973.

Kohen, E., B. Thorell, C. Kohen, and J. M. Solmon: Studies on metabolic events in localized compartments of the living cell by rapid microspectro-fluorometry. *Adv. Biol. Med. Phys.* 15, 271–97, 1974.

Lotka, A.J.: *Elements of Physical Biology*. Williams and Wilkins, Baltimore, 1924 (reprinted as Elements of Mathematical Biology, Dover, New York, 1956).

Mavrovouniotis, M.L., G. Stephanopoulos, and G. Stephanopoulos: Computer-aided synthesis of biochemical pathways. *Biotechn. Bioeng.* 36, 1119–32, 1990.

Michaelis, L., and M.L. Menten: Die Kinetik der Invertinwirkung. *Biochem. Zeitschrift* 49, 333–69, 1913.

Nielsen, J.: Microbial process kinetics. In: C. Ratledge and B. Kristiansen (Eds.), *Basic Biotechnology* (Chapter 6). Cambridge University Press, Cambridge, U.K., 2001.

Ogata, K.: *Modern Control Engineering* (2nd ed.). Englewood Cliffs, NJ: Prentice Hall, 1990.

Papoutsakis, E.T.: Equations and calculations for fermentations of butyric acid bacteria *Biotechnol. Bioeng.* 26, 174–87, 1984.

Papoutsakis, E.T., and C.L. Meyer: Equations and calculations of product yields and preferred pathways for butanediol and mixed-acid fermentations. *Biotechnol. Bioeng.* 27, 50–66, 1985.

Peschel, M., and W. Mende: *The Predator-Prey Model: Do We Live in a Volterra World?* Akademie-Verlag, Berlin, 1986.

Planck, M.: *Treatise on Thermodynamics* (3rd ed.). Dover, New York, 1945 (translated from the 7th German edition with the author's sanction by Alexander Ogg).

Pons, A., C.G. Dussap, C. Pequignot, and J.B. Gros: Metabolic flux distribution in Cornybacterium melassecola ATCC 17965 for various carbon sources. *Biotechnol. Bioeng.* 51, 177–89, 1996.

Pramanik, J., and J.D. Keasling: Stoichiometric model of *Escherichia coli* metabolism: Incorporation of growth-rate dependent biomass composition and mechanistic energy requirements. *Biotechnol. Bioeng.* 56, 398–421, 1997.

Prigogine, I.: *Étude Thermodynamique des Processus Irreversibles*. Desoer, Liège, 1947 (*Introduction to the Thermodynamics of Irreversible Processes*, Thomas, Springfield, IL, 1955).

Reder, C.: Metabolic control theory: A structural approach. *J. Theor. Biol.* 135, 175–201, 1988.

Reiner, J.M.: *Behavior of Enzyme Systems*. Van Nostrand Reinhold, New York, 1969.

Ricard, J.: *Biological Complexity and the Dynamics of Life Processes*. Elsevier, Amsterdam, 1999.

Richter. P.H., I. Procaccia, and J. Ross: Chemical instabilities. *Adv. Chem. Phys.* 43, 217–68, 1980.

Roberts, D.V.: *Enzyme Kinetics*. Cambridge University Press, Cambridge, U.K., 1977.

Sauro, H.M., J.R. Small, and D.A. Fell: Metabolic control and its analysis. Extensions to the theory and matrix methods. *Eur. J. Biochem.* 165, 215–22, 1987.

Savageau, M.A.: Biochemical systems analysis, I. Some mathematical properties of the rate law for the component enzymatic reactions. *J. Theor. Biol.* 25, 365–9, 1969a.

Savageau, M.A.: Biochemical systems analysis, II. The steady-state solutions for an n-pool system using a power-law approximation. *J. Theor. Biol.* 25, 370–9, 1969b.

Savageau, M.A.: Biochemical systems analysis, III. Dynamic solutions using a power-law approximation. *J. Theor. Biol.* 26, 215–26, 1970.

Savageau, M.A.: *Biochemical Systems Analysis. A Study of Function and Design in Molecular Biology*. Addison-Wesley, Reading, MA, 1976.

Savageau, M.A.: Growth of complex systems can be related to the properties of their underlying determinants. *Proc. Natl. Acad. Sci. USA* 76, 5413–17, 1979a.

Savageau, M.A.: Allometric morphogenesis of complex systems: Derivation of the basic equations from first principles. *Proc. Natl. Acad. Sci. USA* 76, 6023–5, 1979b.

Savageau, M.A.: Mathematics of organizationally complex systems. *Biomed. Biochim. Acta* 44, 839–44, 1985.

Savageau, M.A.: Biochemical systems theory: Operational differences among variant representations and their significance. *J. Theor. Biol.* 151, 509–30, 1991.

Savageau, M.A.: Critique of the enzymologist's test tube. In: E.E. Bittar (Ed.), *Fundamentals of Medical Cell Biology* (Vol. 3A, pp. 45–108). JAI Press, Greenwich, CT, 1992a.

Savageau, M.A.: Dominance according to metabolic control analysis: Major achievement or house of cards? *J. Theor. Biol.* 154, 131–6, 1992b.

Savageau, M.A.: Design principles for elementary gene circuits: Elements, methods, and examples. *Chaos II*, 142–59, 2001.

Savageau, M.A., E.O. Voit, and D.H. Irvine: Biochemical systems theory and metabolic control theory. I. Fundamental similarities and differences. *Math. Biosci.* 86, 127–45, 1987a.

Savageau, M.A.: Michaelis-Menten mechanism reconsidered: Implications of fractal kinetics. *J. Theor. Biol.* 176, 115–24, 1995a.

Savageau, M.A.: Enzyme kinetics *in vitro* and *in vivo*: Michaelis-Menten revisited. In: E.E. Bittar (Ed.), *Principles of Medical Biology* (Vol. 4, pp. 93–146). JAI Press, Greenwich, CT, 1995b.

Savageau, M.A., and A. Sorribas: Constraints among molecular and systemic properties: Implications for physiological genetics. *J. Theor. Biol.* 141, 93–115, 1989.

Savageau, M.A., and E.O. Voit: Recasting nonlinear differential equations as S-systems: A canonical form. *Mathem. Biosci.* 87, 83–115, 1987.

Savageau, M.A., E.O. Voit, and D.H. Irvine: Biochemical systems theory and metabolic control theory. II. The role of summation and connectivity relationships. *Math. Biosci.* 86, 147–69, 1987b.

Savinell, J.M.: Analysis of Stoichiometry in Metabolic Networks. Ph.D. dissertation, University of Michigan, 1991.

Savinell, J.M., and B.Ø. Palsson: Network analysis of intermediary metabolism using linear optimization. I. Development of mathematical formalism. *J. Theor. Biol.* 154(4), 421–54, 1992a.

Savinell, J.M., and B.Ø. Palsson: Network analysis of intermediary metabolism using linear optimization. II. Interpretation of *hybridoma* cell metabolism. *J. Theor. Biol.* 154(4), 455–73, 1992b.

Savinell, J.M., and B.Ø. Palsson: Optimal selection of metabolic fluxes for *in vivo* measurement. I. Development of mathematical methods. *J. Theor. Biol.* 155(2), 201–14, 1992c.

Savinell, J.M., and B.Ø. Palsson: Optimal selection of metabolic fluxes for *in vivo* measurement. II. Application to *Escherichia coli* and *hybridoma* cell metabolism. *J. Theor. Biol.* 155(2), 215–42, 1992d.

Schauer, M., and R. Heinrich: Quasi-steady-state approximation in the mathematical modeling of biochemical reaction networks. *Math. Biosci.* 65, 155–70, 1983.

Schilling, C.H., and B.Ø. Palsson: The underlying pathway structure of biochemical reaction networks. *Proc. Natl. Acad. Sci. USA* 95(8), 4193–8, 1998.

Schilling, C.H., and B.Ø. Palsson: Assessment of metabolic capabilities of *Haemophilus influenzae* Rd through a genome-scale pathway analysis. *J. Theor. Biol.* 203, 249–83, 2000.

Schilling, C.H., S. Schuster, B.Ø. Palsson, and R. Heinrich: Metabolic pathway analysis: Basic concepts and scientific applications in the post-genomic era. *Biotechnol. Prog.* 15(3), 296–303, 1999.

Schilling, C.H., D. Letscher, and B.Ø. Palsson: Theory for the systemic definition of metabolic pathways and their use in interpreting metabolic function from a pathway-oriented perspective. *J. Theor. Biol.* 203, 229–248, 2000.

Schnell, S., and P.K. Maini: Enzyme kinetics at high enzyme concentration. *Bull. Math. Biol.* 62, 483–99, 2000.

Schnell, S., and C. Mendoza: Closed-form solution for time-dependent enzyme kinetics. *J. Theor. Biol.* 187, 207–12, 1997.

Schulz, A.R.: *Enzyme Kinetics. From Diastase to Multi-enzyme Systems.* Cambridge University Press, Cambridge, U.K., 1994.

Segel, L.A.: On the validity of the steady state assumption of enzyme kinetics. *Bull. Math. Biol.* 50, 579–93, 1988.

Segel, L.A.: *Biological Kinetics.* Cambridge University Press, Cambridge, U.K., 1991.

Segel, L.A., and M. Slemrod: The quasi-steady-state assumption: A case study in perturbation. *SIAM Rev.* 31, 446–77, 1989.

Seressiotis, A., and J.E. Bailey: MPS: An artificially intelligent software system for the analysis and synthesis of metabolic pathways. *Biotechnol. Bioeng.* 31, 587–602, 1988.

Shiraishi, F., and M.A. Savageau: The tricarboxylic acid cycle in *Dictyostelium discoideum*. I. Formulation of alternative kinetic representations. *J. Biol. Chem.* 267, 22912–18, 1992a.

Shiraishi, F., and M.A. Savageau: The tricarboxylic acid cycle in *Dictyostelium discoideum*. II. Evaluation of model consistency and robustness. *J. Biol. Chem.* 267, 22919–25, 1992b.

Shiraishi, F., and M.A. Savageau: The tricarboxylic acid cycle in *Dictyostelium discoideum*. III. Analysis of steady state and dynamic behaviour. *J. Biol. Chem.* 267, 22926–33, 1992c.

Shiraishi, F., and M.A. Savageau: The tricarboxylic acid cycle in *Dictyostelium discoideum*. IV. Resolution of discrepancies between alterntative methods of analysis. *J. Biol. Chem.* 267, 22934–43, 1992d.

Shiraishi, F., and M. A. Savageau: The tricarboxylic acid cycle in *Dictyostelium discoideum*. V. Systemic effects of including protein turnover in the current model. *J. Biol. Chem.* 268, 16917–28, 1993.

Sols, A., and R. Marco: Concentrations of metabolites and binding sites. Implications in metabolic regulation. *Curr. Top. Cell. Reg.* 2, 227–73, 1970.

Sorribas, A., J. March, and E.O. Voit: Estimating age-related trends in cross-sectional studies using S-distributions. *Stat. Med.* 10(5), 697–713, 2000.

Stephanopoulos, G.N., A.A. Aristidou, and J. Nielsen: *Metabolic Engineering. Principles and Methodologies*. Academic Press, San Diego, CA, 1998.

Thornton, H.G.: On the development of a standardised Agar medium for counting soil bacteria, with especial regard to the repression of spreading colonies. *Ann. Appl. Biol.* IX, 241–74, 1922.

Tsai, S.P., and Y.H. Lee: Application of Gibbs' rule and a simple pathway method to microbial stoichiometry. *Biotechnol. Prog.* 4(2), 82–8, 1988.

Tyson, J.J.: *The Belousov-Zhabotinskii Reaction. Lecture Notes in Biomathematics* (Vol. 10). Springer Verlag, Berlin, 1976.

Vallino, J.J., and G. Stephanopoulos: Metabolic flux distribution in *Corynebacterium glutamicum* during growth and lysine overproduction. *Biotechnol. Bioeng.* 41, 633–46, 1993.

Varma, A., and B.Ø. Palsson: Parametric sensitivity of stoichiometric flux balance models applied to wild-type *Escherichia coli* metabolism. *Biotechnol. Bioeng.* 45, 69–79, 1995.

Varma, A., B.W. Boesch, and B.Ø. Palsson: Biochemical production capabilities of *Escherichia coli*. *Biotechnol. Bioeng.* 42, 59–73, 1993.

Varma, A., B.W. Boesch, and B.Ø. Palsson: Metabolic flux balancing: Basic concepts, scientific and practical use. *Bio/Technol.* 12, 994–8, 1994.

Voit, E.O. (Ed.): *Canonical Nonlinear Modeling. S-System Approach to Understanding Complexity*, (xi + 365 pp.). Van Nostrand Reinhold, New York, 1991.

Voit, E.O.: *Computational Analysis of Biochemical Systems. A Practical Guide for Biochemists and Molecular Biologists* (xii + 530 pp.). Cambridge University Press, Cambridge, U.K., 2000a.

Voit, E.O.: Canonical modeling: A review of concepts with emphasis on environmental health. *Environ. Health Perspect.* 108 (Suppl. 5): (*Mathematical Modeling in Environmental Health Studies*), 895–909, 2000b.

Voit, E.O., and M.A. Savageau: Accuracy of alternative representations for integrated biochemical systems. *Biochem.* 26, 6869–80, 1987.

Waller, K.V., and P.M. Mäkllä: Chemical reaction invariants and variants and their use in reactor modeling, simulation, and control. *Ind. Eng. Chem. Proc. Des. Dev.* 20, 1–11, 1981.

Westerhoff, H.V., and K. van Dam: *Thermodynamics and Control of Biological Free-Energy Transduction*. Elsevier, Amsterdam, 1987.

Woolfolk, C.A., and E.R. Stadtman: Regulation of glutamine synthetase. III. Cumulative feedback inhibition of glutamine synthetase from *Escherichia coli*. *Arch. Biochem. Biophys.* 118, 736–55, 1967.

Wright, B.E., M.H., Butler, and K.R. Albe: Systems analysis of the tricarboxylic acid cycle in *Dictyostelium discoideum*. I. The basis for model construction. *J. Biol. Chem.* 267, 3101–5, 1992.

Wright, E.M.: Solution of the equation $z \exp(z) = a$. *Proc. Roy. Soc. Edinburgh*, A 65, 193–203, 1959.

## CHAPTER TWO

Methods of Biochemical Systems Theory

The appeal of the power-law approximation for representing enzyme-catalyzed reactions has led to the development of an entire modeling framework that is known in the field as *Biochemical Systems Theory* (BST). Hundreds of theoretical treatises and applications concerned with BST have been published in the peer-reviewed literature, in books, abstracts, and presentations. It is not the purpose of this chapter to provide a detailed and comprehensive treatment, and we will only highlight topics that are of relevance for the remainder of the book. The reader interested in further details, hands-on examples, and exercises is referred to a recent, comprehensive textbook on the subject (Voit 2000).

The typical modeling process contains the following components:

- Model Design
- Dynamic Analysis
- Steady-State Analysis
- Evaluation and Interpretation of Results

That looks pretty straightforward and simple. In reality, however, the process is not as direct as it may seem, but cyclic in nature. The first model design is seldom all encompassing and optimal. This usually becomes clear from the dynamic and steady-state analyses of the model or during the evaluation and interpretation of results. Indicators of potential problems arise in the form of discrepancies between model and observations, lack of robustness, or predictions that may not even be testable at the present time, but are suspect for one reason or another.

Diagnosis of these flaws may require further analyses that pinpoint the problems as precisely as possible and ultimately suggest a change in model design. Such a change may be easy to implement, for instance, if it consists of refined parameter values, but it may also be more drastic, requiring alterations in the fundamental structure of the model. Within an effective modeling framework, discrepancies between model and reality provide very valuable clues and can be used to guide the redesign of the initial model. The redesigned model is subjected to the same analyses, as the former model and, if the redesign is successful, should exhibit improved performance. Still, it is to be expected that the redesigned model does not ameliorate all problems, and a third cycle of design, analysis, and reality check is initiated. The hope in this modeling process is that the cycling is finite and that the kth iteration (k small!) yields a model that captures the features of highest interest with sufficient accuracy and reliability, even though it may still not capture every aspect of the investigated phenomenon.

MODEL DESIGN

Paraphrasing W. Somerset Maugham, James Haefner (1996) described model design as follows: "There are three simple rules for creating a model. Unfortunately, nobody knows what they are." Clever and cynical at once, Haefner does have a point in that there are no strict, generally accepted guidelines for mapping reality onto a mathematical model.

Although rules are lacking, it is rather clear where the gaps in the process are, even if these gaps are hard to fill. The first concerns the abstraction of reality, the condensation of what is important, the identification of what may be omitted or ignored without too much loss in validity. The second gap is the translation of observed processes into mathematical terminology. As we discussed in the previous chapter, many mathematical functions may have very similar graphs, and the selection of the optimal model is an unsolved problem that requires a fairly deep understanding of the subject area as well as insight in the feasibility of specific mathematical and computational techniques. Finally, there is a lack of clear rules about the types of analyses to which a model should be subjected. This issue may not seem to be a matter of model design. However, the ultimate structure of a model determines to some degree what types of analyses are readily executed and which are cumbersome, difficult, or even impossible. For instance, it is very informative for the general robustness and validity of a model to compute its sensitivities with respect to changes in parameters or inputs. Yet, many published model analyses do not include this aspect, presumably because the model structure does not facilitate sensitivity analysis with reasonable effort (Shiraishi and Savageau 1992).

In comparison to most ad hoc approaches, the design strategy for a model within BST comes relatively close to a set of guidelines. It involves three major components: the identification of variables and processes and their interactions, the translation of the model constituents and interactions into mathematical functions and equations, and the identification of numerical values for all involved parameters, variables, and factors. The notable difference between BST and ad hoc models is that the translation into equations is a straightforward step.

One might think that the establishment of functions and equations is the most important among the three components, but it is often the first component of identifying model components and their interactions that is most influential. This identification step sets the boundaries for questions that can or cannot be asked and possibly answered. If important variables are missing from the model design or if interactions between variables are misrepresented, all later analyses are affected, and one may not even realize it. It is therefore crucial to dedicate sufficient effort to model design.

The identification of relevant variables is a matter of subject area knowledge and professional judgment. As an illustration, consider a pathway model of glycolysis. The standard biochemistry textbook is very likely to contain a scheme of this central pathway, and the task of identifying the important metabolites seems simple. Nonetheless, judgment is needed in two directions. First, is it really necessary to model all metabolites that are listed? The conversion between glucose 6-phosphate and fructose 6-phosphate is highly reversible and very fast. Could we not simply define a "hexose-6-phosphate pool?" Second, is it desirable or necessary to include more metabolites or branches than shown in the textbook? Glucose 6-phosphate can have different fates. Although the glycolytic branch clearly dominates the overall flux in magnitude and importance, some material is used for the production of glycogen and trehalose, and some eventually enters the pentose-phosphate cycle. The typical scheme of glycolysis does not always depict these branches, even though they obviously exist. Do we need to include them? The textbook organization of metabolism into pathways is useful for learning purposes, but in reality is rather arbitrary. How should we deal with this web of pathways?

The philosophy of "the more the merrier" is not the best strategy for model design. Every component requires the determination of associated parameter values, most of which are to some degree uncertain and variable, or even entirely unknown. The accumulation of assumptions that are necessitated by the lack of full information can lead to model results that are highly variable in themselves and impede the predictive power of the model. Furthermore, if the method of analysis is computer simulation, in which parameters are varied within their alleged ranges, the necessary number of simulations grows multiplicatively with every new parameter. The inclusion of a further variable in a model typically requires an additional equation. If the variable is not really important, the effort of constructing, parameterizing, and implementing the equation increases unnecessarily and slows down the analysis.

These brief arguments in favor and against additional variables may suffice to stress the importance of a judicial choice of variables, which has to be driven by arguments from the subject area and a clear definition of what the model is supposed to accomplish. It is noted that the identification of components is often the most contentious issue in discussions with subject area scientists, who do not like to see details ignored even though they are known to exist. In fact, subject area scientists very seldom challenge the mathematical analysis itself or doubt the mathematical results, but they readily take issue with the abstraction process and its simplifications.

Once all variables are identified, they should be categorized as *dependent* or *independent*. Dependent variables are affected by the dynamics of the model and usually change in value over the course of a mathematical experiment. A typical example is the concentration of ethanol produced in a fermentation process. Independent variables are not affected by the dynamics of the process. They are often – but not always – under the direct control of the experimenter. Two examples are substrates, which the experimenter keeps at constant levels throughout the experiment, and enzyme activities, which often do not change over the time period of an experiment, but are not directly controllable during the experiment. For streamlined bookkeeping, it is customary in BST to list the dependent variables first and the independent variables second; the number of dependent variables is typically called n, and the number of independent variables is called m.

Our terminology of dependent and independent variables is not the only choice. In chemistry, the variables are sometimes called *species* (Chevalier, Schreiber, and Ross 1993), which Clarke (1980) classifies as *internal* and *external*. Other authors have distinguished between *state* and *control* or *input* variables (Richter, Procaccia, and Ross 1980; Mayhan 1983; Chen 1984; Casti 1985), or between *controlled* and *manipulated* variables (Ogata 1990). Lübbert and Simutis (2001) talk about *manipulable* and *actuator* variables, whose alterations lead to changes in a *control* variable. Each terminology has its own advantages and drawbacks. Biochemical Systems Theory uses the categories *dependent* and *independent*, because independent variables are not necessarily outside the system ("external") or under the control of the experimenter, as discussed before. Confusion between an independent variable in this sense and the mathematical notion of *time* as the independent variable is seldom an issue.

The distinction between dependent and independent variables is very important for model design. Because independent variables are constant throughout any given experiment, they do not require formulation through a time-dependent function or differential equation, as dependent variables do. Although the distinction is crucial, the change in status from dependent to independent or vice versa is a typical occurrence throughout the iterative process of model design and redesign. For the first model of a biochemical pathway, one might start by defining some component X_5 as an independent variable, only to learn later that X_5 indeed changes over time. In a model refinement, X_5 is replaced with a dependent variable, which is then characterized by its own differential equation. If X_6 is originally dependent, but its dynamics turns out to be irrelevant, for instance, because it changes extremely slowly, it may be justifiable to eliminate its defining differential equation and to fix X_6 at an appropriate constant value.

Once all relevant variables are identified, it is necessary to establish their interactions. For a metabolic pathway, these interactions have typically one of two manifestations. Either *material is flowing* between two variables, or a variable affects a process by virtue of a *signal*. An example for the former case is the conversion between chemical species through an enzyme-catalyzed reaction or a transport step. The latter may be the inhibition of a pathway by its end product. In this case, the concentration of the end product is not directly affected by the inhibition it exerts. The end product just sends the signal to slow down production of an earlier step in the pathway. Eventually, the decreased flux will have an effect on the end product, but this effect is indirect. Similarly, if an enzyme catalyzes a reaction, one commonly assumes that there is no change in the total concentration of enzyme. The enzyme affects the process without contributing mass to it.

It is customary and very much recommended to construct a graphical diagram or *map* showing the interactions between variables and to distinguish graphically the flow of material from the modulating or signaling effects of a variable. The flow of

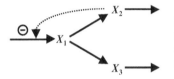

Figure 2.1. Simple branched pathway, in which X_1 is used for the production of X_2 and X_3. X_2 exerts an inhibitory effect on the synthesis of X_1. Note the difference between signal (dashed) and material flow (heavy) arrows.

material is typically represented with solid, heavy arrows, whereas signals are represented with light and/or dashed arrows. Figure 2.1 shows a simple map. Chapters 3 and 7 contain larger, more realistic maps. Also, numerous maps of different degrees of complexity are found in Voit (2000).

Once the construction of the map is completed, it is advisable to double-check the correctness of all interactions. A simple rule is that heavy (material flow) arrows always point from one variable to another (e.g., fluxes from X_1 to X_2 and from X_1 to X_3 in Figure 2.1), with the exception of influxes into the system, which may not have an explicit source (e.g., the influx to X_1 in Figure 2.1), or effluxes out of the system, which may not have an explicit target (e.g., the effluxes from X_2 and X_3 in Figure 2.1). In contrast to material flow arrows, light or dashed (signal) arrows always originate at a variable and point to a heavy (material flow) arrow (e.g., the inhibition of the production of X_1 by X_2 in Figure 2.1). The distinction is important because it affects the construction of the model equations from the map.

DESIGN OF EQUATIONS IN BST

The translation of a map into a symbolic S-system or GMA model is a straightforward process. In fact, some attempts have been made to automate this process with a computer algorithm (Okamoto et al. 1997; http://helios.brs.kyushu-u.ac.jp/~bestkit/). The rules for setting up the equations are very similar for the two types of systems. We begin with the S-system formulation.

Consider any dependent variable X_i, such as X_2 in Figure 2.1. If the map is constructed correctly, none, one, or several heavy (material flow) arrows should point to the variable, and none, one, or several heavy arrows should point away from the variable. The arrows pointing toward X_i represent the various flows of material contributing to increases in X_i. To obtain the S-system form, one lists all variables that *directly* affect any of these flows. It is immaterial for this step whether a variable plays the role of a substrate or affects the production of X_i through a signal. It is also immaterial whether a variable affects the production of X_i in one or several different ways. At this point, it is sufficient simply to list all variables that have a direct effect on the production of X_i. These variables enter the describing power-law term of X_i, each with its own kinetic order. The numerical value of the kinetic order is not known at this point, but we can certainly represent the kinetic order with the symbol g_{ij}. The first index of the kinetic order refers to the dependent variable being produced (X_i) and the second to the dependent or independent variable (X_j) affecting the production, no matter in what role. Once all variables are listed, their product receives a rate constant α_i, where the index again coincides with the index of the dependent variable under consideration (X_i).

Figure 2.2. Branched pathway as in Figure 2.1, except that the substrate for the production of X_1 is explicitly modeled by the independent variable X_4.

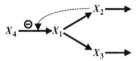

The degradation of X_i is handled in exactly the same fashion. All variables that *directly* affect any of the processes of degradation are collected in a product, along with symbolic kinetic orders. For the degradation terms, one customarily uses the symbols h_{ij} for kinetic orders and β_i for rate constants.

Subtracting the degradation term from the production term, the complete S-system equation for X_i reads

$$\dot{X}_i = \alpha_i \prod_{j=1}^{n+m} X_j^{g_{ij}} - \beta_i \prod_{j=1}^{n+m} X_j^{h_{ij}}. \tag{2.1}$$

This design of equations is repeated for all dependent variables $X_i (i = 1, 2, \ldots, n)$. Independent variables $(X_j; j = n+1, n+2, \ldots, n+m)$ are not modeled with equations, because they are by definition constant throughout any given experiment.

To complete the formulation of model equations, one defines initial values for all dependent variables. These values determine the state at which the system resides at the beginning of the experiment and may be written as $X_i(0)(i = 1, 2, \ldots, n)$.

As an example, consider the branched pathway with inhibition shown in Figure 2.2. It is very similar to that in Figure 2.1, but explicitly models the initial substrate with the independent variables X_4. The system thus has one independent and three dependent variables. The production of X_1 is affected by the independent input X_4 and also by the inhibition exerted by X_2. These two variables enter the production term. The degradation of X_1 is affected only by X_1 itself. Even though material flows toward X_2 and X_3, these two variables are only the passive recipients and have no direct effect on the rate of degradation of X_1. The symbolic S-system formulation of the map is therefore

$$
\begin{aligned}
\dot{X}_1 &= \alpha_1 X_2^{g_{12}} X_4^{g_{14}} - \beta_1 X_1^{h_{11}} & X_1(0) &= \text{initial value of } X_1 \\
\dot{X}_2 &= \alpha_2 X_1^{g_{21}} - \beta_2 X_2^{h_{22}} & X_2(0) &= \text{initial value of } X_2 \\
\dot{X}_3 &= \alpha_3 X_1^{g_{31}} - \beta_3 X_3^{h_{33}} & X_3(0) &= \text{initial value of } X_3 \\
X_4 &= \text{constant.}
\end{aligned} \tag{2.2}
$$

The rules for setting up the symbolic GMA model are essentially the same. The notable exception is that each process of production or degradation is modeled individually with a product of power-law functions and a rate constant. As in the S-system formulation, it does not matter whether it is a dependent or independent variable that affects the process, as long as the influence is direct; both types of variables are treated in the same fashion.

The consequence of modeling every process individually is that a GMA equation contains as many terms on the right-hand side as processes affecting the variable under consideration. In contrast to an S-system equation, a GMA equation thus may have any number of power-law terms.

The kinetic orders of GMA systems are customarily called f_{ijk}. The first index refers to the dependent variable under consideration, the second to the dependent or independent variable affecting the process, no matter in what form, and the third index counts the processes affecting the dynamics of X_i. The rate constants are called γ_{ik}. They are by definition positive. If the kth term represents a production process, the rate constant receives a plus sign, and if it represents a degradation process, the rate constant receives a minus sign.

Using the same example as before (Figure 2.2), the production of X_1 and the degradation processes of X_2 and X_3 are modeled in the GMA representation exactly as they were in the S-system. The only difference lies in the degradation of X_1 and the concomitant conversion into X_2 and X_3. These steps are represented as two terms in the first equation of the GMA model, one for each process. Retaining the S-system terminology for fluxes that are identical between the two forms, the result is

$$\dot{X}_1 = \alpha_1 X_2^{g_{12}} X_4^{g_{14}} - \gamma_1 X_1^{f_{111}} - \gamma_2 X_1^{f_{112}}$$

$$\dot{X}_2 = \gamma_1 X_1^{f_{111}} - \beta_2 X_2^{h_{22}}$$

$$\dot{X}_3 = \gamma_2 X_1^{f_{112}} - \beta_3 X_3^{h_{33}}$$

$$X_4 = \text{constant.}$$

(2.3)

In fact, the new terms are already part of the S-system representation, namely, $\gamma_1 X_1^{f_{111}} = \alpha_2 X_1^{g_{21}}$ and $\gamma_2 X_1^{f_{112}} = \alpha_3 X_1^{g_{31}}$. In spite of these equivalences, the GMA and S-system models do differ mathematically in the way the degradation of X_1 is represented.

Differences between the two forms only exist at branch points, where either two or more independent processes converge or diverge. All other steps are identical, even if they are highly modulated by activators or inhibitors. In particular, models of linear chains of reactions or of cascades are identical in the GMA and S-system forms. The literature on the mutual advantages and disadvantages of GMA and S-system representations is quite extensive.

PARAMETER ESTIMATION

Models of biochemical systems in BST contain only two types of parameters, or maybe three if one counts the initial values at which the system resides at the beginning of the mathematical experiment. The two types of true parameters are kinetic orders and rate constants. In most cases, kinetic orders are estimated first.

Estimation Based on Fluxes

Likely ranges. Before we review some standard methods for identifying values for the kinetic orders, it is useful to explore the range within which such values typically fall. Recall that the kinetic order g_{ij} in an S-system represents the direct effect that variable X_j has on the production of X_i. If the kinetic order is zero, the effect is nil, a positive kinetic order reflects an increasing effect, and a negative kinetic order represents some type of inhibition. Suppose variable X_j has a value of 10 and the kinetic order g_{ij} is 1. A 10% change in the value of X_j changes the production flux V_i^+ of X_i from $\alpha_i X_1^{g_{i1}} X_2^{g_{i2}} \ldots 10^1 \ldots X_{n+m}^{g_{i,n+m}}$ to $\alpha_i X_1^{g_{i1}} X_2^{g_{i2}} \ldots 11^1 \ldots X_{n+m}^{g_{i,n+m}}$. Thus, the flux increases by a factor of 1.1, which again corresponds to 10%. If the kinetic order were 50, the flux would increase to over hundred-fold its original value ($1.1^{50} = 117.39$). Such a strong response to a 10% change would be very unusual. Back-of-the-envelope computations of this nature allow us to establish ranges that kinetic orders might reasonably assume. Combining them with experience from systems that have actually been implemented and analyzed, we suggest that kinetic orders representing biochemical reactions or transport steps are very often in the range between 0 and 1. Inhibitory effects typically call for kinetic orders between 0 and -0.5.

The situation is different in signaling pathways, which have the purpose of strongly amplifying signals. In these cases, the kinetic orders may be much higher. For instance, Segall et al. (1986) describe how a rather weak external signal can trigger a change in the direction of a moving bacterium. If the effect of the signal on the cellular motor were represented in BST, the corresponding kinetic order would have a very high value at the order of 50.

Based on work of Kopelman (1986, 1991), Savageau (1993, 1995a,b) proposed another exception to the relatively small magnitudes of kinetic orders that we discussed previously. He demonstrated that kinetic orders might reach much higher values if enzymatic reactions do not occur in well-mixed, homogeneous media, but at surfaces or within channels. These caveats must be recognized when one put limits on possible values for kinetic orders.

Rate constants are more difficult to box in. First, rates of processes indeed can differ by several orders of magnitude (Clarke 1980), and, secondly, the rate constants depend on the units used for metabolites and time. Clearly, if a rate constant reflects turnover per time unit, then a change from minutes to seconds or from millimoles to micromoles alters the rate constant correspondingly.

Experience with models of biochemical systems has shown that kinetic orders are very important, but that robust models are not much affected by small inaccuracies in their values. This implies that rounding a kinetic order to two significant digits is usually more than sufficient. Again, there are exceptions. Nonlinear systems can exhibit bifurcations in the parameter space, where the system behavior abruptly and qualitatively changes if a parameter crosses some threshold. Close to such a bifurcation point, even the smallest inaccuracies can very significantly alter the predictions of the model. We will present an example later in this chapter. However, most systems in reality are quite tolerant, and small alterations in parameter values are rather ineffectual. Methods of sensitivity analysis can alert us of critical parameters.

If the system is robust, even relatively low accuracy in the parameter values is sufficient to explore typical responses to changes in inputs or slight alterations in parameter values. In fact, supposing that the modeled pathway is robust, one may initiate the analysis with an order-of-magnitude model, in which all kinetic orders are chosen within the brackets proposed previously and rate constants are chosen to reflect the relative magnitudes of fluxes among the system components. If the system is not too big, one might even run Monte-Carlo simulations in which numerous combinations of parameter values are tested. The values for each iteration of such a simulation study may be drawn randomly from within the most likely ranges or one may subdivide all ranges with a finite number of grid points and exhaust all combinations. Alves and Savageau (2000a–d) recently used such simulation methods to discover general design principles of biochemical pathways. Voit and Del Signore (2001) used Monte-Carlo simulations to study the effect of inaccuracies in pathway optimization. Hatzimanikatis, Floudas, and Bailey (1996a,b) used grid point analysis for kinetic orders in an optimization of pathway structure.

Direct estimation from flux data. The ideal situation for estimating kinetic orders is a series of flux measurements in vivo upon variation of one contributing variable at a time. For instance, if the effect of X_j on flux V_i^+ is to be determined, it would be ideal to keep all variables but X_j at their nominal (typical) values, and slightly to vary X_j about its nominal value (Savageau 1976). The flux V_i^+ becomes a univariate function of X_j, and because of the power-law structure of all fluxes in BST, this function is linear when plotted as the logarithm of V_i^+ against the logarithm of X_j. The slope of the function is equivalent to the kinetic order g_{ij}. It may be obtained by inspection of the plot or with formal linear regression. Examples of this type of estimation are found in Curto et al. (1998) (see also Figures 1.8 and 3.3).

Estimation from alleged rate functions. In very many cases, the literature presents kinetic information within the conceptual framework of Michaelis–Menten mechanisms. It is immaterial for the present discussion whether such rate laws are sufficiently close to the truth. If the only information available is given in this form, we better make the most of it. The corresponding power-law representation is computed via partial differentiation in logarithmic space, as demonstrated in Chapter 1. At an operating point of choice, the rate law and its power-law representation are exactly equivalent, and close to this point, the two are very similar. Thus, if the alleged rate law is presented as the function $R(X_1, \ldots, X_{n+m})$ and the corresponding power-law representation is V_i^+, the kinetic order g_{ij} is straightforwardly computed as

$$g_{ij} = \frac{\partial R}{\partial X_j} \cdot \frac{X_j}{R}, \tag{2.4}$$

which is computed and evaluated at the chosen operating point. In practical applications of BST, the majority of kinetic orders have been estimated in this fashion. Examples are given throughout the literature and in Chapters 1, 3, and 7.

Estimation of rate constants. Once all kinetic orders in a flux term have been estimated, the rate constant is computed based on the equivalence of the power-law representation and the approximated rate function at the operating point of choice. Thus, one computes straightforwardly

$$\alpha_i = R(X_1, \ldots, X_{n+m}) X_1^{-g_{i1}} X_2^{-g_{i2}} \ldots X_{n+m}^{-g_{i,n+m}}, \tag{2.5}$$

which again is evaluated at the operating point.

GMA- and S-systems. The methods described previously are equally valid for GMA- and S-systems. The difference between the two modeling strategies rests with the definition of influxes and effluxes for each variable or pool. In the GMA form, each influx is treated individually, which results in a product of power-law functions for every influx and efflux. By contrast, the S-system form aggregates all influxes into one process, which is subsequently represented as one product of power-law functions, as shown previously. Similarly, the S-system form calls for aggregation of all effluxes, before any approximation is executed. These differences and their implications were discussed before.

For purposes of parameter estimation, it is often useful to construct a GMA model first and, if deemed beneficial, to add the aggregation step toward the corresponding S-system model. The aggregation is purely a matter of simple algebra and can be executed in symbolic programs such as Mathematica and to some degree in the BST-freeware PLAS (Ferreira 2000). The advantage of estimating the GMA-system first is that this form is sometimes closer to biochemical intuition and that information from the literature is more often given for individual fluxes rather than overall influxes and effluxes. The aggregation step from the GMA to the S-system form is again a matter of partial differentiation, with the special simplification that the approximated function always consists of a sum of products of power-law functions. For instance, if the ith equation of the GMA-system contains three terms with positive sign,

$$V_{i1} = \gamma_{i1} \prod_{j=1}^{n+m} X_j^{f_{ij1}}, V_{i2} = \gamma_{i2} \prod_{j=1}^{n+m} X_j^{f_{ij2}}, \quad \text{and} \quad V_{i3} = \gamma_{i3} \prod_{j=1}^{n+m} X_j^{f_{ij3}},$$

the kinetic orders of the corresponding S-system equation term are computed as

$$g_{ij} = \frac{\partial (V_{i1} + V_{i2} + V_{i3})}{\partial X_j} \cdot \frac{X_j}{V_{i1} + V_{i2} + V_{i3}}. \tag{2.6}$$

As always, this expression is evaluated at the operating point of choice. Because of the power-law structure of all terms, it is easy to intuit that the kinetic order g_{ij} is a weighted sum of the kinetic orders f_{ijk} and that the weights are the magnitudes of fluxes corresponding with each term of the GMA representation:

$$g_{ij} = \frac{V_{i1} f_{ij1} + V_{i2} f_{ij2} + V_{i3} f_{ij2}}{V_{i1} + V_{i2} + V_{i3}}. \tag{2.7}$$

The rate constant is computed again from the equivalence of the GMA- and S-system

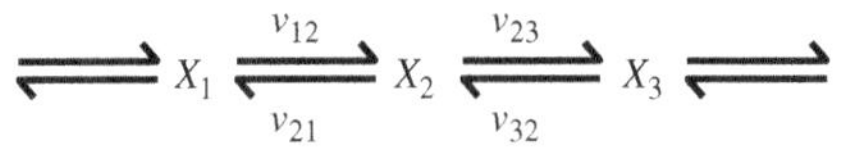

$$\rightleftharpoons X_1 \underset{v_{21}}{\overset{v_{12}}{\rightleftharpoons}} X_2 \underset{v_{32}}{\overset{v_{23}}{\rightleftharpoons}} X_3 \rightleftharpoons$$

Figure 2.3. Map of a generic reversible pathway.

terms at the operating point:

$$\alpha_i = (V_{i1} + V_{i2} + V_{i3}) \cdot \prod_{j=1}^{n+m} X_j^{-g_{ij}}. \tag{2.8}$$

Reversible reactions. The pathway illustrated in Figure 2.3 permits two options for aggregating fluxes. Focusing on X_2, one could consider its production flux as the net influx from X_1 and the degradation flux as the net efflux toward X_3. Thus, the appropriate formulation for the equation of X_2 would be

$$\dot{X}_2 = (v_{12} - v_{21}) - (v_{23} - v_{32}). \tag{2.9}$$

However, there is an alternative. X_2 receives material from X_1 and from X_3 and it loses material used for the production of X_1 and X_3. Thus, one could legitimately formulate the change in X_2 as

$$\dot{X}_2 = (v_{12} + v_{32}) - (v_{21} + v_{23}). \tag{2.10}$$

Equations (2.9) and (2.10) are mathematically equivalent, but suggest different power-law representations. In the first case, $V_2^+ = (v_{12} - v_{21})$ is a function of X_1 and X_2 and $V_2^- = (v_{23} - v_{32})$ is a function of X_2 and X_3. In the second case, both fluxes $V_2^+ = (v_{12} + v_{32})$ and $V_2^- = (v_{21} + v_{23})$ are functions of all three variables. The differences in the numbers of variables may suggest different degrees of freedom. However, these dissipate because of constraints among the parameters.

If the reaction occurs far from the thermodynamic equilibrium, the reverse fluxes become insignificant, and the so-called *irreversible strategy* of Eq. (2.9) is the choice of convenience. However, if the reaction is close to the thermodynamic equilibrium, the *reversible strategy* (Eq. 2.10) is superior with respect to a number of criteria, including accuracy of representation and redirection of flux (Sorribas and Savageau 1989).

Haldane (1930) studied a reversible reaction between X_i and X_j with forward and reverse rates v_F and v_R. He suggested formulating the net rate v_{net} as

$$v_{net} = \frac{(V_{max(F)}/K_{Mi}) \cdot (X_i - X_j/K_{eq})}{1 + X_i/K_{Mi} + X_j/K_{Mj}}, \tag{2.11}$$

where $V_{max(F)}$ is the maximum velocity of the forward reaction, K_{Mi} and K_{Mj} are the Michaelis constants with respect to X_i and X_j, and K_{eq} is the equilibrium constant of the reaction. If the mass action ratio is denoted as Γ, which is defined as

$$\Gamma = \frac{[\text{Reaction Product}]}{[\text{Reaction Substrate}]},$$

the maximum velocity of the reverse reaction as $V_{max(R)}$, and the net flux as J, the kinetic orders for the two reactions in the corresponding power-law representation may be computed as

$$g_{ji} = \frac{1}{1 - \Gamma/K_{eq}} - \frac{v_F}{V_{\max(F)}} = \frac{1}{1 - \Gamma/K_{eq}}\left(1 - \frac{J}{V_{\max(F)}}\right),$$ (2.12)

$$g_{jj} = -\frac{\Gamma/K_{eq}}{1 - \Gamma/K_{eq}} - \frac{v_R}{V_{\max(R)}} = -\frac{\Gamma/K_{eq}}{1 - \Gamma/K_{eq}}\left(1 + \frac{J}{V_{\max(R)}}\right)$$ (2.13)

(Groen et al. 1982; Voit 2000). If the reaction occurs far from equilibrium or if the equilibrium constant is very large, the ratio Γ/K_{eq} is close to zero, and the expressions for g_{jj} and g_{ji} simplify to

$$g_{ji} \approx 1 - \frac{v_F}{V_{\max(F)}},$$ (2.14)

$$g_{jj} = -\frac{v_R}{V_{\max(R)}}.$$ (2.15)

There are a few other methods for estimating kinetic orders from fluxes. Because they are not used all that much, we do not mention them here and refer the interested reader to Voit (2000).

Estimation from Dynamical Data

All previous methods were based on representations of individual fluxes. An alternative, or sometimes a complement, is the estimation of parameter values from dynamical data. These data consist of series of measurements of all dependent variables over a period of time. In principle, the estimation in this case is a straightforward nonlinear regression problem, in which one starts with a trial solution, computes the differences between the solution and the observed data, and updates the solution with some nonlinear search algorithm. (Chapter 4 discusses such algorithms in more detail.) The problem in the context of parameter estimation is that these algorithms are everything but straightforward in practical applications. Even for moderately large systems, the logistic and numerical problem become so significant that this method of parameter estimation is simply unfeasible. A few applications have used this method, but they dealt with rather small models or made use of problem-specific simplifications (Voit and Savageau 1982; Torsella and Bin-Razali 1991; Berg, Voit, and White 1996; Sands and Voit 1996; Voit and Sands 1996).

It might be useful to develop estimation methods for power-law representations based on genetic algorithms, because the structure of these representations is always the same, which allows for efficient standardized representations, and because relatively low accuracy of the estimates is often sufficient. So far, only preliminary studies have pursued this avenue (Zhang, Voit, and Schwacke 1996; Okamoto et al. 1997; Tominaga et al. 1999).

Aware of problems with all methods of estimation from dynamic data, several authors have proposed hybrid methods that use local information of the dynamic response of a steady-state system following a perturbation. These methods are based on the fact that such a response is approximately linear if the perturbation is small. The mathematical reason is that the (uni- or multivariate) linear representation constitutes the first-order terms of a Taylor series expansion, which is sufficient in accuracy if the perturbations are small. There is a long history of studying the linearized system rather than the more complicated nonlinear system. The most prominent example may be the assessment of the local stability of a nonlinear system, which is very often fully characterized by its linearization; stability will be discussed in a later section of this chapter. For the present context, a few articles should be mentioned that specifically deal with the estimation of parameter values or the related topic of determining the structure of a pathway from observed time series data. Chevalier et al. (1993) presented a good review of such methods.

The starting point is a nonlinear dynamical system that is symbolically given in the general form

$$\dot{X}_i = V_i(X_1, \ldots, X_n). \tag{2.16}$$

If the variable X_k is slightly perturbed by the quantity δX_k, the response of the system in terms of deviations δX_i is approximately

$$\delta \dot{X}_i = \frac{\partial V_i}{\partial X_k} \delta X_k. \tag{2.17}$$

This expression has the structure of a linear differential equation that can be solved analytically in terms of exponential functions, from which kinetic parameters may be estimated (Savageau 1976, p. 33). Bar-Eli and Geiseler (1983) used small perturbations in a single variable to identify a chemical network in a continuous stirred tank reactor.

If many variables are perturbed, either one at a time or simultaneously as a "cocktail" (Díaz-Sierra, Lozano, and Fairén 1999), the measurements of slopes can be interpreted as partial derivatives, which constitute the elements of the Jacobian $\mathbf{J}$. The matrix equation corresponding to Eq. (2.17) then reads

$$\frac{d\delta \mathbf{X}}{dt} = \mathbf{J} \cdot \delta \mathbf{X}. \tag{2.18}$$

The matrix $\delta \mathbf{X}$ contains as elements the concentrations X_i, from which the corresponding steady-state values X_{iS} are subtracted: $\delta X_i = X_i - X_{iS}$. These differences between measured and steady-state values are assumed to be caused by experimentally induced changes in other variables X_j. Given that the magnitudes of these induced changes are known and assuming that the changes δX_i can be measured for sufficiently small perturbations, it is theoretically possible to estimate the elements of $\mathbf{J}$, which are the (infinitesimal) quantities $\partial V_i / \partial X_j$.

A variation on this theme is a perturbation in a single chemical species and the measurement of its concentration X_i^k at N time points ($k = 1, \ldots, N$). One may define differences between these concentrations and the corresponding steady-state values

X_{iS}, namely $\delta U_{ik} = X_i^k - X_{iS}$, and formulate the matrix equation

$$\frac{d\delta \mathbf{U}}{dt} = \mathbf{J} \cdot \delta \mathbf{U}, \tag{2.19}$$

from which one can estimate $\mathbf{J}$, at least in principle.

Another possibility for estimating $\mathbf{J}$ is the introduction of artificial, delayed feedback, which is possible in well-controlled systems such as a continuous-flow stirred-tank reactor (Chevalier et al. 1993). Once $\mathbf{J}$ is known, no matter from which types of experiments, it is possible to deduce the chemical network structure with different methods (Chevalier et al. 1993; Sorribas, Lozano, and Fairén 1998; Díaz-Sierra et al. 1999).

Sorribas et al. (1993) did a similar analysis in logarithmic space, where the partial derivatives are called *logarithmic gains*. (Logarithmic gains will be discussed in a later section of this chapter.) They realized that logarithmic gains alone are seldom sufficient to identify all parameters of a system but suggested using them to constrain the set of admissible parameter values. Díaz-Sierra and Fairén (2001) estimated the Jacobian for systems in power-law representation from time series measurements via linear regression. As did Sorribas et al. (1993), they recognized that the Jacobian is not sufficient for parameter estimation, but that it is useful in establishing valuable constraints on the admissible set of parameter values.

It seems that very few actual data have been analyzed using these methods based on measurement of the Jacobian. As Chevalier et al. (1993) and others have pointed out, the challenge with the use of the Jacobian is the accuracy with which its elements can be determined from experimental data. Strictly speaking, Eq. (2.19) is only true for infinitesimally small perturbations, even though analyses with artificial data have shown that it holds fairly well for perturbations of realistic magnitude. However, to obtain changes in metabolites that are accurately measurable, one might need much larger perturbations than justifiable, which casts doubts on the estimation.

A direct generalization from a mathematical point of view is the use not only of the (first-order) Jacobian but also of the (second-order) Hessian matrix. Chevalier et al. (1993) warn that inherent inaccuracies in measurements probably make the Hessian unfeasible in most practical applications.

As an alternative to measuring initial slopes, one may change the input to the system, allow the system to assume a new steady state, and study the relationships between input and metabolite variations. For small "concentration shift measurements" (Chevalier et al. 1993), the changes in metabolites are again approximated by linear functions, and this provides insight in the parameters of the system. The interested reader is referred to Kacser and Burns (1979), Fell (1992, 1997), Chevalier et al. (1993), and Voit (2000) for examples and discussion.

Beyond the numerical challenges associated with nonlinear regression, genetic algorithms and transient responses, parameter estimation from dynamical data has not received much attention because such data used to be scarce. However, it is to be expected that dynamic data will be more readily available in the near future. For instance, Almeida, Reis, and Carrondo (1995) and Neves et al. (1999) have shown that it is possible to record concentrations of metabolites in individual cells in time intervals of a few seconds. Furthermore, microarray experiments are currently still expensive and relatively inaccurate, but it is only a matter of time that they will become a part of the standard repertoire of biotechnology labs around the world. It is no problem in principle to use microarrays or DNA chips for establishing detailed temporal responses to a stimulus. In fact, some data of this type are already available on the web. An example is a large database of yeast responses to a variety of stresses (http://rana.lbl.gov/). The temporal expression patterns of genes over time clearly contain valuable information on the organism's response to the stimulus, and the only missing link in our context is a valid interpretation of this information in terms of estimating metabolic parameters. Rudimentary attempts in this direction are beginning to appear in the literature (Voit and Radivoyevitch 2000; Voit, in press).

DYNAMIC ANALYSIS

Once the model is constructed and all its parameters are assigned numerical values, the model is ready for analysis. The types of analyses can be categorized in different ways. We distinguish dynamic analyses and steady-state analyses, realizing that some important types of assessments have components of both. For instance, the analysis of local stability usually refers to one particular steady state at a time, but the very concept of this analysis is an evaluation of how the system responds to a small perturbation away from the steady state, which clearly has a dynamic component. Similarly, the analysis of structural stability identifies particular values of parameters, at which the behavior of the system changes qualitatively. The methods of such an analysis are mainly algebraic, focusing on critical thresholds in the parameter space, but the consequences of actually moving a parameter across a critical threshold are often seen most clearly in the dynamics of the system.

As a simplified categorization, we consider an analysis as dynamic if it focuses on temporal responses. We may distinguish two types of analyses, which again have some overlap. The first type studies in detail the temporal responses to a single change in a system feature. Such a change may be an alteration in input or in some parameter value. The second type is a simulation study, in which some system features are randomly changed within some ranges. The random changes evoke variations in system responses, and the collections of responses form statistical distributions that provide insight in the likely, worst, and best behaviors the system may be able to yield.

Alterations in System Characteristics

Imagine a real system such as an organism. It is subject to two types of alterations. The first is a change in its environment, to which the organism reacts. Typical examples include fluctuations in the amount of substrate or food, the availability of particular cofactors, or the external temperature. Under normal conditions, the organism reacts to environmental changes by turning on or off appropriate genes or maybe by shivering or sweating. The reactions do not change the internal structure of the organism and are easily reversed once the stimulus ends. The second change is an alteration in structure. A typical example is a (permanent) mutation. A gene is knocked out, and the organism must rewire its physiological mechanisms to cope with the new circumstances. An alteration of this type is often irreversible for the remainder of the organism's life.

Both types of alterations are readily implemented in a mathematical model. A change in input often corresponds to a change in initial conditions or independent variables, whereas a change in structure usually affects kinetic orders and rate constants. Again, there is some overlap in implementation between the two scenarios: A permanent change in an independent variable is mathematically equivalent to a numerical change in the corresponding rate constant(s), even though the interpretation of the two alterations is somewhat different. As an illustration, suppose there is a constant supply of substrate, which is modeled as an independent variable. A change in the constant value of this independent variable would be classified as an environmental change; it usually leads the system to assume a new steady state. Similarly, the activity of an enzyme is normally modeled as an independent variable. A mutation affecting the activity is clearly a structural change but also leads to an adjustment of the value of the independent variable. From a mathematical point of view it is more convenient simply to distinguish temporary from permanent changes.

The generic scenario for studying a temporary change is the following. A system is residing at a steady state, that is, material is flowing in and out, but the influxes and effluxes of all variables are in balance, so that the actual concentration or quantity of the variable is constant. At time point t_0, one or several of the inputs or of the dependent variables are instantly changed, and one studies how the system responds. An example is shown in Figure 2.4. It depicts (in the bottom left panel) a situation in which the system receives a unique bolus of input and the three dependent variables respond with some damped oscillations, before they return to their original steady states. A second example would be an instantaneous change in a dependent variable. In this case, it is immaterial why the variable changed, but it is of interest to study how the system absorbs the perturbation.

For a generic example of a permanent change, we may use the same example. In contrast to the previous scenario, the increase in the independent variable X_4 at time 1 is now irreversible. The graph of the response demonstrates the differences between temporary and permanent changes (Figure 2.4, bottom right panel). Both alterations evoke strong responses in the form of overshoots. However, the dynamic is different in extent and time scale. Furthermore, the metabolites in the first scenario return to their original steady state values of 1, while they approach a new steady state $(X_1 = X_2 = 9, X_3 = 3)$ in the second scenario.

A permanent change in one of the parameter values may have any number of consequences. The system may approach a new steady state, it may become unstable, enter a pattern of sustained oscillations, or it may "cease to exist," in a sense that one or more variables become zero.

Linear Pathway with Feedback Inhibition

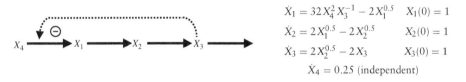

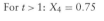

$$\dot{X}_1 = 32 X_4^2 X_3^{-1} - 2 X_1^{0.5} \quad X_1(0) = 1$$
$$\dot{X}_2 = 2 X_1^{0.5} - 2 X_2^{0.5} \quad X_2(0) = 1$$
$$\dot{X}_3 = 2 X_2^{0.5} - 2 X_3 \quad X_3(0) = 1$$
$$\dot{X}_4 = 0.25 \text{ (independent)}$$

At $t = 1$: $X_4 = 0.75$

At $t = 1.05$: $X_4 = 0.25$

For $t > 1$: $X_4 = 0.75$

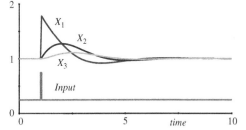

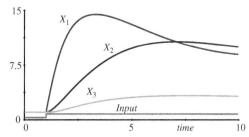

Figure 2.4. The same pathway is subjected to two types of alterations. One consists of a very brief, temporary increase in input (bottom left panel), whereas the other consists of a permanent change in input (bottom right panel). Note the difference in scale of the y axis.

Monte-Carlo Simulation

The technique of exploring possible and likely behaviors of systems through massive simulations arose during the development of the first hydrogen bombs (Rugen and Callahan 1996). The name "Monte-Carlo" was chosen, because the random generation of input reminded the developers of gambling in the famous casino at the French Riviera. The great strength of the simulation technique is that it often allows assessments that are impossible to execute with purely algebraic means. Although there is an art to successful Monte-Carlo simulation, the method is really a brute-force approach, in which the computational speed of an algorithm replaces the rigor and completeness that a purely algebraic approach would have, if it were feasible. A good introduction to Monte-Carlo simulation is available on a website of the United States Environmental Protection Agency (U.S. EPA 1997).

The typical situation for a Monte-Carlo simulation is a complex system that contains features or inputs that are variable or uncertain. Variability may be the reflection of true differences among individuals: not all organisms – even of the same strain – have exactly the same size, the same profile of enzyme activities, or the same concentrations of all metabolites. Of course, there is also variability in input, especially if organisms reside in the outside world. Uncertainties may stem from a lack of knowledge about parameters, variables, or input features. Measurements may be associated with some experimental error, data from the literature may provide only indirect information, or some kinetic or physiological information may only be available

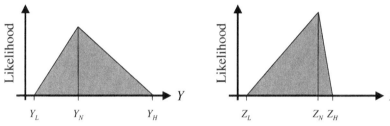

Figure 2.5. Triangular distributions of input variables Y and Z.

from a different organism. Uncertainties of this type are a fact of life. In principle, though not always in practice, they could be reduced or even eliminated through further experimentation, whereas true variability is intrinsic and usually cannot be changed.

For the logistics of implementing a Monte-Carlo simulation, the differentiation between uncertainty and variability is not all that important. What is significant is the identification of variable or uncertain parameters or features, a quantification of their possible ranges, and an estimation of the probabilities with which a given feature has a given value.

For simplicity of argument, suppose a system has ten dependent variables $X_1, \ldots,$ X_{10} and two inputs Y and Z with "normal" values Y_N and Z_N. With the normal inputs, the system has a steady state $(X_{1N}, \ldots, X_{10N})$. The inputs are subject to variability, and it is known that their values could possibly be as low as Y_L and Z_L or as high as Y_H and Z_H. An assumption of simplicity may be that their potential values form triangular distributions as shown in Figure 2.5.

A Monte-Carlo simulation with the system consists of hundreds or thousands of iterations. For each iteration, a random number generator draws a value of Y and a value of Z in such a fashion that all Ys and all Zs collectively form the distributions shown in Figure 2.5. With each randomly drawn pair of Y and Z, the differential equations describing the system are solved, so that the overall simulation result consists of a collection of system responses.

Suppose there are three distributed input variables, say $X_{n+1} - X_{n+3}$, and four dependent variables, say $X_1 - X_4$. The simulation with randomly drawn $X_{n+1}, X_{n+2},$ and X_{n+3} then yields four output distributions, one each of $X_1, X_2, X_3,$ and X_4 values (Figure 2.6). Typical questions asked with such an approach are: What are the most likely outcomes for $X_1, X_2, X_3,$ and X_4? What are the statistical features of the distributions of $X_1, X_2, X_3,$ and X_4? What are the extreme outcomes (best- or worst-case scenarios) for $X_1, X_2, X_3,$ and X_4? Are the ranges of the input variables amplified or attenuated in $X_1, X_2, X_3,$ and X_4? In addition to studying dependent variables at steady state, about any other feature of the system may be assessed. Examples are fluxes, peak concentrations, and transition times. We demonstrate the use of Monte-Carlo simulations in Chapter 6.

As mentioned previously, some of the output features of a Monte-Carlo simulation would be very difficult to obtain with any other means. At the same time, simulations lack the completeness and rigor of a purely mathematical analysis, and it may happen

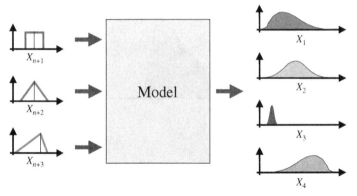

Figure 2.6. Schematic illustration of a Monte-Carlo simulation. Three independent (input) variables X_{n+1}, X_{n+2}, and X_{n+3}, are distributed as shown on the left. Their distributions lead to output, such as X_1, X_2, X_3, and X_4, that is also distributed. The output distributions may have a variety of shapes that depend on the model and are often difficult to predict without an actual execution of the simulation. (See www.cambridge.org/9780521177481 for color version.)

that some rare situation never comes up in a simulation, even though it is possible and potentially relevant.

STEADY-STATE ANALYSIS

The great advantage of representing a pathway with an S-system rather than some other nonlinear model – or even a GMA model – is the fact that its steady state can be expressed explicitly in the form of a system of linear equations (Savageau 1969). One important consequence is that the steady state itself, as well as derived features like sensitivities and logarithmic gains, are easily computed with the free-ware PLAS (Ferreira 2000), which was developed specifically for analyses of GMA and S-systems. Thus, even if the following theoretical derivation might look complicated, actual analyses are readily executed numerically; full examples are given in Chapters 3, 6, and 7.

At steady state, none of the variables changes in value, even though material is flowing through the system. Mathematically, this situation requires that all derivatives are zero:

$$0 = \alpha_i \prod_{j=1}^{n+m} X_j^{g_{ij}} - \beta_i \prod_{j=1}^{n+m} X_j^{h_{ij}} \qquad i = 1, \ldots, n. \tag{2.20}$$

Given that none of the rate constants α_i and β_i is zero, one may bring all terms with negative signs (the "β-terms") to the left-hand side. Taking logarithms, one obtains equations that are linear in the logarithms of the original variables. In detail, one obtains

$$\ln\left(\beta_i \prod_{j=1}^{n+m} X_j^{h_{ij}}\right) = \ln\left(\alpha_i \prod_{j=1}^{n+m} X_j^{g_{ij}}\right), \tag{2.21}$$

which becomes

$$\ln(\beta_i) + \sum_{j=1}^{n+m} [h_{ij} \ln(X_j)] = \ln(\alpha_i) + \sum_{j=1}^{n+m} [g_{ij} \ln(X_j)] \qquad i = 1, 2, \dots, n. \quad (2.22)$$

For streamlined notation, we define $y_i = \ln(X_i)$ and move all terms including y_is on the left side and all terms without y_i on the right side. The result is

$$\sum_{j=1}^{n+m} g_{ij} y_j - \sum_{j=1}^{n+m} h_{ij} y_j = \ln(\beta_i) - \ln(\alpha_i) \qquad i = 1, 2, \dots, n. \quad (2.23)$$

Finally, we rename $a_{ij} = g_{ij} - h_{ij}$ and $b_i = \ln(\beta_i) - \ln(\alpha_i) = \ln(\beta_i/\alpha_i)$ for all i and all j. This notation makes it evident that an S-system with n-dependent and m-independent variables has a steady state that is characterized by a set of n linear equations of the form

$$a_{i1} y_1 + a_{i2} y_2 + \cdots + a_{i,n} y_n + a_{i,n+1} y_{n+1} + \cdots + a_{i,n+m} y_{n+m} = b_i. \quad (2.24)$$

It is useful to separate the terms that contain dependent variables from those that do not. If $\mathbf{A}_D$ denotes the matrix containing coefficients $a_{ij} = g_{ij} - h_{ij}$ that are composed of kinetic orders associated with dependent variables ($j = 1, \dots, n$), and if $\mathbf{A}_I$ refers to the corresponding quantities $a_{ij} = g_{ij} - h_{ij}$ associated with the independent variables ($j = n+1, \dots, n+m$), the steady-state equations can be succinctly written in matrix form as

$$\mathbf{A}_D \cdot \vec{y}_D + \mathbf{A}_I \cdot \vec{y}_I = \vec{b}. \quad (2.25)$$

The vectors $\vec{y}_D$ and $\vec{y}_I$ contain the (logarithms of the) dependent and independent variables, respectively, and $\vec{b}$ consists of the logarithms of the ratios of the rate constants: $b_j = \ln(\beta_j/\alpha_j)$ (Savageau 1969).

Linear algebra allows us to solve for the dependent variables via multiplication with the inverse of the matrix $\mathbf{A}_D$, which exists if the system has a non-zero steady-state point (Savageau 1976). The result is

$$\vec{y}_D = \mathbf{A}_D^{-1} \cdot \vec{b} - \mathbf{A}_D^{-1} \cdot \mathbf{A}_I \cdot \vec{y}_I = \mathbf{M} \cdot \vec{b} + \mathbf{L} \cdot \vec{y}_I. \quad (2.26)$$

The matrix $\mathbf{M} = \mathbf{A}_D^{-1}$ contains as elements the *sensitivities* of the system with respect to changes in rate constants, whereas the product $\mathbf{L} = -\mathbf{A}_D^{-1} \cdot \mathbf{A}_I$ contains as elements the logarithmic gains $L_{j,n+k} = L(X_j, X_{n+k})$ (Savageau 1971a,b, 1972). Each sensitivity quantifies the relative change in a dependent variable X_i that is evoked by a relative change in a rate constant, such as α_j. Similarly, each logarithmic gain $L(X_j, X_{n+k})$ quantifies the relative change in a dependent variable X_j that is evoked by a relative change in the independent variable X_{n+k}. Sensitivities and logarithmic gains connect the local properties of a pathway to its overall systemic responses (Savageau 1971b, 1976). It is useful to study these local properties in some more detail, because they are valuable diagnostic tools if something with the model is wrong.

Strictly speaking, a sensitivity is an infinitesimal quantity, which is mathematically expressed as

$$S(X_i, \alpha_j) = \frac{\partial \ln(X_i)}{\partial \ln(\alpha_j)} = \frac{\partial y_i}{\partial \ln(\alpha_j)}. \tag{2.27}$$

To see its relationship with the components of the steady-state Eq. (2.26) most clearly, it is useful to study a system without independent variables. Not much changes by the presence of independent variables, but the derivation is less cluttered. The steady-state matrix equation for a system without independent variables is

$$\mathbf{A} \cdot \vec{y} = \vec{b}, \tag{2.28}$$

where $\mathbf{A} = \mathbf{A}_D$. According to Cramer's rule, each dependent variable y_i can be expressed as

$$y_i = \frac{\det(\mathbf{A}_i)}{\det(\mathbf{A})}, \tag{2.29}$$

where $\mathbf{A}_i$ is obtained from $\mathbf{A}$ by substituting the ith column with the vector $\vec{b}$ (Lang 1968). The rate constants α_j and β_j appear on the right-hand side as contributors to the parameter b_j, which by definition equals $b_j = \ln(\beta_j/\alpha_j) = \ln(\beta_j) - \ln(\alpha_j)$. Replacement of b_j with this expression yields

$$y_i = \frac{\begin{vmatrix} a_{11} & a_{12} & a_{13} & \cdots & b_1 & \cdots & a_{1n} \\ a_{21} & a_{22} & a_{23} & \cdots & b_2 & \cdots & a_{2n} \\ \cdot & \cdot & \cdot & \cdots & \cdot & \cdots & \cdot \\ a_{j1} & a_{j2} & a_{j3} & \cdots & \ln(\beta_j) - \ln(\alpha_j) & \cdots & a_{jn} \\ \cdot & \cdot & \cdot & \cdots & \cdot & \cdots & \cdot \\ \cdot & \cdot & \cdot & \cdots & \cdot & \cdots & \cdot \\ a_{n1} & a_{n2} & a_{n3} & \cdots & b_n & \cdots & a_{nn} \end{vmatrix}}{|\mathbf{A}|} \tag{2.30}$$

Without actually executing the operation, suppose one would expand the determinant $|\mathbf{A}_i|$ in the numerator. The result would be complicated, but the expression $\ln(\beta_j) - \ln(\alpha_j)$ would always remain intact. Each term in the end would either contain the entire term $\ln(\beta_j) - \ln(\alpha_j)$ as a multiplicative factor or not at all. Because all computations in the expansion of the determinant are linear operations, the symbolical result is of the form

$$y_i = C_1 + (\ln(\beta_j) - \ln(\alpha_j)) \cdot C_2, \tag{2.31}$$

where C_1 and C_2 are sums of products of the coefficients a_{ij} and of solution coefficients $b_i \neq b_j$, which are all independent of $\ln(\beta_j)$ and $\ln(\alpha_j)$. Differentiation of y_i with respect to $\ln(\alpha_j)$ or $\ln(\beta_j)$ thus identifies the sensitivity of X_i with respect to α_j or β_j as

$$S(X_i, \alpha_j) = -C_2$$
$$S(X_i, \beta_j) = +C_2. \tag{2.32}$$

The sensitivities of a variable with respect to the rate constants α_j and β_j are always the same in magnitude but opposite in sign. They are functions of the entire system, that is, of potentially all kinetic orders and all rate constants. This is important to remember, because if one is interested in the effect of changes in the rate of one particular process, it is not sufficient to focus on just this process or on processes that are "close by" in the pathway, but the entire system must be included in the analysis.

A word of caution is in order. The sensitivities computed here assume independence between the rate constants. If the change in one rate constant is accompanied by a corresponding change in another rate constant, the dependency must be accounted for. This is not difficult in principle, but requires additional algebra (Voit 2000).

Effects of alterations in independent variables are quantified by logarithmic gains. In mathematical terms, "log gains" of metabolites are defined as

$$L(X_j, X_k) = \frac{\partial \ln(X_j)}{\partial \ln(X_k)} = \frac{\partial y_j}{\partial y_k}. \tag{2.33}$$

Like sensitivities, they are readily derived from the steady-state Eq. (2.26). Because alterations in independent variables do not affect the kinetic orders and rate constants, the term $\mathbf{M} \cdot \vec{b}$ and the matrix $\mathbf{L}$ are unchanged, and all effects are limited to the vectors $\vec{y}_D$ and $\vec{y}_I$. Log gains are computed in logarithmic coordinates. In general systems theory, the linearization is typically executed in Cartesian space, where the log gain matrix corresponds to the Jacobian. Chevalier et al. (1993) give a succinct interpretation of the Jacobian elements in terms of chemical networks (see also Cascante, Franco, and Canela 1991).

In addition to log gains of metabolites, one defines log gains of fluxes with respect to independent variables. Because the system is considered at a steady state, the influx and the efflux of a variable are in balance. It is therefore immaterial which flux is used for calculation of flux gains. In the S-system form, the steady-state flux through any dependent variable $X_i (i = 1, 2, \ldots, n)$ is given as

$$V_i^+ = \alpha_i \prod_{j=1}^{n+m} X_j^{g_{ij}} = \beta_i \prod_{j=1}^{n+m} X_j^{h_{ij}} = V_i^-. \tag{2.34}$$

Taking logarithms, the equation becomes a linear equation in logarithmic variables $y_j = \ln(X_j) (j = 1, 2, \ldots, n+m)$. For instance, in terms of V_i^+ one obtains

$$\ln(V_i^+) = \ln(\alpha_i) + \sum_{j=1}^{n} g_{ij} y_j + g_{ik} y_k + \sum_{j=n+1, j \neq k}^{n+m} g_{ij} y_j, \tag{2.35}$$

where we have split up the right-hand side in four terms. The first two terms contain the rate constant and features strictly related to dependent variables. If some independent variable (e.g., y_k), is altered, these terms are unaffected. The third term represents the *direct* effect of y_k on the flux V_i^+, whereas the fourth term quantifies *indirect* effects that are caused by responses of other metabolites to changes in y_k.

In strict analogy to metabolite gains, the logarithmic gain of flux V_i^+ with respect to X_k is defined as the change in $\ln(V_i^+)$ or $\ln(V_i^-)$ with respect to a change in y_k.

This change is again obtained through partial differentiation, which yields

$$L(V_i^+, X_k) = g_{ik} + \sum_{j=1}^{n} g_{ij} L_{jk}, \qquad (2.36)$$

where $L_{jk} = L(X_j, X_k)$ is the metabolite gain $\partial y_j / \partial y_k$, as it was described in Eq.(2.33). The gain of flux V_i^- with respect to X_k is similarly given as

$$L(V_i^-, X_k) = h_{ik} + \sum_{j=1}^{n} h_{ij} L_{jk}, \qquad (2.37)$$

and because V_i^+ and V_i^- are equivalent at steady state, the gains are equivalent as well:

$$L(V_i^-, X_k) = L(V_i^+, X_k). \qquad (2.38)$$

This makes intuitive sense, because the system is in a steady state before and after the perturbation. Therefore, any changes in V_i^+ and V_i^- must be of the same magnitude.

Sensitivities and log gains are very useful tools for the diagnosis of a model. High sensitivities and gains indicate that small changes in a parameter or in an independent variable have very strong effects on the response of the system to perturbations. In many cases, such strong responses may not be realistic, which indicates that something may be wrong with the model.

Log gains are of prominent importance for questions of optimization. Because they indicate how much each dependent variable or flux is affected by a change in an independent variable, log gains provide a screening tool for independent variables that might be useful for optimization. If an independent variable is associated with log gains close to zero, even substantial changes in this variable will have little effect. On the other hand, if the log gain of a desired metabolite or flux with respect to a particular independent variable is high, this variable might be a good candidate for manipulation. We will return to this type of assessment in later chapters.

STABILITY ANALYSIS

The assessment of local stability has components of dynamic and steady-state analysis. On one hand, locally stability is a feature of a given steady state, and a system may simultaneously have several steady states with different stability properties. On the other hand, local stability is concerned with responses to perturbations, and these clearly have a dynamic component.

The steady state of a system is said to be locally stable if the system returns to it after a small perturbation. In most relevant cases, local stability is assessed upon linearization of the system at the steady state. This is justified because an old theorem of Hartman and Grobman (see Guckenheimer and Holmes 1983) proves that the nonlinear and the linearized system have locally the same stability behavior.

Although linearization is a great simplification, it is still is not a trivial matter, because it calls for the characterization of the complex-valued eigenvalues of the

system. Their exact computation would require the solution of an nth-order polynomial, which is seldom possible even for moderately large n. Fortunately, the precise
values of the eigenvalues are not needed, and it turns out that a simpler criterion is
sufficient. This criterion states that all eigenvalues must have negative real parts for
the system to be stable at the tested steady state. The existence of even one eigenvalue
with a positive real part is sufficient to make the steady state unstable.

The existence of negative real parts among the eigenvalues can be tested with
iterative algorithms that can theoretically be performed by hand. One is called the
Routh–Hurwitz criterion (Routh 1930; Savageau 1976; Clarke 1980); another one
is called the Jury test (Jury 1971; Edelstein-Keshet 1988). Very detailed treatises of
stability and instabilities in chemical networks are found in Gavalas (1968), Clarke
(1980), and Richter et al. (1980). Jou and Llebot (1990) presented a discussion of
different types of stability from a thermodynamic point of view. Examples of stability
analyses in S-systems can be found in Savageau (1975, 1976) and Voit (1991, 2000).

Other properties of the eigenvalues contain interesting information as well. In
particular, imaginary parts of the eigenvalues that are different from zero occur in
complex conjugate pairs. They imply the potential for oscillations in transient responses of the system. Furthermore, the magnitudes of the eigenvalues are inversely
related to the relaxation times with which the various variables return to the steady
state after a perturbation (for specific examples in the context of biochemical systems,
see Palsson and Lightfoot 1984; Palsson, Jamier, and Lightfoot 1984, 1985; Shiraishi
and Savageau 1992).

For most practical purposes, stability in S-systems will be addressed numerically
in PLAS (Ferreira 2000), which produces all eigenvalues automatically along with
the steady-state solution. In the display, the positive or negative real parts, the orders
of magnitude of all real parts, and the existence of nonzero imaginary parts are
immediately visible.

Structural stability analysis identifies combinations of parameter values at which
the system behavior experiences drastic changes. As Clarke (1980) pointed out, chemical networks can exhibit a variety of "exotic" behaviors, including the switching
between multiple steady states, explosions, sustained oscillations, and chaos. For a
parameter value below some threshold, the system may have a stable steady state,
but as soon as the parameter exceeds the threshold, stability is lost or the system
begins to oscillate. Criteria for such thresholds and their evaluation in a particular
case require a bit of algebra. For the most common case of a Hopf bifurcation, small
S-systems provide shortcuts that can be executed with a pocket calculator (Lewis
1991).

As an example for a "structural mutation" (Jou and Llebot 1990), that is, for a
switch in qualitative behavior, consider the simple S-system

$$\dot{X}_1 = \alpha \cdot \left(X_1^4 - X_1^3 X_2^2 \right) \qquad X_1(0) = 1$$
$$\dot{X}_2 = X_1^3 - X_1^2 X_2 \qquad X_2(0) = 1 \qquad (2.39)$$

(Lewis 1991). For clarity, all parameters are fixed, except for α. If α has a value

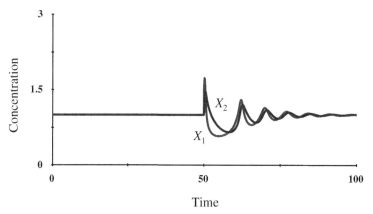

Figure 2.7. Dynamical responses of the system in Eq. (2.39) with $\alpha = 0.8$. At $t = 50$, X_1 is changed to 1.5. X_1 and X_2 immediately respond with (further) increases, before oscillating back to the steady state at (1, 1).

below 1, the system responds to a perturbation in X_1 or X_2 with damped oscillations; after some while, the system returns to the original steady state $(X_{1S}, X_{2S}) = (1, 1)$. Figure 2.7 shows the response to a change in X_1 from 1 to 1.5 at time $t = 50$, when the critical parameter α has the value 0.8. If α exceeds the threshold of 1, the behavior is drastically different. In the example of Figure 2.8, $\alpha = 1.2$. The same change in X_1 now leads to the onset of stable oscillations (note the different scales of the x axes). The differences in responses are not due to the relatively large change in X_1 or to the relatively large difference in the values of α of the two scenarios. The threshold at $\alpha = 1$ is mathematically crisp. Any value below corresponds to a stable steady state and leads to a response in the form of damped oscillations, whereas any value above 1 leads to stable oscillations about an unstable steady state. For α exactly at 1, the responses depend on the specific properties of the perturbation.

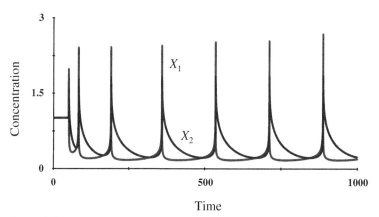

Figure 2.8. Dynamical responses of the system in Eq. (2.39) with $\alpha = 1.2$. At $t = 50$, X_1 is changed to 1.5. After some transient period, X_1 and X_2 begin and continue to oscillate with constant frequency and amplitude. Apparent slight differences in peak height are artifacts of the numerical solution.

As a second example, consider a GMA model proposed by Steinmetz and Larter (1991), which represents one of the first chemical systems shown to exhibit chaos in vitro. The experimental system, proposed originally by Yamazaki, Yokota, and Nakajima (1965) and by Olsen and Degn (1977, 1978), is called the peroxidase–oxidase reaction. It consists of an aerobic oxidation of NADH, which is catalyzed by peroxidase and involves dissolved O_2. The overall reaction scheme is

$$2NADH + 2H^+ + O_2 \rightarrow 2NAD^+ + 2H_2O.$$

Based on a more detailed reaction scheme proposed by Degn, Olsen, and Perram (1979), Steinmetz and Larter (1991) modeled the reaction with four dependent variables: A, B, X, and Y, which represent O_2, NADH, and two intermediates that were suspected to be the NAD· radical and a radical form of the enzyme, respectively. With reasonable initial values, the GMA model has the form

$$\dot{A} = 0.89 - kABX - 0.046875\,ABY - 0.1175\,A \qquad A(0) = 4$$
$$\dot{B} = 0.5 - kABX - 0.046875\,ABY \qquad B(0) = 15$$
$$\dot{X} = 0.001 + kABX + 0.09375\,ABY - 20X - 2500\,X^2 \qquad X(0) = 0.02$$
$$\dot{Y} = 2500\,X^2 - 0.046875\,ABY - 1.104\,Y \qquad Y(0) = 0.25.$$

$$(2.40)$$

The rate constant k is kept in symbolic form, because it constitutes the parameter of interest, which we intend to vary from one experiment to the next. As Steinmetz and Larter demonstrated, the system allows for primary and secondary Hopf bifurcations and, as a consequence, different values of k lead to dramatically different responses of the system.

For $k = 1$, the system has a stable steady state, to which it returns from perturbations with strongly damped oscillations. If k is decreased, the oscillatory transients span longer time periods. Between $k = 0.27$ and $k = 0.2699$, the real parts of the pair of complex conjugate eigenvalues cross zero, and the stable steady state is supplanted with a stable limit cycle. Figure 2.9a shows the time courses of A and B for $k = 0.2$ (note that A and B were scaled to $A/4$ and $B/30$ to facilitate plotting). Further decreases in k destabilize the formerly regular oscillations. For $k = 0.1$, the limit cycle oscillations are interrupted by irregular fluctuations (Figure 2.9b), and for yet smaller values (e.g., $k = 0.05$), the behavior becomes rather exotic (Figure 2.9c). Steinmetz and Larter provide a detailed analysis of these transitions.

EVALUATION AND INTERPRETATION

Modeling is a combination of art and mathematical technique. The initial step of identifying variables and processes requires judgment about which observations and other pieces of information are important enough to warrant inclusion in the model, and this judgment is based on a mix of subject area knowledge and modeling experience. Once the variables and processes are defined and parameter values determined, the model analysis itself is, by and large, a matter of applied mathematics

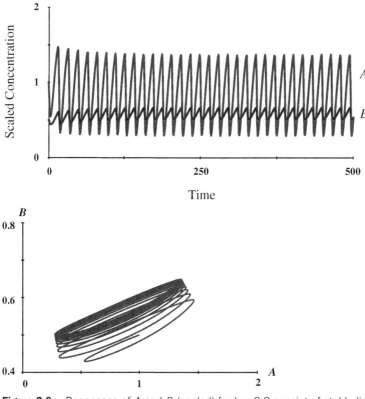

Figure 2.9a. Responses of A and B (scaled) for $k = 0.2$ consist of stable limit cycle oscillations. Top panel: A and B plotted against time. Bottom panel: The "phase plane" plot of B against A shows how the system, after a transient phase, approaches the limit cycle.

and computation that does not really depend on biology, except for the selection of questions to be asked and scenarios to be tested.

After all mathematical analyses are completed, a final step of the current modeling cycle is needed, namely, making sense of the mathematical results. These consist of collections of numbers and graphs, for instance, exhibiting time courses and sensitivity profiles, which have a clearly defined mathematical meaning. The mathematical implication of an eigenvalue with a positive real part is instability at some steady state. The crossing of a critical threshold by a parameter may mean the onset of oscillation. The last part of the analysis requires translation of these mathematical insights into the terminology of biology. Only biological understanding can determine whether some mathematical result is even relevant. Maybe the critical threshold will never be crossed in reality. Maybe the actual constellation of parameter values prevents the system from becoming unstable. Conversely, maybe the crossing of a threshold is the key to understanding a drastic morphological or physiological change during development and differentiation. Mathematics cannot answer these questions, and it is up to the modeler, maybe in consultation with a biologist, to decipher and interpret what the numbers mean, what is relevant, and what is important. This evaluation

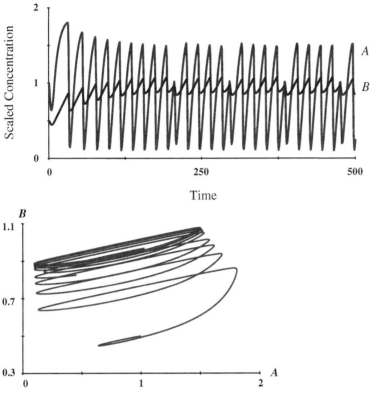

Figure 2.9b. For $k = 0.1$, the oscillations in A and B (scaled) become irregular. This is more clearly seen in the time course (top panel) than in the phase plane plot (bottom panel).

and interpretation step is critically important, because without translation of the mathematical key results into the terminology of biology, the entire model analysis will have little impact within the biological community.

A part of the evaluation and interpretation phase is the identification of further scenarios to be tested and of amendments to the model. New scenarios usually require additional sets of simulations or analyses with untested sets of parameter values. They do require time and effort, but are typically executed with relative ease. However, the evaluation of results may also make it clear that important questions cannot be answered with the current model or that some observations are simply not modeled correctly, which becomes apparent from qualitative or significant quantitative discrepancies between model results and reality. In this case, it is likely that the model structure needs to be revised, which may require the redefinition or addition of variables or a closer look at the governing processes. Structural changes are almost always accompanied with new or refined parameter values, and these may necessitate a new series of analyses and simulations. As a consequence, a second iteration of the modeling cycle begins, as is indicated in Figure 2.10. The graph is somewhat deceiving in that it implies that one goes through modeling cycles only to return to the same point. A more encouraging and indeed more realistic representation would

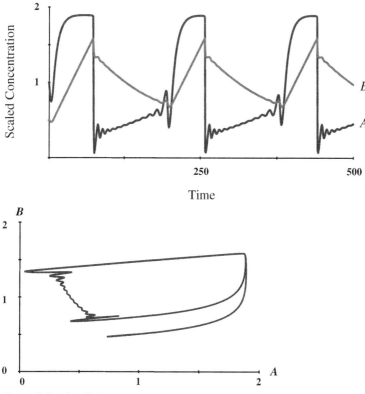

Figure 2.9c. Oscillations in A and B (scaled) for $k = 0.05$ are highly irregular (top panel), but still exhibit a limit cycle (bottom panel).

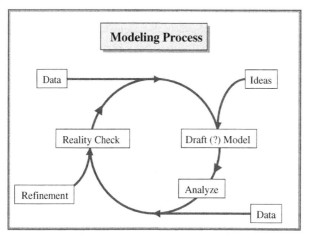

Figure 2.10. Iterative, cyclic nature of the modeling process. Each cycle is hoped to provide additional insight and understanding.

be that of a spiral that indicates the increase in insight and understanding during each modeling cycle.

REFERENCES

Almeida, J.S., M.A.M. Reis, and M.J.T. Carrondo: Competition between nitrate and nitrite reduction in denitrification by Pseudomonas fluorescence. *Biotechnol. Bioeng.* 46, 476–84, 1995.

Alves, R., and M. Savageau: Comparing systemic properties of ensembles of biological networks by graphical and statistical methods. *Bioinformatics* 16, 527–33, 2000a.

Alves, R., and M. Savageau: Systemic properties of ensembles of metabolic networks: Application of graphical and statistical methods to simple unbranched pathways. *Bioinformatics* 16, 534–47, 2000b.

Alves, R., and M. Savageau: Extending the method of mathematically controlled comparison to include numerical comparisons. *Bioinformatics* 16, 786–98, 2000c.

Alves, R., and M. Savageau: Effect of overall feedback inhibition in unbranched biosynthetic pathways. *Biophys. J.* 79, 2290–304, 2000d.

Bar-Eli, K., and W. Geiseler: Perturbations around steady states in a continuous stirred tank reactor. *J. Phys. Chem.* 87, 1352–7, 1983.

Berg, P.H., E.O. Voit, and R. White: A pharmacodynamic model for the action of the antibiotic Imipenem on *Pseudomonas in vitro*. *Bull. Math. Biol.* 58(5), 923–38, 1996.

Cascante, M., R. Franco, and E. Canela: Sensitivity analysis: A common foundation of theories for the quantitative study of metabolic control. In E.O. Voit (Ed.), *Canonical Nonlinear Modeling. S-System Approach to Understanding Complexity* (Chapter 4). Van Nostrand Reinhold, New York, 1991.

Casti, J.L.: *Nonlinear System Theory*. Academic Press, Orlando, FL, 1985.

Chen, C.-T.: *Linear System Theory and Design*. Holt, Rinehart and Winston, New York, 1984.

Chevalier, T., I. Schreiber, and J. Ross: Toward a systematic determination of complex reaction mechanisms. *J. Phys. Chem.* 97, 6776–87, 1993.

Clarke, B.L.: Stability of complex reaction networks. *Adv. Chem. Phys.* 43, 1–215, 1980.

Curto, R., E.O. Voit, A. Sorribas, and M. Cascante: Mathematical models of purine metabolism in man. *Math. Biosci.* 151, 1–49, 1998.

Degn, H., L.F. Olsen, and J.W. Perram: Bistability, oscillation, and chaos in an enzyme reaction. *Ann. N.Y. Acad. Sci.* 316, 622–37, 1979.

Díaz-Sierra, R., and V. Fairén: Simplified method for the computation of parameters of power-law equations from time-series. *Math. Biosci.* 171, 1–19, 2001.

Díaz-Sierra, R., J.B. Lozano, and V. Fairén: Deduction of chemical mechanisms from the linear response around steady state. *J. Phys. Chem.* 103, 337–43, 1999.

Edelstein-Keshet, L.: *Mathematical Models in Biology*. Birkhäuser Mathematics Series. McGraw-Hill, New York, 1988.

Fell, D.A.: Metabolic control analysis – a survey of its theoretical and experimental development. *Biochem. J.* 286, 313–30, 1992.

Fell, D.A.: *Understanding the Control of Metabolism*. Portland Press, London, 1997.

Ferreira, A.E.N.: PLAS©: http://correio.cc.fc.ul.pt/~aenf/plas.html, 2000.

Gavalas, G.R.: *Nonlinear Differential Equations of Chemically Reacting Systems*. Springer-Verlag, Berlin, 1968.

Groen, A.K., R. van den Meer, H.V. Westerhoff, R.J.A. Wanders, T.P.M. Akerboom, and J.M. Tager: Control of metabolic fluxes. In H. Sies (Ed.), *Metabolic Compartmentation* (pp. 9–37). Academic Press, New York, 1982.

Guckenheimer, J., and P. Holmes: *Nonlinear Oscillations, Dynamical Systems, and Bifurcations of Vector Fields*. Springer-Verlag, New York, 1983.

Haefner, J.A.: *Modeling Biological Systems. Principles and Applications*. Chapman and Hall, New York, 1996.

Haldane, J.B.S: *Enzymes* (pp. 28–64). Longmans, Green and Co., London, 1930.

Hatzimanikatis, V., C.A. Floudas, and J.E. Bailey: Optimization of regulatory architectures in metabolic reaction networks. *Biotechnol. Bioeng.* 52(4), 485–500, 1996a.

Hatzimanikatis, V., C.A. Floudas, and J.E. Bailey: Analysis and design of metabolic reaction networks via mixed-integer linear optimization. *AIChE J.* 42(5), 1277–92, 1996b.

Jou, D., and J.E. Llebot: *Introduction to the Thermodynamics of Biological Processes*. Prentice Hall, Englewood Cliffs, NJ, 1990.

Jury, E.I.: The inners approach to some problems of systems theory. *IEEE Trans. Automatic Contr.* AC-16, 233–40, 1971.

Kacser, H., and J.A. Burns: Molecular democracy: Who shares the controls? *Biochem. Soc. Trans.* 7, 1149–60, 1979.

Kopelman, R.: Rate processes on fractals: Theory, simulations, experiments. *J. Stat. Phys.* 42, 185–200, 1986.

Kopelman, R.: Reaction kinetics in restricted spaces. *Israel J. Chem.* 31, 147–57, 1991.

Lang, S.: *A Second Course in Calculus*. Addison-Wesley Publishing Company, Reading, MA, 1968.

Lewis, D.C.: A qualitative analysis of S-systems: Hopf bifurcations. In E.O. Voit (Ed.), *Canonical Nonlinear Modeling. S-System Approach to Understanding Complexity* (Chapter 16). Van Nostrand Reinhold, New York, 1991.

Lübbert, A., and R. Simutis: Measurement and control. In: C. Ratledge and B. Kristiansen (Eds.), *Basic Biotechnology* (Chapter 10). Cambridge University Press, Cambridge, U.K., 2001.

Mayhan, R.J.: *Discrete-Time and Continuous-Time Linear Systems*. Addison-Wesley, Reading, MA, 1983.

Neves, A.R., A. Ramos, M.C. Nunes, M. Kleerebezem, J. Hugenholtz, W.M. de Vos, J. Almeida, and H. Santos: In vivo nuclear magnetic resonance studies of glycolytic kinetics in *Lactococcus lactis. Biotechnol. Bioeng.* 64(2), 200–12, 1999.

Ogata, K.: *Modern Control Engineering* (2nd ed.). Englewood Cliffs, NJ: Prentice Hall, 1990.

Okamoto, M., Y. Morita, D. Tominaga, K. Tanaka, N. Kinoshita, J.-I. Ueno, Y. Miura, Y. Maki, and Y. Eguchi: Design of virtual-labo-system for metabolic engineering: Development of biochemical engineering system analyzing tool-kit (BEST KIT). *Comp. Chem. Eng.* 21(Suppl.), S745–50, 1997.

Olsen, L.F., and H. Degn: Chaos in an enzyme reaction. *Nature* 267, 177–8, 1977.

Olsen, L.F., and H. Degn: Oscillatory kinetics of the peroxidase-oxidase reaction in an open system. Experimental and theoretical studies. *Biochem. Biophys. Acta* 523(2), 321–34, 1978.

Palsson, B.Ø., and E.N. Lightfoot: Mathematical modelling of dynamics and control in metabolic networks. I. On Michaelis-Menten kinetics. *J. Theor. Biol.* 111, 273–302, 1984.

Palsson, B.Ø., R. Jamier, and E.N. Lightfoot: Mathematical modelling of dynamics and control in metabolic networks. II. Simple dimeric enzymes. *J. Theor. Biol.* 111, 303–21, 1984.

Palsson, B.Ø., H. Palsson, and E.N. Lightfoot: Mathematical modelling of dynamics and control in metabolic networks. III. Linear reaction sequences. *J. Theor. Biol.* 113, 231–59, 1985.

Richter, P.H., I. Procaccia, and J. Ross: Chemical instabilities. *Adv. Chem. Phys.* 43, 217–68, 1980.

Routh, E.J.: *Advanced Part of Dynamics of a System of Rigid Bodies* (6th edition, Vol. II). Macmillan, London, 1930.

Rugen, P., and B. Callahan: An overview of Monte-Carlo, a fifty year perspective. *Human Ecological Risk Assess.* 2(4), 671–80, 1996.

Sands, P.J., and E.O. Voit: Flux-based estimation of parameters in S-systems. *Ecol. Model.* 93, 75–88, 1996.

Savageau, M.A.: Biochemical Systems Analysis, II. The steady-state solutions for an n-pool system using a power-law approximation. *J. Theor. Biol.* 25, 370–79, 1969.

Savageau, M.A.: Parameter sensitivity as a criterion for evaluating and comparing the performance of biochemical systems. *Nature* 229(5286), 542–4, 1971a.

Savageau, M.A.: Concepts relating behaviour of biochemical systems to their underlying molecular properties. *Arch. Biochem. Biophys.* 145, 612–21, 1971b.

Savageau, M.A.: The behavior of intact biochemical control systems. *Curr. Top. Cell. Reg.* 6, 63–129, 1972.

Savageau, M.A.: Optimal design of feedback control by inhibition: Dynamic considerations. *J. Mol. Evol.* 5, 199–222, 1975.

Savageau, M.A.: *Biochemical Systems Analysis. A Study of Function and Design in Molecular Biology.* Addison-Wesley, Reading, MA, 1976.

Savageau, M.A.: Influence of fractal kinetics on molecular recognition. *J. Mol. Recogn.* 6, 149–157, 1993.

Savageau, M.A.: Michaelis-Menten mechanism reconsidered: Implications of fractal kinetics. *J. Theor. Biol.* 176, 115–24, 1995a.

Savageau, M.A.: Enzyme kinetics in vitro and in vivo: Michaelis-Menten revisited. In: E.E.Bittar (Ed.), *Principles of Medical Biology* (Vol. 4, pp. 93–146). JAI Press, Greenwich, CT 1995b.

Segall, J.E, S.M. Block, and H.C. Berg: Temporal comparisons in bacterial chemotaxis. *Proc. Natl. Acad. Sci.* U.S.A. 83, 8987–91, 1986.

Shiraishi, F., and M.A. Savageau: The tricarboxylic acid cycle in *Dictyostelium discoideum*. II. Evaluation of model consistency and robustness. *J. Biol. Chem.* 267, 22919–25, 1992.

Sorribas, A., and M.A. Savageau: Strategies for representing metabolic pathways within biochemical systems theory: Reversible Pathways. *Math. Biosci.* 94, 239–69, 1989.

Sorribas, A., S. Samitier, E.I. Canela, and M. Cascante: Metabolic pathway characterization from transient response data obtained in situ: Parameter estimation in S-system models. *J. Theor. Biol.* 162, 81–102, 1993.

Sorribas, A., J.B. Lozano, and V. Fairén: Deriving chemical and biochemical model networks from experimental measurements. *Rec. Devel. Phys. Chem.* 2, 553–73, 1998.

Steinmetz, C.G., and R. Larter: The quasiperiodic route to chaos in a model of the peroxidase-oxidase reaction. *J. Chem. Phys.* 94(2), 1388–96, 1991.

Tominaga, D., M. Okamoto, Y. Maki, S. Watanabe, and Y. Eguchi: Nonlinear numerical optimization technique based on genetic algorithm for inverse problem: towards the inference of genetic networks. *Proc. German Conf. on Bioinformatics*, 127–140, 1999.

Torsella, J., and A.M. Bin Razali: An analysis of forestry data. In: E.O. Voit (Ed.), *Canonical Nonlinear Modeling. S-System Approach to Understanding Complexity* (Chapter 10). Van Nostrand Reinhold, New York, 1991.

U.S. EPA: Guiding Principles for Monte-Carlo Analysis (EPA/630/R-97/001), http://www.epa.gov/ncea/monteabs.htm, 1997.

Voit, E.O. (Ed.): *Canonical Nonlinear Modeling. S-System Approach to Understanding Complexity,* (xi + 365 pp.). Van Nostrand Reinhold, New York, 1991.

Voit, E.O.: *Computational Analysis of Biochemical Systems. A Practical Guide for Biochemists and Molecular Biologists* (xii + 532 pp.). Cambridge University Press, Cambridge, U.K., 2000.

Voit, E.O.: Models-of-data and models-of-processes in the post-genomic era (*Mathem. Biosci.* (in press)).

Voit, E.O., and M. Del Signore: Assessment of effects of experimental imprecision on optimized biochemical systems. *Biotechnol. Bioeng.* 74(5), 443–8, 2001.

Voit, E.O., and T. Radivoyevitch: Biochemical systems analysis of genome-wide expression data. *Bioinformatics* 16(11), 1023–37, 2000.

Voit, E.O., and P.J. Sands: Modeling forest growth. II. Biomass partitioning in Scots pine. *Ecol. Model.*86, 73–89, 1996.

Voit, E.O., and M.A. Savageau: Power-law approach to modeling biological systems. III. Methods of analysis. *J. Ferment. Technol.* 60(3), 233–41, 1982.

Yamazaki, I., K. Yokota, and R. Nakajima: Oscillatory oxidations of reduced pyridine nucleotide by peroxidase. *Biochem. Biophys. Res. Commun.* 21, 582–6, 1965.

Zhang, Z., E.O. Voit, and L.H. Schwacke: Parameter estimation and sensitivity analysis of S-systems using a genetic algorithm. In: T. Yamakawa and G. Matsumoto (Eds.), *Methodologies for the Conception, Design, and Application of Intelligent Systems.* World Scientific, Singapore, 1996.

## CHAPTER THREE

A Model of Citric Acid Production in the Mold *Aspergillus niger*

INTRODUCTION

This chapter addresses the question of how mathematical modeling can use detailed and localized biochemical information to deepen our understanding of biological functioning at the level of integrated metabolic pathways. It describes the cyclical process of mathematical modeling with the example of citric acid metabolism in *Aspergillus niger*. This system has some nice features that facilitate mathematical analysis and provide a good example for demonstrating the various steps of model design, analysis, and refinement. Although citric acid metabolism is quite complex and involves several biochemical pathways in two different cellular compartments, the system is relatively closed. Moreover, under conditions where *A. niger* synthesizes citric acid, this synthesis is essentially the only metabolic pathway of quantitative importance, the substrate is almost entirely glucose and/or fructose, and citric acid is the sole final product. The system is rich in external and internal controls, and the governing biochemical mechanisms are well documented with a comprehensive body of kinetic information. The intricate regulatory structure is a source of nonlinear responses that make nonmathematical approaches unsuitable.

Citric acid production in *A. niger* was first approached with mathematical modeling techniques by Meyrath (1967). He used a macroscopic approach based on mass and energy balances and concluded that up to 85% of the consumed hexoses could be converted into citric acid. Verhoff and Spradlin (1976) also presented a mass balance model with the aim of unraveling the metabolic pathways involved in citric acid biosynthesis. In 1994, we presented the first structured mathematical model dealing with the first part of the biosynthetic pathway, from glucose up to pyruvate and the anaplerotic reactions (Torres 1994a,b). Alvarez-Vasquez (2000) and Alvarez-Vasquez, González-Alcón, and Torres (2000) recently proposed extensions and refinements of this model.

To demonstrate the iterative nature of mathematical modeling, we begin with our original model, which involved all cellular processes that at the time were considered important for citric acid production in *A. niger*. This information is first presented as a description of the main regulatory features of the system and subsequently summarized in the form of a biochemical map. The map is translated into symbolic S-system equations, with methods shown in Chapter 2. We discuss standard techniques of parameter estimation that allow us to implement the symbolic S-system structure as a numerical model. Next, the numerical S-system model is used for steady-state and stability analyses and for a first exploration of dynamic features. A particularly important component of the steady-state analysis will be the exploration of sensitivity and gain profiles, which will give us a first impression of which system constituents may be most promising for improvements in yield. The assessment of the system's dynamic shows the evolution of the system after different perturbations and provides first information for transient optimizations. We will conclude the analysis of the original model with the identification of limitations and shortcomings, which historically prompted us to develop a more refined and comprehensive model of the pathway. The subsequent section will introduce this revised model, which is described in greater detail in the dissertation research of Alvarez-Vasquez (2000). It represents the culmination of a ten-year effort of integrating biological information, data analysis, and experimental and mathematical optimization studies on citric acid production in *A. niger*.

One may ask why we would begin the discussion with a model that ultimately will be discarded. We follow this historical development for didactic reasons. It hardly ever happens that the first model is satisfactory in all aspects, and it is therefore beneficial to demonstrate what types of diagnostics may be employed to detect specific causes of discrepancies between reality and model, and how features of the diagnosis can be converted into amendments and refinements of the initial model. These types of thought processes become evident only if the first model is shown with its strengths, but also with its limitations.

Once the quality of the refined model has been assured, we will be ready to proceed to the optimization of enzyme profiles, with the goal of maximizing citric acid flux production under conditions of cell viability. This optimization is the topic of Chapter 6. The biotechnological implementation of these profiles is an application of modern techniques of metabolic engineering. It requires artificial over- or underexpression of some of the *A. niger* genes, many of which are already available for experimental manipulation.

BASIC CHEMISTRY AND BIOCHEMISTRY OF CITRIC ACID

Citric acid (2-hydroxy-1,2,3-propane tricarboxylic acid) is widely distributed throughout nature, occurring in both plants and animals. Its chemical structure is shown in Figure 3.1. Citric acid is the initial compound for numerous chemical processes of industrial relevance, because it possesses one hydroxy and three carboxyl groups, which facilitate transformation into a wide variety of complex molecules. It is furthermore a very valuable industrial compound because of its ability to form complexes with metal ions, most notably the alkaline earths calcium and magnesium, but also the transition metals copper, nickel, zinc, and cobalt. As a metal complexing

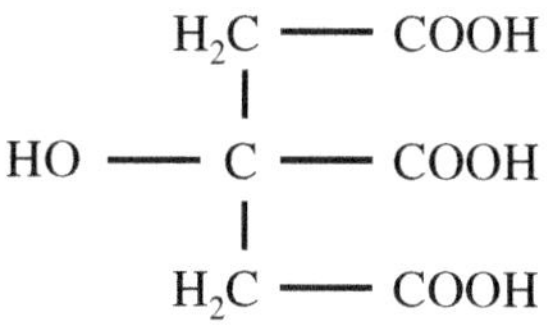

$$
\begin{array}{c}
H_2C \!-\!\!-\!\!-\! COOH \\
| \\
HO \!-\!\!-\!\!-\! C \!-\!\!-\!\!-\! COOH \\
| \\
H_2C \!-\!\!-\!\!-\! COOH
\end{array}
$$

Figure 3.1. Chemical structure of citric acid.

agent, it is becoming increasingly important for metal cleansing and for stabilizing oils and fats that are used against oxidative deterioration. However, citric acid plays its most prominent role in food preservation. Citric acid and its sodium, potassium, and calcium salts are classified by the U.S. Food and Drug Administration as "generally recognized as safe" (GRAS) for sequestrant, miscellaneous, and general uses in foods. As a consequence of this classification and of its cheap production price, citric acid is heavily used as an acidulant and preservative in the soft drink and canning industry. In fact, of the total production of citric acid almost half is used in the beverage industry, about 20% for food and confections, 8% for the pharmaceutical industry, and about 20% for detergents and cleaning agents (Kubicek 2001).

Citric acid was originally obtained through physical extraction from lemon juice. Today, most citric acid worldwide is commercially produced by fermentation processes using the mold *Aspergillus niger* as the fermenting organism and either dextrose or beet molasses as raw material for substrate. Within the fermentation industry, citric acid is a key player, with a global production capacity estimated at several hundred thousand tons per year. The overall demand is expected to increase exponentially, and so is the demand for higher productivity and product yields. According to Kubicek (2001), high-yield citric acid-producing *A. niger* strains are "among the most secretly kept organisms in biotechnology." In fact, this microbe is so important, it has its own website: http://www.aspergillus.man.ac.uk/.

Among organisms for citric acid fermentation, strains of the mold *A. niger* have played a major role since the very beginnings of biotechnological production. The first studies date back to the turn of the last century. Of particular importance was the work of Wehmer (1903) and Zahorsky (1913), who pioneered the investigation of citric acid formation. They immediately recognized the commercial potential of a fermentation process that was based on microorganisms, which were able to yield product in very large amounts. Since these early beginnings, more than two thousand scientific reports and several hundred patents related to the biotechnological exploitation of the mold have been presented (Röhr 1998). In fact, this fermentation process can be considered a paradigm for applying knowledge about a metabolic pathway to a practically relevant commercial application.

In spite of its industrial importance and the considerable, long-term attention devoted to the physiology and biochemistry of citric acid production in *A. niger*, the fermentation process itself is still not completely understood. Numerous variables have an effect on the process, but cannot be fully controlled, thereby causing potential impairment of yield. Modern strategies of yield and process improvement therefore concentrate not only on process control, but also on breeding more efficient strains of *A. niger*. These improvements are guided by the current understanding of microbial metabolism and molecular genetics. Although strain manipulations with methods of molecular biology and recombinant DNA technology have been initiated only recently, a severe limitation is already becoming apparent – in this as well as in other biotechnological processes. The available biological information in itself is not sufficient for prescribing specific alterations that would effectively redirect metabolic fluxes toward the desired final product. The biological techniques must be complemented with mathematical strategies of systems analysis that are able to pinpoint the steps most relevant for flux enhancement.

The process of citric acid production in *A. niger* begins when conidia suspensions are inoculated in a culture medium that contains sucrose as substrate, together with inorganic phosphate, Mg^{2+}, Ca^{2+}, and Zn^{2+}. The culture medium should be set up first in the absence of Mn^{2+}, which is a determining factor for the accumulation process (Kubicek and Röhr 1978). Under these conditions, the culture enters a phase of rapid growth (tropophase), and between 96 to 288 hours attains a steady state in citric acid mycelium level. At this stage, called *idiophase*, or *nongrowing, fermentation* stage, the pH in the medium is about 2. During the time period between 70 and 170 hours of cultivation, citrate production is the only metabolic activity of quantitative importance (Kubicek, Hampel, and Röhr 1979). The system of interest here refers to the fermentation stage, and most kinetics and steady-state data correspond specifically to measurements at 150 hours.

Five metabolic processes play a leading role in the citric acid biosynthesis of *A. niger*: (1) breakdown of hexoses to pyruvate and acetyl-CoA during glycolysis; (2) citric acid formation within the tricarboxylic acid cycle; (3) anaplerotic formation of oxalacetate from pyruvate and CO_2; (4) transport processes throughout the cytoplasmatic and mitochondrial membranes; and (5) oxygen consumption. In our first model, only the first and third metabolic processes were considered, along with some transport steps; the map of the system is depicted in Figure 3.2.

Carbohydrate metabolism. The fungus possesses an extracelullar, mycelium-bound invertase. It is active under conditions of citric acid fermentation and capable of rapidly hydrolyzing sucrose that is supplied in the medium (Kubicek and Röhr 1986). The action of this invertase in combination with hydrolysis during autoclaving means that the effective carbon source is glucose. The metabolism of glucose involves joint participation of the glycolytic pathway and the pentose phosphate pathway. During the initial, rapid growth phase, the relative involvement of the two pathways is 2:1, but during citric acid fermentation, it is 4:1. This observation was made already in some early studies (Cleland and Johnson 1954; Kubicek and Röhr 1977).

Once glucose has entered the cytoplasm, it is quickly phosphorylated by the action of hexokinases (Schreferl-Kunar et al. 1989; Steinbock 1993). Glycolysis is regulated at the phosphofructokinase step. Under normal conditions, citrate inhibits this enzyme when it reaches an average concentration in the mycelium of about 9 mM. However, NH_4^+, whose concentration in the culture may increase up to 2–3 mM, counteracts this inhibitory effect (Habison, Kubicek, and Röhr 1983; Arts, Kubicek, and Röhr 1987).

Anaplerosis of the tricarboxylic acid cycle. The CO_2 formed from pyruvate in its transformation to acetyl-CoA is used by the cytosolic pyruvate carboxylase for the formation of oxalacetate. The importance of this anaplerotic reaction for high yields of citric acid accumulation has been recognized early on, because it supplies the tricarboxylic acid cycle with oxalacetate (Cleland and Johnson 1954). Pyruvate carboxylase is a constitutive cytoplasmic enzyme (Osmani and Scrutton 1983) that has been characterized in detail (Feir and Suzuki 1969). The oxalacetate so formed is converted to malate by means of cytoplasmatic malate dehydrogenase and then transported to the mitochondrion, where it acts as a countersubstrate for the tricarboxylate carrier (Evans, Scragg, and Ratledge 1981, 1983).

Transport processes. A complete characterization of mitochondrial transport processes in *A. niger* has so far remained elusive, and most available evidence comes from studies in yeast. This organism apparently possesses a transport carrier for pyruvate (Halestrap, Scott, and Thomas 1980) and a countertransport system, in which citrate and isocitrate leave the mitochondrial compartment by a specific carrier, with malate as counter-ion (Evans et al. 1983). This form of transport is feasible because malate dehydrogenase is present in both the mitochondrial and cytosolic compartments of *A. niger* (Ma, Kubicek, and Röhr 1981).

The facts summarized here, together with available data on metabolite pools, fluxes, and enzyme levels at the idiophase steady state (at 150 hours of cultivation), were arranged in the metabolic map shown in Figure 3.2. The metabolite concentrations and enzyme activities in Tables 3.1 and 3.2 also refer to 150 hours of fermentation. Supposing average distribution in the mycelium, concentrations were calculated under the assumption that 1 μmol/mg is equivalent to 0.26 mM (Kubicek and Röhr 1978).

A FIRST MATHEMATICAL MODEL

Mass Balance and S-System Equations

The fundamental equations characterizing the dynamics of the dependent concentrations are determined by the concept of mass balance (see Chapters 1 and 2). They are written generically in the following form for each dependent variable or pool:

$$\dot{X}_i = \sum_{j=1}^{n+m} \gamma j v_{ji} - \sum_{j=1}^{n+m} v_{ij} = V_i^+ - V_i^- \qquad i = 1, \ldots, n. \tag{3.1}$$

Each individual metabolite concentration is represented by a variable X_i, and each individual flux from a pool X_i to a pool X_j is denoted as v_{ij}; stoichiometric relationships are taken into account by the parameters γj. The collection of all fluxes contributing to the synthesis of X_i is denoted as V_i^+, and the net transformation flux of X_i is collectively denoted as V_i^-. The numbers of dependent and independent variables are traditionally called n and m, respectively.

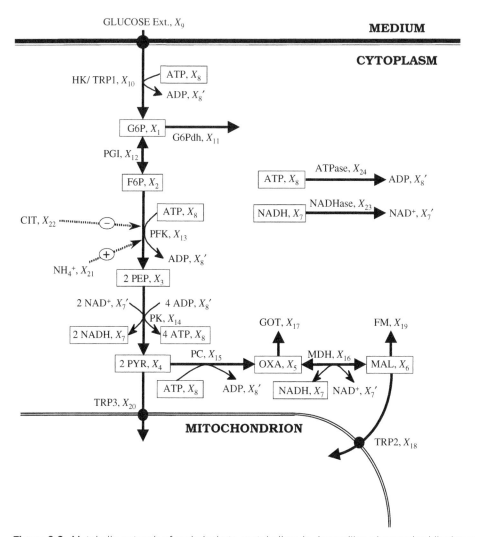

Figure 3.2. Metabolic network of carbohydrate metabolism in *Aspergillus niger* under idiophase conditions of citric acid accumulation. Metabolites are shown in boxes and reactions by arrows. Solid lines indicate reactions or transport processes whereas dashed lines show regulatory interactions. Dots represent transport systems. Dependent variables are numbered from 1 to 8: G6P (X_1, glucose 6-phosphate); F6P (X_2, fructose 6-phosphate); PEP (X_3, phosphoenol pyruvate); PYR (X_4, pyruvate); OXA (X_5, oxalacetate); MAL (X_6, malate); NADH (X_7); ATP (X_8). Independent variables are numbered from 9 to 24: Glucose ext. (X_9, glucose in medium); HK/TRP1 (X_{10}, hexokinase/glucose carrier); G6Pdh (X_{11}, glucose 6-phosphate dehydrogenase); PGI (X_{12}, phosphoglucose isomerase); PFK (X_{13}, phosphofructokinase); NH$_4^+$(X_{21}, ammonium); PK (X_{14}, pyruvate kinase); PC (X_{15}, pyruvate carboxylase); GOT (X_{17}, aspartate aminotransferase); MDH (X_{16}, malate dehydrogenase); TRP2 (X_{18}, mitochondrial malate transport system); TRP3 (X_{20}, mitochondrial pyruvate transport system); FM (X_{19}, fumarase); ATPase (X_{24}); NADHase (X_{23}, dehydrogenases).

Table 3.1. Metabolite Concentrations During Idiophase Steady State (at 150 Hours of Cultivation)

Metabolite	Abbreviation	X_i	Concentration (mM)	References
Glucose in medium	GLUCOSE, Ext	9	290	Kubicek and Röhr 1986; Xu et al. 1989
Glucose 6-phosphate	G6P	1	0.1	Habison et al. 1983
Fructose 6-phosphate	F6P	2	0.025	Habison et al. 1983
Phosphoenol pyruvate	PEP	3	0.01	Reich and Sel'kov 1981
Pyruvate	PYR	4	0.12	Kubicek and Röhr 1978
Oxalacetate	OXA	5	0.00058	Kubicek and Röhr 1978
Malate	MAL	6	2.6	Kubicek and Röhr 1978
Nicotinamide adenine dinucleotide, reduced form	NADH	7	0.035	Führer, Kubicek, and Röhr 1980
Nicotinamine adenine dinucleotide	NAD^+	7'	0.12	Führer et al. 1980
Adenosine triphosphate	ATP	8	0.5	Habison, Kubicek, and Röhr 1979
Adenosine diphosphate	ADP	8'	0.075	Habison et al. 1979
Ammonium	NH_4^+	21	2.35	Habison et al. 1979
Citrate	CIT	22	5.62	Meixner-Monori et al. 1986

For the mechanistic model in Figure 3.2, the mass balance equations take the form

$$\dot{X}_1 = v_{9,1} - [(v_{1,2} - v_{2,1}) + v_{1,0}] = V_1^+ - V_1^-$$
$$\dot{X}_2 = [v_{1,2} - v_{2,1}] - v_{2,3} = V_2^+ - V_2^-$$
$$\dot{X}_3 = 2v_{2,3} - v_{3,4} = V_3^+ - V_3^-$$
$$\dot{X}_4 = v_{3,4} - [v_{4,5} + v_{4,0}] = V_4^+ - V_4^-$$
$$\dot{X}_5 = v_{4,5} - [v_{5,0} + (v_{5,6} - v_{6,5})] = V_5^+ - V_5^-$$
$$\dot{X}_6 = [v_{5,6} - v_{6,5}] - [v_{6,0} + v_{6,0'}] = V_6^+ - V_6^-$$
$$\dot{X}_7 = v_{3,4} - [(v_{5,6} - v_{6,5}) + v_{7,7'}] = V_7^+ - V_7^-$$
$$\dot{X}_8 = v_{3,4} - [v_{0,1} + v_{2,3} + v_{4,5} + v_{8,8'}] = V_8^+ - V_8^-.$$

(3.2)

To facilitate the transition from individual fluxes v_{ij} to aggregated influxes and effluxes V_i^+ and V_i^-, the grouping of fluxes is identified in Eq. (3.2) with brackets. The grouping in this case may actually happen in several combinations of two alternatives, because several steps of the pathway are reversible (see Chapter 2). The first alternative is the so-called *irreversible strategy*, in which one replaces the forward and reverse fluxes between two pools with the resultant net flux. The second

Table 3.2. Enzyme Activities (μmol/min mg Protein) and Fluxes (nmol/min mg Protein) Through Reactions and Transport Steps During Idiophase (at 150 Hours of Cultivation), as Well as Relevant Kinetic and Thermodynamic Constants[a]

Enzyme	Symbol	X_i	Activity	Net Flux	Kinetic and Thermodynamic Constants	References
Hexokinase/substrate transport	HK/Subs	10	0.016	6.3	K_m (ATP) 0.15 K_m (Subs) 0.27	Mischak, Kubicek, and Röhr 1984 Röhr et al. 1987 Schreferl-Kunar et al. 1989
Glucose-6-P dehydrogenase	G6Pdh	11	0.017	1.26	K_m (G6P) 0.045	Jaklitsch, Kubicek, and Scrutton 1991 Legisa and Mattey 1988 Mattey 1992 Kubicek 1987
Phosphoglucose isomerase	PGI	12	15.3	5.04	K_{eq} 0.4	Newsholme and Start 1973
Phosphofructokinase	PFK	13	0.075	5.04	$S_{0.5}$ (F6P) 0.25	Habison et al. 1983 Arts et al. 1987
Pyruvate kinase	PK	14	0.15	10.08	$S_{0.5}$ (ATP) 0.03 K_m (PEP) 0.1 K_m (ADP) 0.07	Meixner-Monori, Kubicek, and Röhr 1983
Pyruvate carboxylase	PC	15	0.015	5.18	K_m (PYRc) 0.28 K_m (ATP) 0.23	Feir and Suzuki 1969 Jaklitsch et al. 1991 Jaklitsch et al. 1991
Malate dehydrogenase	MDH	16	5.6	4.9	K_{eq} 35714.2	Führer et al. 1980 Henson and Cleland 1964
Aspartate aminotransferase	GOT	17	0.048	0.28	K_m (OXAc) 0.088	Osmani and Scrutton 1983
Malate transport	TRP2	18	0.54	4.886	K_m (MALc) 11	Perkins et al. 1973
Fumarase	FM	19	0.02	0.1	K_m (MALc) 0.052	Osmani and Scrutton 1983
Pyruvate transport	TRP3	20	0.045	4.9	K_m (PYRc) 6.2	Briquet 1977 Halestrap et al. 1980
Dehydrogenases	NADHase	23	1	10.068	K_m (NADH) 0.5	—
ATPase	ATPase	24	1	6.83	K_m (ATP) 0.8	Supply, Wach, and Goffeau 1993

[a] K_m and $S_{0.5}$ are given in mM.

alternative is the *reversible strategy* of aggregating all fluxes entering a pool and of all fluxes leaving a pool, independent of whether these fluxes are indeed irreversible or represent one direction of a reversible step (Sorribas and Savageau 1989).

For the initial model, and under the most relevant conditions, all steps were found to be essentially irreversible, and the S-system model was set up according to the irreversible strategy, which is indicated with brackets in Eq. (3.2). The result is shown in Eq. (3.3).

$$\dot{X}_1 = \alpha_1 X_8^{g_{1,8}} X_9^{g_{1,9}} X_{10}^{g_{1,10}} - \beta_1 X_1^{h_{1,1}} X_{11}^{h_{1,11}} X_{12}^{h_{1,12}}$$

$$\dot{X}_2 = \alpha_2 X_1^{g_{2,1}} X_2^{g_{2,2}} X_{12}^{g_{2,12}} - \beta_2 X_2^{h_{2,2}} X_8^{h_{2,8}} X_{13}^{h_{2,13}} X_{21}^{h_{2,21}} X_{22}^{h_{2,22}}$$

$$\dot{X}_3 = \alpha_3 X_2^{g_{3,2}} X_8^{g_{3,8}} X_{13}^{g_{3,13}} X_{21}^{g_{3,21}} X_{22}^{g_{3,22}} - \beta_3 X_3^{h_{3,3}} X_7^{h_{3,7}} X_8^{h_{3,8}} X_{14}^{h_{3,14}}$$

$$\dot{X}_4 = \alpha_4 X_3^{g_{4,3}} X_7^{g_{4,7}} X_8^{g_{4,8}} X_{14}^{g_{4,14}} - \beta_4 X_4^{h_{4,4}} X_8^{h_{4,8}} X_{15}^{h_{4,15}} X_{20}^{h_{4,20}}$$

$$\dot{X}_5 = \alpha_5 X_4^{g_{5,4}} X_8^{g_{5,8}} X_{15}^{g_{5,15}} - \beta_5 X_5^{h_{5,5}} X_6^{h_{5,6}} X_7^{h_{5,7}} X_{16}^{h_{5,16}} X_{17}^{h_{5,17}}$$

$$\dot{X}_6 = \alpha_6 X_5^{g_{5,5}} X_6^{g_{6,6}} X_7^{g_{6,7}} X_{16}^{g_{6,16}} - \beta_6 X_6^{h_{6,6}} X_{18}^{h_{6,18}} X_{19}^{h_{6,19}}$$

$$\dot{X}_7 = \alpha_7 X_3^{g_{7,3}} X_7^{g_{7,5}} X_8^{g_{7,8}} X_{14}^{g_{7,14}} - \beta_7 X_5^{h_{7,5}} X_6^{h_{7,6}} X_7^{h_{7,7}} X_{16}^{h_{7,16}} X_{23}^{h_{7,23}}$$

$$\dot{X}_8 = \alpha_8 X_3^{g_{8,3}} X_7^{g_{8,5}} X_8^{g_{8,8}} X_{14}^{g_{8,14}} - \beta_8 X_2^{h_{8,2}} X_4^{h_{8,4}} X_8^{h_{8,8}} X_9^{h_{8,9}} X_{10}^{h_{8,10}} X_{13}^{h_{8,13}} X_{15}^{h_{8,15}}$$
$$\cdot X_{21}^{h_{8,21}} X_{22}^{h_{8,22}} X_{24}^{h_{8,24}}.$$

$$(3.3)$$

Parameter Estimation and S-System Representation

The transition from the symbolic S-system representation in Eq. (3.3) to a numerically complete model requires the quantitative determination of 70 kinetic orders (30 $g_{i,j}$ and 40 $h_{i,j}$), as well as 16 rate constants (8 α_i and 8 β_i). These parameter values can be estimated with a variety of methods that are discussed in detail in the literature (Savageau 1976; Fell 1992; Voit 2000; see also Chapter 2). Many of these methods are based on four fundamental experimental measurements, namely: the steady-state concentrations of dependent variables; the levels of independent variables, such as enzyme activities; the flux measurements at steady state; and traditional kinetic studies, from which power-law rate laws can be derived through partial differentiation (see Chapter 2). In the present case, the kinetic orders and rate constants were indeed calculated from experimental data that had previously been published within the conceptual framework of Michaelis–Menten rate functions. The values used here are summarized in Tables 3.1 and 3.2; sample estimations are presented in the next section.

Kinetic Orders and Rate Constants
Pyruvate kinase/glyceraldehyde phosphate dehydrogenase reaction. As a first, relatively simple example, let us consider the conversion of X_3 into X_4, which is quantified

by the rate function $v_{3,4}$. The reaction transforms the combined pool of phosphoenolpyruvate plus fructose-1,2-bisphosphate and 1,3-bisphosphoglycerate, X_3, into pyruvate, X_4, through the activity of the pyruvate kinase/glyceraldehyde phosphate dehydrogenase reaction. Because there is no branchpoint at X_3, the rate for this reaction in the S-system model, V_3^-, directly corresponds to $v_{3,4}$ (Figure 3.2 and Eq. 3.2). The power-law representation V_3^- involves those variables that have a direct effect on this process and has the symbolic form

$$V_3^- = \beta_3 X_3^{h_{3,3}} X_7^{h_{3,7}} X_8^{h_{3,8}} X_{14}^{h_{3,14}}.$$

If suitable experimental data were available, one could determine each kinetic order from a plot of V_3^- against one of the contributing variables. However, such direct data are not at hand, and we compute instead each kinetic order by partial differentiation of a published rate function with respect to the corresponding variable. As described in Chapter 2, the partial derivative is evaluated at the nominal steady state, where the traditional and the power-law rate laws have exactly the same value. The relevant quantities of this state are given below (see also Tables 3.1 and 3.2).

X_{3S} (phosphoenol pyruvate):	0.01 mM
$X_{7S'}$ (NAD$^+$):	0.12 mM
$X_{8S'}$ (ADP):	0.075 mM
X_{14} (pyruvate kinase):	0.15 units/mg
V_3^- (overall degradation flux of X_3):	10.08 nmol/min mg

The kinetic order $h_{3,3}$ quantifies the effect of phosphoenol pyruvate (PEP), X_3, on its own degradation through the pyruvate kinase (PK) activity. In freshly prepared extracts of *A. niger*, PK exhibits Michaelis–Menten kinetics toward PEP with a Michaelis constant K_M of 0.1 mM (Meixner–Monori et al. 1983). Direct (partial, logarithmic) differentiation of the Michaelis-Menten rate law (MMRL) and evaluation with the steady-state values presented here yields the kinetic order $h_{3,3}$ as

$$h_{33} = \frac{\partial \text{MMRL}}{\partial X_3} \cdot \frac{X_{3S}}{\text{MMRL}_S} = \frac{K_M}{K_M + X_{3S}} = \frac{0.1}{0.1 + 0.01} = 0.9$$

(Chapter 2; Voit 2000, Eq. 5.25).

The kinetic orders $h_{3,7}$ and $h_{3,8}$ are -0.028 and -3.196, respectively, as we will show after we have discussed constraints on NAD$^+$ and ADP. The kinetic order $h_{3,14}$ reflects the influence of the PK activity on the PEP pool tranformation. Assuming direct proportionality between velocity and enzyme activity, the numerical value of the kinetic order is $h_{3,14} = 1$.

The rate constant is computed from equivalence of V_3^- and v_{34} at steady state. Thus,

$$v_{34} = \beta_3 X_3^{h_{3,3}} X_7^{h_{3,7}} X_8^{h_{3,8}} X_{14}^{h_{3,14}} \qquad \text{at steady state}$$

$$10.08 = \beta_3 \cdot 0.01^{0.9} \cdot 0.035^{-0.028} \cdot 0.5^{-3.196} \cdot 0.15.$$

Accounting for the differences in units (metabolites in mM; fluxes in nM per minute per milligram dry weight), we obtain $\beta_3 = 0.42$.

Glucose 6-phosphate degradation. As the second example, consider the kinetic orders and rate constant in the reaction that removes glucose 6-phosphate (X_1). The aggregated rate for the reaction that removes glucose 6-phosphate from the pool X_1 is given by

$$V_1^- = [(v_{1,2} - v_{2,1}) + v_{1,0}]$$

(see Eq. 3.2). The corresponding power-law representation in symbolic form can be written down immediately and without any further information. It involves those and only those variables that have a direct effect on the process. In other words, it includes all variables that contribute to one or more of the steps $v_{1,2}$, $v_{2,1}$, or $v_{1,0}$. The result is

$$V_1^- = \beta_1 X_1^{h_{1,1}} X_2^{h_{1,2}} X_{11}^{h_{1,11}} X_{12}^{h_{1,12}}.$$

Each kinetic order is computed via partial differentiation of the alleged traditional rate function(s) with respect to the variable of interest, and the partial derivative is evaluated at the nominal steady state, at which the traditional and the power-law rate laws have exactly the same value. The relevant quantities of this state were measured as

X_1 (glucose 6-phosphate):	0.1 mM
X_2 (fructose 6-phosphate):	0.025 mM
X_{11} (glucose 6-phosphate dehydrogenase activity):	0.017 units/mg
X_{12} (phosphoglucose isomerase activity):	15.3 units/mg
V_1^- (overall degradation flux of X_1):	0.0063 (μM/min/mg dr. wt.)

(see Tables 3.1 and 3.2). The kinetic orders are computed first, followed by the rate constant. The computations are not quite as straightforward as in the first example and in Chapter 2, but still strictly adhere to the strategy of differentiation and evaluation at the chosen operating point. We discuss them one at a time.

The kinetic order $h_{1,1}$ quantifies the effect of glucose 6-phosphate, X_1, on its own degradation, which involves three fluxes, namely, v_{12}, v_{21}, and v_{10}. X_1 affects all three, and the partial differentiation must therefore include them all. The computation proceeds as follows:

$$h_{1,1} = \frac{\partial V_1^-}{\partial X_1} \frac{X_1}{V_1^-} = \frac{\partial (v_{1,2} - v_{2,1})}{\partial X_1} \frac{X_1}{V_1^-} + \frac{\partial v_{1,0}}{\partial X_1} \frac{X_1}{V_1^-} = \frac{\partial v_n}{\partial X_1} \frac{X_1}{v_n} \frac{v_n}{V_1^-} + \frac{\partial v_{1,0}}{\partial X_1} \frac{X_1}{v_{1,0}} \frac{v_{1,0}}{V_1^-}.$$

In the last terms of the equation, v_n codes for the net velocity of the reversible step; that is, $v_n = (v_{1,2} - v_{2,1})$. Using this quantity, the equation for $h_{1,1}$ can also be formulated as

$$h_{1,1} = h'_{1,1} \frac{v_n}{V_1^-} + h''_{1,1} \frac{v_{1,0}}{V_1^-},$$

where

$$h'_{1,1} = \frac{\partial v_n}{\partial X_1} \frac{X_1}{v_n}$$

$$h''_{1,1} = \frac{\partial v_{1,0}}{\partial X_1} \frac{X_1}{v_{1,0}}.$$

This particular formulation demonstrates that the value of $h_{1,1}$ is composed of the weighted contributions of two terms, $h'_{1,1}$ and $h''_{1,1}$. The first quantifies the influence of glucose 6-phosphate on the synthesis of fructose 6-phosphate through the phospho-glucose isomerase reaction. The second expression quantifies the influence of glucose 6-phosphate on its own transformation via the glucose 6-phosphate dehydrogenase activity.

Groen et al. (1982) demonstrated that the kinetic order for the substrate of a reversible enzyme with Michaelis–Menten kinetics is given by

$$h'_{1,1} = \frac{1}{1 - \frac{\Gamma}{K_{eq}}} \left(1 - \frac{J}{V^f_{max}}\right),$$

where V^f_{max} is the maximum velocity of the forward reaction, that is, of substrate consumption; K_{eq}, the equilibrium constant; J, the net flux throughout the enzyme; and Γ is the mass action ratio. Substitution of the corresponding values yields a value of $h'_{1,1} = 2.66$.

$h''_{1,1}$ is obtained from direct differentiation of the alleged Michaelis–Menten kinetics of glucose 6-phosphate dehydrogenase toward glucose 6-phosphate at the steady-state substrate concentrations, yielding the value $h''_{1,1} = 0.31$. Substituting these values in the expression of $h_{1,1}$, one obtains $h_{1,1} = 2.19$.

The expression for the kinetic order $h_{1,2}$ is

$$h_{1,2} = \frac{\partial V^-_1}{\partial X_2} \frac{X_2}{V^-_1} = \frac{\partial(v_{1,2} - v_{2,1})}{\partial X_2} \frac{X_2}{V^-_1} + \frac{\partial v_{1,0}}{\partial X_1} \frac{X_2}{V^-_1}$$

$$= h'_{1,2} \frac{v_n}{V^-_1} + h''_{1,2} \frac{v_{1,0}}{V^-_1}$$

$$h'_{1,2} = \frac{\partial v_n}{\partial X_2} \frac{X_2}{v_n}$$

$$h''_{1,2} = \frac{\partial v_{1,0}}{\partial X_2} \frac{X_2}{v_{1,0}}.$$

Groen et al. (1982) showed that the kinetic order for the product of a reversible enzyme with Michaelis–Menten kinetics may be computed as

$$h'_{1,2} = -\frac{\frac{\Gamma}{K_{eq}}}{1 - \frac{\Gamma}{K_{eq}}} \left(1 - \frac{J}{V^r_{max}}\right),$$

where J, Γ, and K_{eq} are defined as in the previous paragraphs, and V^r_{max} is the maximum velocity of the reverse reaction. After substitution of the corresponding values, one directly obtains $h'_{1,2} = -1.66$. Additionally, $h''_{1,2}$ is zero, because $V_{1,0}$ is not affected by X_2. The result is therefore $h_{1,2} = -1.333$.

The calculation of $h_{1,11}$ begins with the same *Ansatz*

$$h_{1,11} = \frac{\partial V_1^-}{\partial X_{11}} \frac{X_{11}}{V_1^-} = \frac{\partial (v_{1,2} - v_{2,1})}{\partial X_{11}} \frac{X_{11}}{V_1^-} + \frac{\partial v_{1,0}}{\partial X_{11}} \frac{X_{11}}{V_1^-},$$

which may be reformulated as

$$h_{1,11} = h'_{1,11} \frac{v_{1,0}}{V_1^-}$$

$$h'_{1,11} = \frac{\partial v_{1,0}}{\partial X_{11}} \frac{X_{11}}{v_{1,0}}.$$

Assuming direct proportionality between velocity and enzyme activity suggests $h'_{1,11} = 1$, which results in $h_{1,11} = 0.2$. The same procedure applies to $h_{1,12}$, yielding a value of 0.8.

The value of the rate constant β_1 is determined from the set of values for the rate, the concentrations, and the kinetic orders we just determined. The result is

$$\beta_1 = \frac{V_1^-}{X_1^{h_{1,1}} X_2^{h_{1,2}} X_{11}^{h_{1,11}} X_{12}^{h_{1,12}}} = 1.833 \cdot 10^{-3}.$$

Fructose 6-phosphate degradation. As a final example, consider the inhibitory effect of citrate on the phosphofructokinase I activity, which catalyzes the conversion of X_2 to X_3 and therefore appears in the form of the two kinetic orders $h_{2,22}$ and $g_{3,22}$. Because the degradation of X_2 and the synthesis of X_3 in fact constitute the same process, the two kinetic orders may be coded with different symbols, but have the same numerical value.

The kinetic order $h_{2,22}$ is determined directly from published experimental results of Habison et al. (1983), who studied the regulatory properties of phosphofructo-kinase in *Aspergillus niger*. Among a variety of features, the authors investigated the inhibition of *A. niger* phosphofructokinase by citrate under different conditions.

Figure 3.3 shows phosphofructokinase activity over the relevant range of citrate concentrations. The study producing these results was carried out under conditions resembling the molecular medium during citric acid accumulation that is relevant here. Specifically, the authors used 0.5 mM ATP, 10 mM inorganic phosphate, and 1 mM of fructose 6-phosphate, which rather well approximates the steady-state conditions in our scenario. The kinetic order $h_{2,22}$ measures the slope of the process as a function of citrate, in logarithmic coordinates (Figure 3.3B). In the range close to the nominal concentration of interest, 5.6 mM citrate, the natural logarithm of phosphofructokinase activity is 1.7, and the measured slope and $h_{2,22}$, is -0.3. The value is negative, which is to be expected, because citrate has an inhibitory effect.

Constraints

In addition to kinetic orders and rate constants that are computed with straightfor-ward differentiation, the system also contains some constraints for which the model

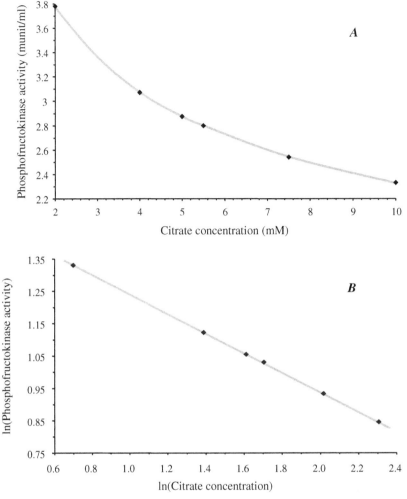

Figure 3.3. Phosphofructokinase activity decreases with increasing citrate concentration (Habison et al. 1983). Panel A shows the original data in Cartesian coordinates. Representation in logarithmic coordinates (panel B) provides the corresponding kinetic $h_{2,22}$ order directly as the slope of the graph.

and its parameter values must account. These constraints deal with the total amounts of NADH and NAD^+ and of ATP and ADP, which are essentially constant and coded as X_{TN} and X_{TA}, respectively. Because one metabolite in each of these pairs is a dependent variable, the other one also changes over time. To avoid requiring four independent variables and subsequently assuring constancy of the total pools, one may use the constraint equations and approximate them with power-law functions, as it is done with other rate functions (see also Appendix).

The constraints in their original form read

$$X_7 + X_7' = X_{TN}$$
$$X_8 + X_8' = X_{TA}.$$

As functions of the constant amounts and the explicit system variables, the variables $X_7'(NAD^+)$ and X_8' (ADP) are expressed in the power-law formalism as

$$X_7' = \gamma_7 X_{TN}^{f_{7',TN}} X_7^{f_{7',7}}$$

$$X_8' = \gamma_8 X_{TA}^{f_{8',TA}} X_8^{f_{8',8}}.$$

The kinetic orders $f_{7',TN}$, $f_{7',7}$, $f_{8',TA}$, and $f_{8',8}$ are computed like other $g_{i,j}$ or $h_{i,j}$ parameters. For instance,

$$f_{7',TN} = \frac{\partial X_7'}{\partial X_{TN}} \cdot \frac{X_{TN}}{X_7'} = \frac{\partial(X_{TN} - X_7)}{\partial X_{TN}} \cdot \frac{X_{TN}}{X_7'} = 1 \cdot \frac{X_{TN}}{X_7'} = \frac{X_{TN}}{X_7'}.$$

In the same fashion, we obtain

$$f_{7',7} = \frac{\partial X_7'}{\partial X_7} \frac{X_7}{X_7'} = -\frac{X_7}{X_7'}$$

$$f_{8',TA} = \frac{\partial X_8'}{\partial X_{TA}} \frac{X_{TA}}{X_8'} = \frac{X_{TA}}{X_8'}$$

$$f_{8',8} = \frac{\partial X_8'}{\partial X_8} \frac{X_8}{X_8'} = -\frac{X_8}{X_8'}.$$

The rate constant parameters γ_7 and γ_8 are analogous to the standard rate constants α and β and are calculated in the same fashion. For example, one has

$$\gamma_7 = \frac{X_7'}{X_{TN}^{f_{7',TN}} X_7^{f_{7',7}}},$$

which is evaluated at the steady state. With $X_7 = 0.035$, $X_7' = 0.12$, and $X_{TN} = 0.155$, we thus obtain

$$\gamma_7 = \frac{0.12}{0.155^{0.155/0.12} \cdot 0.035^{0.035/0.12}} = 3.55.$$

Wherever X_7' and X_8' appear in the S-system representation, they are now replaced by their corresponding power-law terms in X_7, X_T, X_8, and $X_{T'}$.

The manipulation of constraints allows us to compute the kinetic orders $h_{3,7}$ and $h_{3,8}$, which we used for the power-law representation of v_{34}.

Rate v_{34} revisited. The kinetic order $h_{3,7}$ quantifies the influence of the dependent system variable X_7 (NADH) on the degradation of PEP, v_{34}. The assessment of this effect is slightly more complicated because the degradation of PEP depends only indirectly on NADH, namely through the constraint NADH + NAD$^+$ = X_{TN} = constant. If NAD$^+(X_7')$ were an explicit variable, we would straightforwardly compute its kinetic order as

$$h_{3,7'} = \frac{\partial v_{34}}{\partial X_7'} \frac{X_7'}{v_{34}}.$$

The constraint equations allow us to compute $h_{3,7}$ from this expression in the following fashion. First, according to its definition, we have

$$h_{3,7} = \frac{\partial v_{34}}{\partial X_7} \frac{X_7}{v_{34}}.$$

Applying the chain rule, we find

$$h_{3,7} = \frac{\partial v_{34}}{\partial X_7'} \frac{\partial X_7'}{\partial X_7} \frac{X_7}{v_{34}}.$$

The constraint $X_7' = X_{TN} - X_7$ leads directly to

$$\frac{\partial X_7'}{\partial X_7} = -1.$$

Substitution of this result into the expression for $h_{3,7}$ and multiplication and division with X_7' yield

$$h_{3,7} = \frac{\partial v_{34}}{\partial X_7'} \cdot \frac{\partial X_7'}{\partial X_7} \cdot \frac{X_7}{v_{34}} \cdot \frac{X_7'}{X_7'} = -1 \cdot \frac{\partial v_{34}}{\partial X_7'} \cdot \frac{X_7'}{v_{34}} \cdot \frac{X_7}{X_7'} = -\frac{X_7}{X_7'} \cdot h_{3,7'}.$$

In general, one obtains for constraints of this type a relationship like

$$h_{3,7} = f_{7',7} \cdot h_{3,7'},$$

where $f_{7',7}$ is defined as previously shown. Thus, the dependence of the flux on the explicit system variable X_7 is readily computed from the direct dependence of the flux on the implicit variable X_7'.

In our particular example, the kinetic order $f_{7',7}$ has a value of -0.29 at the reference steady state. NAD^+ influences the PEP transformation through the activity of the glyceraldehyde phosphate dehydrogenase. This enzyme has not yet been characterized in *A. niger*, but data are available from other systems, where a Michaelis constant of about 0.013 mM was reported (Crow and Wittenberger 1979; Sakai, Hasumi, and Endo 1990). Accordingly, we obtain $h_{3,7'} = 0.097$ and subsequently $h_{3,7} = -0.028$. The negative value of $h_{3,7}$ is a consequence of the fact that the sum $NAD^+ + NADH$ is constant. Thus, any increase in NAD^+ means a concomitant decrease in NADH and consequently a loss of rate through the reaction.

The kinetic order $h_{3,8}$ describes the influence of ADP on the transformation of PEP. As before, the model constraint translates into

$$h_{3,8} = f_{8',8} \cdot h_{3,8'}.$$

Evaluation of $f_{8',8}$ at the reference steady state yields -6.66. PK may be described with a Michaelis–Menten rate law (Meixner-Monori et al. 1983), with a K_M value for ADP of 0.07 mM. Differentiation yields $h_{3,8'} = 0.48$. Accordingly, $h_{3,8}$ is equal to -3.196. The negative magnitude of this kinetic order is to be interpreted as in the previous case.

First Numerical Model

Once all parameter values are computed from measurements, partial differentiation, consideration of constraints, and evaluation at the nominal point of interest, the numerical S-system representation of citric acid metabolism in *A. niger* is complete. It has the numerical specifications:

$$\dot{X}_1 = 0.459 X_8^{0.23} X_9^{0.0009} X_{10} - 1.833 \cdot 10^{-3} X_1^{2.194} X_2^{-1.333} X_{11}^{0.2} X_{12}^{0.8}$$

$$\dot{X}_2 = 3.25 \cdot 10^{-4} X_1^{2.665} X_2^{-1.667} X_{12} - 1.861 X_2^{0.9} X_8^{0.056} X_{13} X_{21}^{0.65} X_{22}^{-0.3}$$

$$\dot{X}_3 = 3.722 X_2^{0.9} X_8^{0.056} X_{13} X_{21}^{0.65} X_{22}^{-0.3} - 0.42 X_3^{0.9} X_7^{-0.028} X_8^{-3.196} X_{14}$$

$$\dot{X}_4 = 0.42 X_3^{0.9} X_7^{-0.028} X_8^{-3.196} X_{14} - 2.538 X_4^{0.836} X_8^{0.161} X_{15}^{0.509} X_{20}^{0.486}$$

$$\dot{X}_5 = 1.895 X_4^{0.7} X_8^{0.315} X_{15} - 4.23 \cdot 10^{14} X_5^{3.2} X_6^{-2.2} X_7^{3.788} X_{16}^{0.945} X_{17}^{0.054}$$

$$\dot{X}_6 = 2.04 \cdot 10^{15} X_5^{0.7} X_6^{-2.33} X_7^{4.005} X_{16} - 0.005 X_6^{0.813} X_{18}^{0.979} X_{19}^{0.02}$$

$$\dot{X}_7 = 0.204 X_3^{0.873} X_7^{-0.028} X_8^{-4.06} X_{14} - 868.709 X_5^{1.618} X_6^{-1.136} X_7^{2.203} X_{16}^{0.468} X_{23}^{0.513}$$

$$\dot{X}_8 = 0.841 X_3^{0.9} X_7^{-0.028} X_8^{-3.196} X_{14} - 0.308 X_2^{0.225} X_4^{0.179} X_8^{0.257} X_9^{0.294}$$

$$\cdot X_{10}^{0.312} X_{13}^{0.25} X_{15}^{0.256} X_{21}^{0.162} X_{22}^{-0.075} X_{24}^{0.18}.$$

$$(3.4)$$

For computational analyses, one also needs initial values. These are typically determined as perturbations about the steady state and not very critical.

Steady State

The S-system representation (Eq. 3.4) of carbohydrate metabolism in *Aspergillus niger* has a nonzero steady-state solution that, in principle, could be determined analytically (Savageau 1969). Without doubt, however, it is easier to obtain the solution numerically with the software package PLAS (Ferreira 2000). The numerical steady-state values coincide with those chosen or observed from experiments. This is to be expected, because the observed steady-state values were used to determine the kinetic orders and rate constants for the S-system representation.

Local Stability

Local stability is confirmed by examination of the eigenvalues of the local representation of the system at its steady state (Savageau 1975). Table 3.3 shows that the real parts of all eigenvalues are negative, which indicates that the nominal steady state is locally stable and that the system will return to this steady state following small perturbations. One also notes that not all imaginary parts are zero, which allows for the possibility of damped oscillatory responses to perturbations. Finally, the magnitudes of the real parts of all eigenvalues indicate the relative time scales of system responses. The smallest values are related to the slowest processes, whereas the larger values are associated with the faster processes.

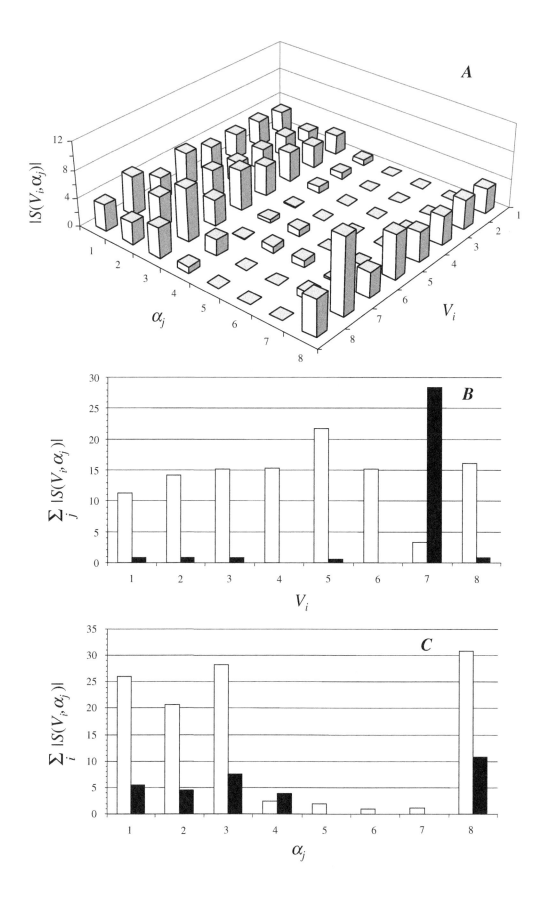

**Table 3.3. Eigenvalues of the First
S-System Model of Citric Acid
Metabolism in *A. niger* During Idiophase**

$$-1.84 \cdot 10^{-3}$$
$$-2.13 \cdot 10^{-3}$$
$$-4.48 \cdot 10^{-2}$$
$$-7.59 \cdot 10^{-2} + 3.89 \cdot 10^{-3}i$$
$$-7.59 \cdot 10^{-2} - 3.89 \cdot 10^{-3}i$$
$$-0.61$$
$$-1.07$$
$$-57.97$$

Robustness

The term *robustness* addresses the question of how much the system is affected by small, structural changes, that is, by permanent variations in one or some of its independent variables or parameter values. Robustness in this sense is typically assessed by computing partial derivatives of steady-state concentrations and fluxes with respect to all parameters, one at a time. These derivatives quantify *systemic responses* and are called *sensitivities* and *(logarithmic) gains*. Assessments of effects of simultaneous changes in two parameters (so-called *synergism coefficients*) are somewhat more complex, requiring tensor algebra (Salvador 2000a,b). The algebraic assessment of large changes in parameters is in general an unsolved problem.

Sensitivities

The influences of small changes in rate constant or kinetic order parameters on metabolites and fluxes of the system are distributed as shown in Figures 3.4–3.7. The histograms represent both types of sensitivities (effects on fluxes and metabolite concentrations) with respect to both types of parameters (rate constants and kinetic orders). They provide a quick impression of how the sensitivities are distributed throughout the model.

Rate constant sensitivities. These sensitivities are presented in Figures 3.4 and 3.5. Although there are sixteen rate constants, results are presented only with respect to the eight αs. This is sufficient, because the sensitivities for the β rate constants can be obtained from the sensitivities for the corresponding α rate constants. With respect to

Figure 3.4. Sensitivities of fluxes V_i with respect to changes in rate constants α_j. Panel A: Most, but not all, sensitivities have magnitudes between 0 and 10, indicating small or moderate effects of alterations in rate constants. Panel B: Magnitudes for a given flux V_i are summed over all the rate constant parameters α_j. Panel C: Magnitudes for a particular flux rate constant parameter α_j are summed over all fluxes. The solid bars in each projection represent the sum of the negative sensitivities, and the open bars represent the sum of the positive sensitivities. Nomenclature of dependent variables is given in Figure 3.2.

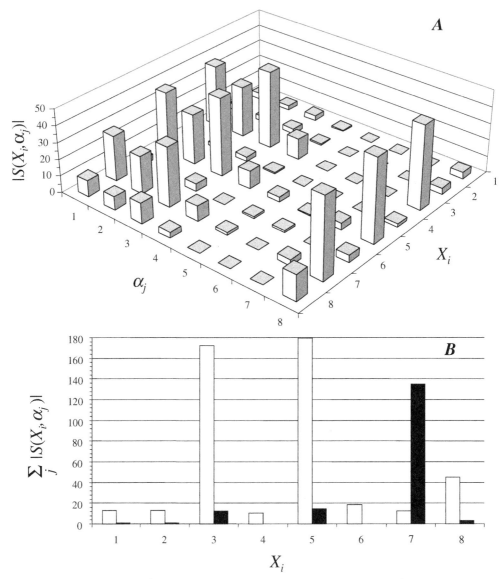

Figure 3.5. Sensitivities of metabolite concentrations with respect to changes in rate constants. Panel A: A quarter of the sensitivities have magnitudes exceeding 10, which may be a reason for concern. Panel B: The two-dimensional projection shows the magnitudes for a particular metabolite concentration X_i, summed over all the rate constant parameters α_j. Panel C: The projection exhibits the magnitudes for a particular rate constant parameter α_j summed over all the metabolite concentrations X_i. The solid bars in each projection represent the sum of the negative sensitivities, and the open bars represent the sum of the positive sensitivities. Nomenclature of dependent variables is given in Figure 3.2.

metabolites, each β_i sensitivity is simply the negative of the α_i sensitivity (see Chapter 2). For fluxes, the sensitivity for a β rate constant can again be obtained by negating the value for the α parameter, except for the sensitivities on the diagonal, for which one first subtracts 1 and then negates the values.

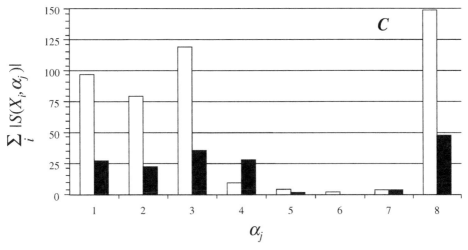

Figure 3.5 *(continued)*

Figure 3.4A demonstrates that almost all rate constants have some influence. Of the total sixty-four sensitivities, thirty-one (48.4%) are below 1, meaning attenuation of change; thirty-two (50%) are between 1 and 10, meaning modest amplification; and one (1.6%) is above 10. The largest sensitivities are associated with fluxes through NADH (X_7) and oxalacetate (X_5). This can be seen more clearly in Figure 3.4B, where the magnitudes for a given flux V_i have been summed over all the independent variables X_k. In this and other plots of this type, the solid portion of each bar represents the sum of the negative sensitivities (or logarithmic gains), whereas the open portion represents the sum of the positive sensitivities (or logarithmic gains). In panel C, the magnitudes for a given independent variable X_i are summed over all fluxes.

The effects of the rate constants on the metabolite concentrations are distributed as shown in Figure 3.5. Here, the largest sensitivities are associated with fructose 6-phosphate (X_3), oxalacetate (X_5), and NADH (X_7), and to a lesser extend with ATP (X_8). These pools are highly influenced by most of the rate constants. Furthermore, the sensitivities are about one order of magnitude higher than those through other pools. Of the sixty-four sensitivities total, twenty-two (34.4%) are below 1, twenty-six (40.6%) are between 1 and 10, and sixteen (25%) are above 10.

Kinetic order sensitivities for fluxes and metabolite pools. The influences of the kinetic orders g and h on fluxes are presented in Figure 3.6. The system contains thirty gs, forty hs, and eight fluxes, thus leading to a total of $70 \times 8 = 560$ sensitivities. The three-dimensional plot indicates that all fluxes are influenced to some degrees by changes in any of the kinetic orders. Almost half of the sensitivities (279, 49.8%) are below 1, 185 (33%) are between 1 and 10, and the remaining 96 (17.1%) are above 10.

The influence of the kinetic orders on the metabolite concentrations is shown in Figure 3.7. Essentially the same parameters that influence the fluxes have also a significant effect on the pools (forty-eight kinetic orders out of seventy). The more sensitive pools are fructose 6-phosphate (X_3), oxalacetate (X_5), NADH (X_7), and to a

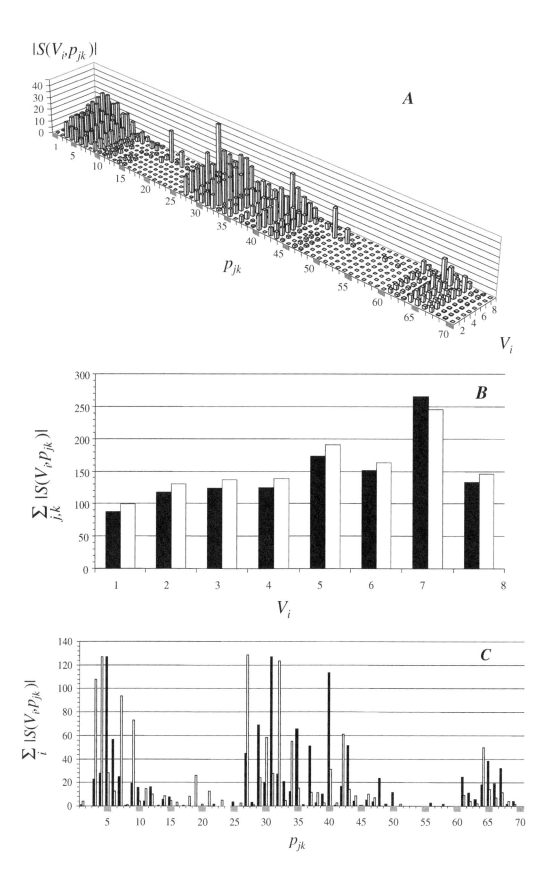

lesser extent ATP (X_8). Of the 560 kinetic order sensitivities, 216 (38.6%) are below 1, 170 (30.4%) are between 1 and 10, and 174 (31.1%) are above 10.

One must note that all sensitivities discussed here are unconstrained. In other words, it is assumed that the change in one parameter is independent of all other parameters. This is clearly not always the case. For instance, the degradation term of X_3 is identical to the production term of X_4, and the parameters in the two terms are equivalent. Nonetheless, the unconstrained sensitivities provide a simple and very indicative diagnostic tool. Issues of constraints among sensitivities are discussed in Salvador (2000a,b) and Voit (2000).

Logarithmic Gains

The numerical characterization of the steady state clearly depends on the values of the independent variables, and changes in the latter usually affect the numerical features of the steady state. These effects are quantified by logarithmic gains (see Chapter 2). As in the case of sensitivities, they are distinguished with respect to fluxes and metabolite concentrations.

Flux logarithmic gains. Figure 3.8 shows the influence of alterations in independent variables on the fluxes through the pools of the pathway. All fluxes are almost equally affected; an exception is the NADH flux (V_7), which reacts more strongly to changes. The impact of perturbations in independent variables can be seen more clearly in Figure 3.8B, where the magnitudes for a given flux V_i are summed over all independent variables X_k. In panel C, the magnitudes for a given independent variable

#	1	2	3	4	5	6	7	8	9	10
g	1,8	1,9	1,10	2,1	2,2	2,12	3,2	3,8	3,13	3,21
#	11	12	13	14	15	16	17	18	19	20
g	3,22	4,3	4,7	4,8	4,14	5,4	5,8	5,15	6,5	6,6
#	21	22	23	24	25	26	27	28	29	30
g	6,7	6,16	7,3	7,7	7,8	7,14	8,3	8,7	8,8	8,14
#	31	32	33	34	35	36	37	38	39	40
h	1,1	1,2	1,11	1,12	2,2	2,8	2,13	2,21	2,22	3,3
#	41	42	43	44	45	46	47	48	49	50
h	3,7	3,8	3,14	4,4	4,8	4,15	4,20	5,5	5,6	5,7
#	51	52	53	54	55	56	57	58	59	60
h	5,16	5,17	6,6	6,18	6,19	7,5	7,6	7,7	7,16	7,23
#	61	62	63	64	65	66	67	68	69	70
h	8,2	8,4	8,8	8,9	8,10	8,13	8,15	8,21	8,22	8,24

Figure 3.6. Sensitivities of fluxes with respect to changes in kinetic orders. Panel A: Many of the sensitivities have small magnitudes, but 17% exceed 10, which may be problematic. Panel B: Two-dimensional projection of the magnitudes for a particular flux V_i, summed over all kinetic order parameters p_{jk}. Panel C: Projection of the magnitudes for a particular kinetic parameter p_{jk}, summed over all the fluxes V_i. The open and solid bars in each projection represent the sum of the positive and negative logarithmic gains, respectively. The kinetic orders are lexicographically presented according to the assignments in the table. For instance, parameter #31 is $h_{1,1}$.

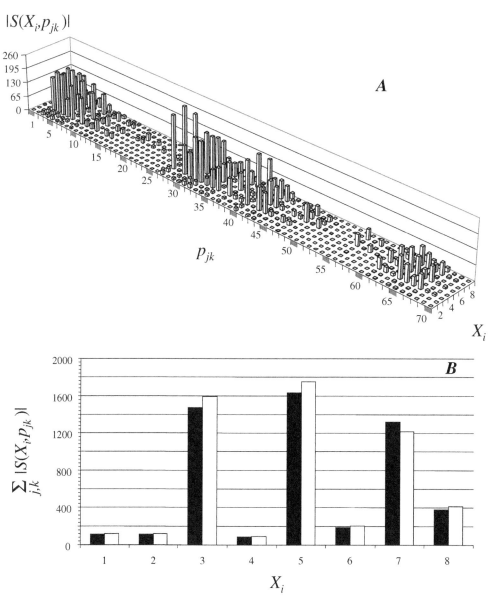

Figure 3.7. Sensitivities of metabolite concentrations with respect to changes in kinetic orders. Panel A: The same kinetic orders that have a strong effect on fluxes (Figure 3.6) also affect the metabolite concentrations in a significant fashion. Panels B and C show two-dimensional projects as in Figure 3.6. Parameters in Panels A and C are numbered as in Figure 3.6.

X_i are summed over all the fluxes. The almost entirely positive control profile of hexokinase/substrate transport (X_{10}) suggests that this step is most important for controlling – and especially, increasing – the fluxes throughout the system.

Quite a few experimental findings correspond well with these results. Mischak et al. (1984) showed that sugar transport is a regulatory step in this pathway.

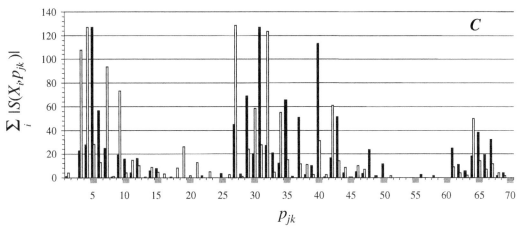

Figure 3.7 *(continued)*

Schreferl-Kunar et al. (1989) investigated mutants with increased hexokinase activity and found that they produced citric acid with better yields. Supporting this result, Steinbock et al. (1991) found that mutants with reduced hexokinase activity produced considerably less citric acid.

Glucose 6-phosphate dehydrogenase (X_{11}) has a negative impact on the flux. This finding is easy to interpret, given that its flux diverts material away from the main glycolytic branch.

Also significant for the control of fluxes is the role of the pyruvate carboxylase (X_{15}) and of the pyruvate transport system (X_{20}). Their activity levels seem to be critical for citrate synthesis, an observation that has been pointed out previously by several authors (e.g., Bloom and Johnson 1962).

Another result is the lacking influence of pyruvate kinase on the fluxes through the pathway. This result confirms previous suggestions by Meixner-Monori et al. (1983), who concluded that this enzyme is not a control point of glycolysis.

Taken together, the model results on flux gains are in good agreement with experimental studies on the control of the system, and indirectly provide some justification for the underlying model assumptions.

Logarithmic gains with respect to metabolite concentrations. Figure 3.9 shows the control structure of the system with respect to metabolite concentrations, as measured by concentration logarithmic gains. In panel B, the magnitudes for a given metabolite concentration X_i are summed over all independent variables X_k. It is evident that four of the eight pools, namely phosphoenolpyruvate (X_3), oxalacetate (X_5), NADH (X_7), and ATP (X_8), are significantly affected by changes in the independent variables, whereas the others are relatively unchanged.

Figure 3.9C shows the overall influence of each independent variable on all metabolite concentrations. The magnitudes of influence suggest three groups of independent variables. The group with the largest influence consists of the substrate (X_9), the activity at the initial hexokinase-substrate transport step (X_{10}), and ATPase (X_{24}). As in the previous case of flux gains, the hexokinase step has a positive

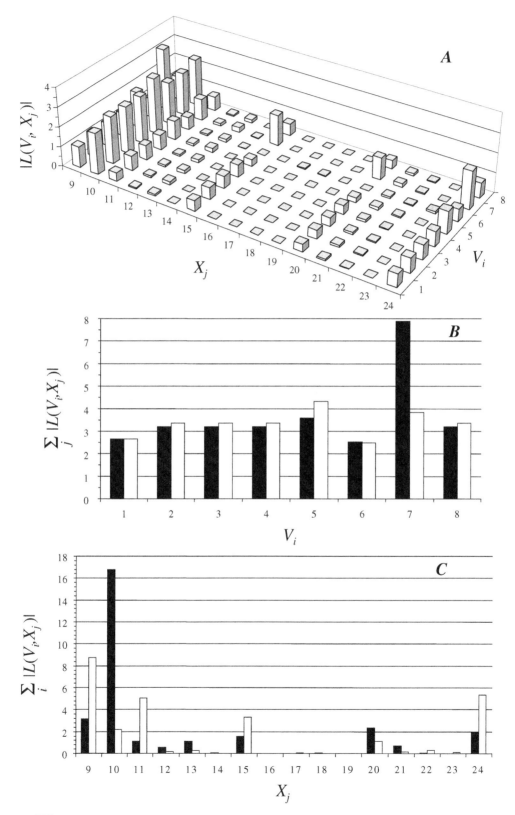

influence on most of the pools, particularly on phosphoenolpyruvate (X_3) and ox-alacetate (X_5); the influence of ATPase (X_{24}) is mainly negative. The second group is formed by glucose 6-phosphate dehydrogenase (X_{11}), pyruvate carboxylase (X_{15}), and pyruvate transport (X_{20}). Glucose 6-phosphate dehydrogenase and pyruvate carboxylase exert a mostly negative impact on the pathway, whereas the impact of pyruvate transport is mainly positive. The rest of the independent variables have a negligible influence on the pools.

Dynamic Behavior

Dynamic responses are most easily assessed with PLAS. As an example, Figure 3.10 shows the results of a simulated twofold increase in the dependent variable X_1 (glucose 6-phosphate). The twofold bolus is partially transformed into fructose 6-phosphate (X_2) and subsequently into phosphoenylpyruvate (X_3). Both X_2 and X_3 overshoot the basal steady-state concentrations by 35–50% over a time period of five minutes, before returning very slowly to the initial steady state. Subsequently, phosphoenylpyruvate is transformed into pyruvate (X_4), initially (for about twenty minutes) at the expense of synthesis of oxalacetate (X_5), which exhibits fluctuations of about 20% around the nominal steady state. From this point on, the perturbation propagates through the anaplerotic oxalacetate synthesis branch, causing 10–20% deviations from the basal steady state. The transient behavior is characterized by damped oscillations that reflect the nonzero imaginary parts of some of the eigenvalues. The time courses show that it takes quite some time (more than seventy minutes) after the initial transition to return to the vicinity of the original steady state. This sluggish response is a cause for concern. Another dynamic analysis of this system furthermore demonstrated that changes in glucose input could lead to severe imbalances between the moieties of ATP and ADP and of NADH and NAD (Voit and Ferreira 1998). Thus, although the steady-state and robustness analyses produced acceptable results, the dynamic system responses to simple changes in input are severe enough to require a second look at the present model.

Limitations and Shortcomings of the Original Model

The model of carbohydrate metabolism in *A. niger* under conditions of citric acid production, as we have discussed it so far, has a stable steady-state solution that coincides with observations and is reasonably robust. Nevertheless, it has also become

---

Figure 3.8. Logarithmic gains of fluxes with respect to changes in independent variables. Panel A: Except for the NADH flux (V_7), all fluxes are affected to a similar degree. Panel B: Magnitudes for a given flux V_i are summed over all independent variables X_k. Panel C: Magnitudes for a given independent variable X_i are summed over all fluxes. The solid bars in each projection represent the sum of negative logarithmic gains and the open bars represent the sum of positive logarithmic gains.

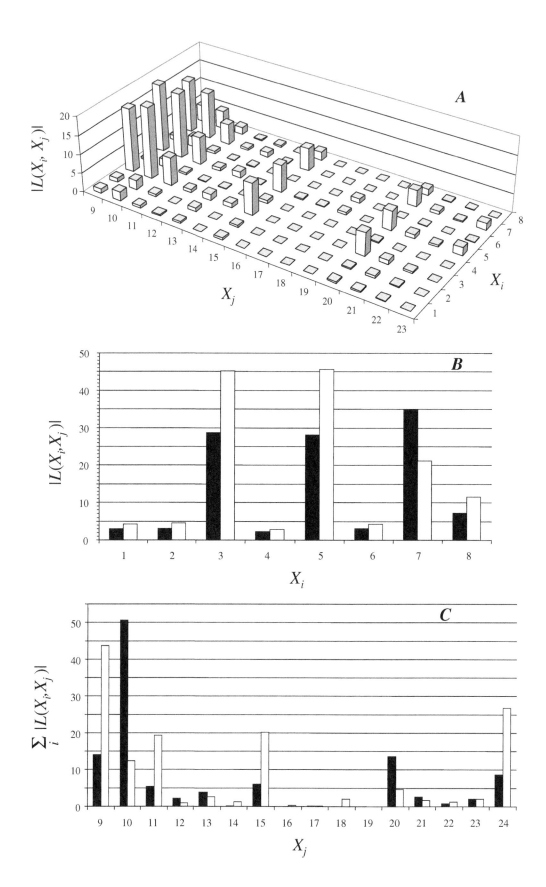

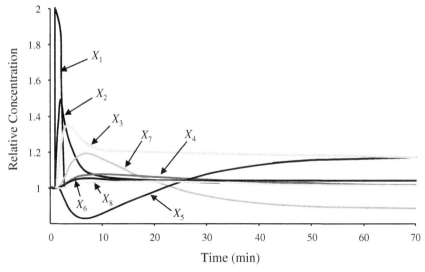

Figure 3.10. Dynamic system response to a twofold increase in glucose 6-phosphate (X_1). At time 1, the glucose 6-phosphate pool (X_1) was increased to twice the steady state. The eight metabolites represented are glucose 6-phosphate (X_1), fructose 6-phosphate (X_2), phosphoenol pyruvate (X_3), pyruvate (X_4), oxalacetate (X_5), malate (X_6), NADH (X_7), and ATP (X_8). Each concentration is normalized with respect to its nominal steady-state value. The long time scale of the response is cause for concern. (See www.cambridge.org/9780521177481 for color version.)

evident that the model is flawed in some respects. Specifically, reactions affecting the fructose 6-phosphate (X_2), oxalacetate (X_5), NADH(X_7), and ATP (X_8) pools have rather high sensitivities, which tend to be a warning sign. Also, some of the dynamic responses do not seem to be acceptable. Although these problems may be cause for disappointment, the results have the strong advantage that they indicate specifically which parts of the pathway should be studied in greater detail. The gain and sensitivity profiles in particular form the basis for model refinements.

Because the model needs to be amended, we might as well ask whether it would be advisable to add other metabolic components. For instance, the current pathway model is limited to carbohydrate metabolism, while other metabolic processes are ignored, even though they could be important for a proper understanding of citric acid synthesis and regulation. Furthermore, since construction of the initial model, new data have become available, and it seems prudent to include them in the map. The following aspects are to be considered in a revision of the model.

Figure 3.9. Logarithmic gains of metabolites with respect to changes in independent variables. Panel A: The three-dimensional plot displays the magnitudes of the logarithmic gains, measuring how a dependent metabolite concentration X_i is affected by an independent variable X_j. Panel B: Magnitudes for a given metabolite X_i are summed over all independent variables X_k. Panel C: Magnitudes for a given independent variable X_i are summed over all metabolites. The solid bars in each projection represent the sum of negatives logarithmic gains and the open bars represent the sum of positive logarithmic gains.

1. The system exhibits some high sensitivities. These are often related to unusually large kinetic orders and rate constants. Indeed, the model contains some kinetic orders with values close to 4 and also some very high rate constants (see Eq. 3.4). Experience with other systems has shown that the inclusion of more biochemical detail in the model often leads to a decrease in the numerical values of power-law parameters and to a consequent increase in robustness. This suggests closer scrutiny of the dynamics of pyruvate and oxalacetate.

2. The combined glucose transport/hexokinase step (X_{10}) seems to be of great importance in controlling the fluxes throughout the system. The current implementation does not allow us to assess the importance of each of the two components separately. This suggests the introduction of an additional independent variable representing specifically the carrier activity.

3. Steinbock et al. (1991) demonstrated that activation by fructose-2,6-bisphosphate makes phosphofructokinase (X_{13}) insensitive to feedback inhibition by citrate. They also observed a pronounced rise in the level of 2,6-biphosphate under conditions of citric acid accumulation. The current model suggests that control of phosphofructokinase lies with the steps involved in the flux through the fructose-2,6-bisphosphate pool, namely synthesis catalyzed by phosphofructokinase II and degradation by fructose bisphosphatase. This suggests better representation of these processes in the revised model.

4. Pyruvate carboxylase (X_{15}) and the pyruvate transport system (X_{20}) exert significant control over the system (see Figures 3.7 and 3.8). However, because these steps are at a branch point, their absolute influence can only be assessed with a model that includes those reactions leading to the synthesis of citric acid within mitochondria. This suggests inclusion at least of the initial reactions of the citric acid cycle.

5. Some of the large sensitivities of the model are associated with fluxes through oxalacetate, X_5 (see Figures 3.3 and 3.4). Although originally thought to be represented well, the dynamics of X_5 needs to be revisited. Woronick and Johnson (1960) described a cytosolic phosphoenolpyruvate carboxylase in *A. niger*. This enzyme allows the cell to transform cytosolic oxalacetate, which is formed by pyruvate carboxylase, into phosphoenolpyruvate. The inclusion of this reaction might alleviate the high sensitivities in this part of the system.

6. Early observations suggested that transport processes other than glucose intake and the mitochondrial pyruvate carrier (X_{20}) could have a significant impact on the control of citric acid production (Kubicek and Röhr 1986). This suggests that the revised model should account for mitochondrial and cytosolic citrate export.

7. Looking a step ahead, one purpose of the model will be the optimization of the rate of citric acid production. This again suggests inclusion of mitochondrial reactions involved in citric acid synthesis through the Krebs cycle and of the respiration systems that are operative in *A. niger* under conditions of citric acid accumulation.

8. Another concern with the model presents itself in the high sensitivities associated with fluxes and concentrations of NADH (X_7) and ATP (X_8). Their dynamics should be revisited.

9. Finally, a revised model should show more appropriate responses to changes in input. In particular, it should avoid the severe imbalances between the moieties of ATP and ADP and NADH and NAD^+ that were observed in the initial model (Voit and Ferreira 1998).

One may ask why these issues were not addressed in the original model. To some degree, not all of the information known now was available at the time of the original model construction. More significant, however, is that the model in its original form seemed to be very reasonable and that only detailed analysis discovered and pinpointed the problems discussed. As pointed out in Chapter 2, the cycling between modeling and refinement is the norm and often requires numerous iterations. Here we consider only two steps; another, more detailed example is discussed in Voit (2000, Chapter 10).

REVISED MODEL

New Processes and Biochemical Data

The amended model retains the basic structure of the original model. As with its predecessor, the backbone consists of the major steps of the glycolytic pathway, namely the breakdown of hexoses to pyruvate and acetyl-CoA and the formation of oxaloacetate from pyruvate and CO_2. The key additions are motivated by the limitations and shortcomings of the initial model, as they were discussed previously. Specifically, the new model will take account of citric acid formation within the tricarboxylic acid cycle and the formation of phosphoenolpyruvate from oxalacetate through the phosphoenol pyruvate carboxylase kinase system. Moreover, transport processes throughout the cytosplasmatic and mitochondrial membranes and the influence of oxygen will be included.

With respect to carbohydrate metabolism, Torres et al. (1996) characterized a glucose-specific carrier that is of relevance here. Panneman et al. (1996, 1998) presented new data on specific hexokinases. Moreover, it was demonstrated that trehalose-6-phosphate might be a potent inhibitor of some hexokinases (Blázquez et al. 1993; Arisan-Atac, Wolschek, and Kubicek 1996; Panneman et al. 1998, Boy-Marcotte et al. 1999). This effect was shown not to increase the product yield or the rate of citrate synthesis, but to reduce the time needed to reach half-maximal and maximal concentration. The inclusion of this regulatory interaction is thus relevant for the search of mechanisms responsible for the onset of the citric acid accumulation, rather than for the characterization of the steady state during idiophase. Kubicek-Pranz et al. (1990) investigated the role of $F-2,6-P_2$ for the regulation of the glycolytic flux at the step catalyzed by PFKI. In particular, they studied the substrate cycle involving the phosphofructokinase II catalyzed synthesis of $F-2,6-P_2$ and its fructose bisphosphatase II catalyzed transformation into F-6-P.

The model will also consider the secondary production of polyols at the flux-diverting branch from the glucose 6-phosphate/fructose 6-phosphate pool. A related potential inclusion would be the process of lipid accumulation. However, this diverting flux will not be modeled, because studies based on metabolic flux analysis of the system showed that fluxes leading to lipids were not significant in comparison with those through the main metabolic pathways (Guebel and Torres 2001).

Oxalacetate may be transformed back into phosphoenolpyruvate through the reaction catalyzed by the phosphoenolpyruvate carboxylase (Woronick and Johnson 1960). Later sensitivity analysis will show that inclusion of this activity significantly improves the robustness of the system (see Section "Robustness" and Figures 3.6 and 3.7). The same is the case with glutamate dehydrogenase, which is located in the cytosole (Pedersen, Carlsen, and Nielsen 1999).

Experimental observations have shown that the enzymes of the glyoxylate shunt are almost inactive (Röhr and Kubicek 1981). In this respect, *A. niger* is similar to other citric acid-producing systems (e.g., *Candida lipolytica*) where the glyoxylate cycle is apparently not relevant for the functioning of the citric acid cycle (Aiba and Matsuoka 1979). Early evidence has also suggested that the tricarboxylic acid cycle is inhibited during citric acid accumulation (Cleland and Johnson 1954) and that the fluxes of the 2-oxoglutarate dehydrogenase, succinate hydrogenase, and fumarate steps are negligible (Kubicek and Röhr 1978; Röhr and Kubicek 1981). Once inside the mitochondria, pyruvate undergoes a transformation to acetyl-CoA by virtue of the pyruvate dehydrogenase reaction, releasing CO_2. The 2-oxoglutarate dehydrogenase is inhibited via feedback loops from oxalacetate and NADH (Meixner-Monori et al. 1985). Because this enzyme catalyzes an irreversible step, the implication is that, under accumulating conditions, malate, fumarate, and succinate are formed by reduction of oxaloacetate. In facultative anaerobic bacteria, this process has been termed the "horse-shoe" version of the tricarboxylic acid cycle (Amaranisingham and Davis 1965).

Aspergillus niger contains only very low amounts of NAD-specific ISDH (Mattey 1977) and an NADP-ISDH that is distributed in the cytosolic and mitochondrial compartments. Feedback inhibition of NADP-ISDH by citrate and isocitrate is well documented (Meixner-Monori et al. 1986). However, inhibition by citrate is relevant only at later stages of citric acid accumulation, when the citrate concentration is high (Meixner-Monori et al. 1986).

Glucose intake has been characterized and identified as a potential regulatory step (Torres et al. 1996). No further advances have been made in the characterization of mitochondrial transport processes in *A. niger*, but the export system of citrate has been characterized (Netik et al. 1997).

The necessity of a high aeration rate as a factor stimulating citric acid accumulation in *A. niger* has been recognized. The biochemical basis of this effect was studied by Kubicek et al. (1980), Zehentgruber, Kubicek, and Röhr (1980), and later by Wallrath, Schmidt, and Weiss (1991), Schmidt et al. (1992), and Prömper, Schneider, and Weiss (1993). Besides the standard respiratory chain, an alternative respiratory system consists of the cyanide and antimycin A-insensitive respiratory and the salicylhydroxamic acid-sensitive pathways. This alternative NADH-oxidase system has been described in *A. niger* (Kirimura, Hirowatari, and Usami 1987; Kirimura

Table 3.4. Metabolite Concentrations During Idiophase Steady State (150 Hours of Cultivation) in the Revised Citric Acid Model

Metabolite	Abbreviation	X_i	Concentration (mM)
Cytosolic glucose	GLUCOSE, Int.	1	0.00274
Glucose 6-phosphate	G6P	2	0.1
Fructose 6-phosphate	F6P	3	0.025
Fructose 2,6-bisphospate	$F2,6P_2$	4	0.0015
Phosphoenolpyruvate	PEP	5	0.01
Cytosolic pyruvate	PYRc	6	0.12
Cytosolic oxalacetate	OXAc	7	0.00058
Cytosolic malate	MALc	8	2.6
Mitochondrial malate	MALm	9	13
Mitochondrial oxalacetate	OXAm	10	0.0022
Mitochondrial pyruvate	PYRm	11	0.56
Coenzyme A	CoA	12	0.108
Acetyl-CoA	AcCoA	13	0.012
Mitochondrial citrate	CITm	14	31.25
Isocitrate	ISC	15	1.21
Cytosolic citrate	CITc	16	5.62
Adenosine triphosphate	ATP	17	0.5
Adenosine diphosphate	ADP	—	0.075
Adenosine monophosphate	AMP	—	0.143
Nicotinamide adenine dinucleotide	NADH	18	0.035
Cytosolic nicotinamine adenine dinucleotide	$NADc^+$	—	0.12
Mitochondrial nicotinamine adenine dinucleotide	$NADm^+$	—	0.6
Glucose in medium	GLUCOSE, Ext.	19	290
Ammonium	NH_4^+	27	2.35
Oxygen	OXYG	41	150^a
Total adenosine phosphate		48	0.718
Total nicotinamine adenine dinucleotide		49	0.755
2-Oxoglutarate	OXO	—	0.26
Alanine	ALA	—	5.75
Carbon dioxide	CO_2	—	0.0012
Glutamate	GLUT	—	10.63

[a] Concentration expressed in milibar.

et al. 2000) and other microorganisms (Henry and Nyns 1975), where it has been suggested to function in the reoxidation of cytoplasmic NADH without concomitant ATP production, thus favoring glycolytic energy production.

Mass Balance and S-System Equations

The new data and processes (Tables 3.4 and 3.5) were arranged in the metabolic map of Figure 3.11. The map alone is sufficient to formulate the corresponding mass

Table 3.5. Enzyme Activities (μmol/min mg Protein), Fluxes (nmol/min mgr Protein) Through Reactions and Transport Steps at Idiophase Steady State (150 Hours Cultivation Time), as Well as Relevant Kinetic and Thermodynamic Constants in the Revised Model[a]

Enzyme	Symbol	X_i	Activity	Net Flux	Kinetic and Thermodynamic Constants	
Glucose transport	TRP1	20	0.032	6.3	K_m (glucose ext.)	3.67
Hexokinase	HK	21	0.06	6.3	K_m (ATP)	0.69
					K_m (glucose int.)	0.32
Glucose-6P-dehydrogenase	G6Pdh	22	0.017	1.26	K_m (G6P)	0.045
Phosphoglucose isomerase	PGI	23	15.3	5.04	$K_{eq}\frac{F6P}{G6P}$	0.4
Fructose 2,6-bisphos-phatase	FBPase	24	0.62	1	K_m (F2,6P)	0.1
6-Phosphofructo-2-kinase	PFK2	25	0.21	1	K_m (F6P)	0.6
					K_m (ATP)	0.7
Phosphofructo-1-kinase	PFK1	26	0.075	5.04	$S_{0.5}$(F6P)	0.25
					$S_{0.5}$(ATP)	0.03
					$S_{0.5}$(F2,6P)	0.128
Pyruvate kinase	PK	28	0.15	10.222	K_m (PEP)	0.1
					K_m (ADP)	0.07
Glyceraldehyde Dehydrogenase	—	—	—	10.222	K_m (NADc$^+$)	0.013
					K_m (GAP)	0.09
Pyruvate carboxylase	PC	29	0.015	5.308	K_m (PYRc)	0.28
					K_m (ATP)	0.23
Phosphoenol-pyruvate carboxykinase	PEPc	30	0.5[b]	0.142	$K_{eq}\frac{PEPc \cdot ADP \cdot CO_2}{OXAc \cdot ATP}$	0.049
Aspartate aminotransferase	GOT	31	0.048	0.280	K_m (OXAc)	0.088
Cytosolic malate dehydrogenase	MDHc	32	5.6	4.886	$K_{eq}\frac{MALc \cdot NADc^+ \cdot CO_2}{OXAc \cdot NADH}$	35714.2
Malate transport	TRP3	33	0.54	4.886	K_m (MALc)	11
Pyruvate transport	TRP2	34	0.045	4.914	K_m (PYRc)	6.2
Alanine transaminase	ALAt	35	0.5[b]	0.028	$K_{eq}\frac{OXO \cdot ALA}{GLUT \cdot PYRm}$	1.5
Pyruvate dehydrogenase	PDH	36	0.217	4.886	K_m (NADm$^+$)	0.061
					K_m (Pyruvate)	0.038
					K_i (NADH)	0.04
					K_m (CoA)	0.01
Citrate synthase	CS	37	0.007	4.886	K_m (OXAm)	0.005
					K_m (AcCoA)	0.01
					K_i (CoA)	0.15
Mitochondrial malate dehydrogenase	MDHm	38	9.4	4.886	$K_{eq}\frac{OXAm \cdot NADH}{MALm \cdot NADm}$	0.00028
Aconitase	ACN	39	0.028	0.25	$K_{eq}\frac{ISC}{CITm}$	0.077
NADP-isocitrate dehydrogenase	ISCDH	40	0.018	0.25	K_m (ISC)	0.077

Table 3.5 (*Continued*)

Enzyme	Symbol	X_i	Activity	Net Flux	Kinetic and Thermodynamic Constants	
Alternative respiratory system	RESP	42	0.2	5.04	K_m (NADH)	0.002
Mitochodrial citrate transport	TRP4	43	0.009	4.636	K_m (CITm)	8
Cytosolic citrate transport	TRP5	44	0.005	4.636	K_m (CITc)	0.72
Adenylate kinase	AK	45	3.06	4.177	$K_{eq} \frac{\text{ADP}^2}{\text{ATP} \cdot \text{AMP}}$	1
ATPase	ATPase	46	5.1	6.83	K_m (ATP)	0.8
Dehydrogenases	NADHase	47	1^c	10.068	K_m (NADH)	0.5^b

[a] K_m and $S_{0.5}$ are given in mM.
[b] Medium size magnitude order of enzyme activities and K_m values have been selected.
[c] Average magnitude of enzyme activities selected.

balance equations:

$$\dot{X}_1 = v_{19,1} - v_{1,2} = V_1^+ - V_1^-$$

$$\dot{X}_2 = (v_{1,2} + v_{3,2}) - (v_{2,3} + v_{2,0}) = V_2^+ - V_2^-$$

$$\dot{X}_3 = (v_{2,3} + v_{4,3}) - (v_{3,2} + v_{3,5} + v_{3,4}) = V_3^+ - V_3^-$$

$$\dot{X}_4 = v_{3,4} - v_{4,3} = V_4^+ - V_4^-$$

$$\dot{X}_5 = (2v_{3,5} + v_{7,5}) - (v_{5,6} + v_{5,7}) = V_5^+ - V_5^-$$

$$\dot{X}_6 = v_{5,6} - (v_{6,7} + v_{6,11}) = V_6^+ - V_6^-$$

$$\dot{X}_7 = (v_{6,7} + v_{5,7} + v_{8,7}) - (v_{7,5} + v_{7,8} + v_{7,0}) = V_7^+ - V_7^-$$

$$\dot{X}_8 = v_{7,8} - (v_{8,7} + v_{8,9}) = V_8^+ - V_8^-$$

$$\dot{X}_9 = (v_{8,9} + v_{10,9}) - v_{9,10} = V_9^+ - V_9^-$$

$$\dot{X}_{10} = v_{9,10} - (v_{10,9} + v_{10,14}) = V_{10}^+ - V_{10}^-$$

$$\dot{X}_{11} = (v_{6,11} + v_{0,11}) - (v_{11,13} + v_{11,0}) = V_{11}^+ - V_{11}^-$$

$$\dot{X}_{12} = v_{13,12} - v_{12,13} = V_{12}^+ - V_{12}^-$$

$$\dot{X}_{13} = (v_{11,13} + v_{12,13}) - (v_{13,12} + v_{13,14}) = V_{13}^+ - V_{13}^-$$

$$\dot{X}_{14} = (v_{10,14} + v_{15,14}) - (v_{14,15} + v_{14,16}) = V_{14}^+ - V_{14}^-$$

$$\dot{X}_{15} = v_{14,15} - (v_{15,14} + v_{15,0}) = V_{15}^+ - V_{15}^-$$

$$\dot{X}_{16} = v_{14,16} - v_{16,0} = V_{16}^+ - V_{16}^-$$

$$\dot{X}_{17} = (2v_{5,6} + v_{5,7} + v_{\text{AKr}}) - (v_{1,2} + v_{3,4} + v_{3,5} + v_{6,7} + v_{7,5}$$
$$+ v_{\text{ATPase}} + v_{\text{AKf}}) = V_{17}^+ - V_{17}^-$$

$$\dot{X}_{18} = (v_{5,6} + v_{9,10} + v_{8,7} + v_{11,13}) - (v_{7,8} + v_{\text{Resp}} + v_{\text{NADHase}}) = V_{18}^+ - V_{18}^-.$$

$$(3.5)$$

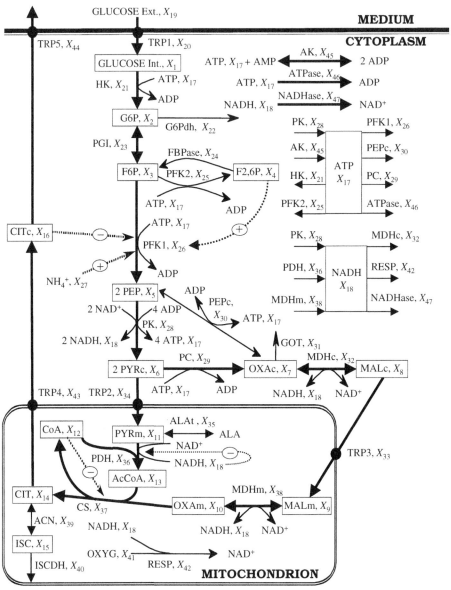

Figure 3.11. Revised map of citric acid metabolism in *A. niger* during idiophase. Metabolites are shown in boxes and reactions as arrows. Solid arrows indicate reactions or transport processes whereas dashed arrows show regulatory interactions. Dots represent transport systems. Dependent variables are numbered from 1 to 18: Glucose int. (X_1, cytosolic glucose); G6P (X_2, glucose 6-phosphate); F6P (X_3, fructose 6-phosphate); F2,6P (X_4, fructose bisphosphate); PEP (X_5, phosphoenol pyruvate); PYRc (X_6, cytosolic pyruvate); OXAc (X_7, cytosolic oxalacetate); MALc (X_8, cytosolic malate); MALm (X_9, mitochondrial malate); OXAm (X_{10}, mitochondrial oxalacetate); PYRm (X_{11}, mitochondrial pyruvate); CoA (X_{12}); AcCoA, (X_{13}, acetyl-CoA); CITm (X_{14}, mitochondrial citrate); ISC (X_{15}, isocitrate); CITc (X_{16}, cytosolic citrate); ATP (X_{17}); NADH(X_{18}). Independent variables are numbered from 19 to 49: Glucose ext. (X_{19}, glucose in medium); TRP1 (X_{20}, glucose carrier); HK (X_{21}, hexokinase); G6Pdh (X_{22}, glucose 6-phosphate dehydrogenase); PGI (X_{23}, phosphoglucose isomerase); FBPase (X_{24}, fructose 2,6-bisphosphatase); PFK2 (X_{25}, 6-phosphofructo-2-kinase); PFK1 (X_{26}, phosphofructokinase); NH_4^+(X_{27}, ammonium); PK (X_{28}, pyruvate kinase);

For intuitive clarity, we have grouped with parentheses all fluxes that will collectively form aggregate influxes and effluxes. For the reversible steps, we chose the reversible modeling strategy, which requires a little bit more modeling effort but has better characteristics if forward and reverse reactions are of a similar magnitude (Sorribas and Savageau 1989; see also Chapter 2).

The symbolic S-system representation that directly derives from the mass balance Eq. (3.5) has the following form:

$$\dot{X}_1 = \alpha_1 X_{19}^{g_{1,19}} X_{20}^{g_{1,20}} - \beta_1 X_1^{h_{1,1}} X_{17}^{h_{1,17}} X_{21}^{h_{1,21}}$$

$$\dot{X}_2 = \alpha_2 X_1^{g_{2,1}} X_2^{g_{2,2}} X_3^{g_{2,3}} X_{17}^{g_{2,17}} X_{21}^{g_{2,21}} X_{23}^{g_{2,23}} - \beta_2 X_2^{h_{2,2}} X_3^{h_{2,3}} X_{22}^{h_{2,22}} X_{23}^{h_{2,23}}$$

$$\dot{X}_3 = \alpha_3 X_2^{g_{3,2}} X_3^{g_{3,3}} X_4^{g_{3,4}} X_{23}^{g_{3,23}} X_{24}^{g_{3,24}}$$
$$- \beta_3 X_2^{h_{3,2}} X_3^{h_{3,3}} X_4^{h_{3,4}} X_{16}^{h_{3,16}} X_{17}^{h_{3,17}} X_{23}^{h_{3,23}} X_{25}^{h_{3,25}} X_{26}^{h_{3,26}} X_{27}^{h_{3,27}}$$

$$\dot{X}_4 = \alpha_4 X_3^{g_{4,3}} X_{17}^{g_{4,17}} X_{25}^{g_{4,25}} - \beta_4 X_4^{h_{4,4}} X_{24}^{h_{4,24}}$$

$$\dot{X}_5 = \alpha_5 X_3^{g_{5,3}} X_4^{g_{5,4}} X_5^{g_{5,5}} X_7^{g_{5,7}} X_{16}^{g_{5,16}} X_{17}^{g_{5,17}} X_{26}^{g_{5,26}} X_{27}^{g_{5,27}} X_{30}^{g_{5,30}}$$
$$- \beta_5 X_5^{h_{5,5}} X_7^{h_{5,7}} X_{17}^{h_{5,17}} X_{18}^{h_{5,18}} X_{28}^{h_{5,28}} X_{30}^{h_{5,30}} X_{48}^{h_{5,48}} X_{49}^{h_{5,49}}$$

$$\dot{X}_6 = \alpha_6 X_5^{g_{6,5}} X_{17}^{g_{6,17}} X_{18}^{g_{6,18}} X_{28}^{g_{6,28}} X_{48}^{g_{6,48}} X_{49}^{g_{6,49}} - \beta_6 X_6^{h_{6,6}} X_{17}^{h_{6,17}} X_{29}^{h_{6,29}} X_{34}^{h_{6,34}}$$

$$\dot{X}_7 = \alpha_7 X_5^{g_{7,5}} X_6^{g_{7,6}} X_7^{g_{7,7}} X_8^{g_{7,8}} X_{17}^{g_{7,17}} X_{18}^{g_{7,18}} X_{29}^{g_{7,29}} X_{30}^{g_{7,30}} X_{32}^{g_{7,32}} X_{48}^{g_{7,48}} X_{49}^{g_{7,49}}$$
$$- \beta_7 X_5^{h_{7,5}} X_7^{h_{7,7}} X_8^{h_{7,8}} X_{17}^{h_{7,17}} X_{18}^{h_{7,18}} X_{30}^{h_{7,30}} X_{31}^{h_{7,31}} X_{32}^{h_{7,32}}$$

$$\dot{X}_8 = \alpha_8 X_7^{g_{8,7}} X_8^{g_{8,8}} X_{18}^{g_{8,18}} X_{32}^{g_{8,32}} - \beta_8 X_7^{h_{8,7}} X_8^{h_{8,8}} X_{18}^{h_{8,18}} X_{32}^{h_{8,32}} X_{33}^{h_{8,33}} X_{49}^{h_{8,49}}$$

$$\dot{X}_9 = \alpha_9 X_8^{g_{9,8}} X_9^{g_{9,9}} X_{10}^{g_{9,10}} X_{18}^{g_{9,18}} X_{33}^{g_{9,33}} X_{38}^{g_{9,38}} - \beta_9 X_9^{h_{9,9}} X_{10}^{h_{9,10}} X_{18}^{h_{9,18}} X_{38}^{h_{9,38}} X_{49}^{h_{9,49}}$$

$$\dot{X}_{10} = \alpha_{10} X_9^{g_{10,9}} X_{10}^{g_{10,10}} X_{18}^{g_{10,18}} X_{38}^{g_{10,38}} X_{49}^{g_{10,49}}$$
$$- \beta_{10} X_9^{h_{10,9}} X_{10}^{h_{10,10}} X_{12}^{h_{10,12}} X_{13}^{h_{10,13}} X_{18}^{h_{10,18}} X_{37}^{h_{10,37}} X_{38}^{h_{10,38}}$$

$$\dot{X}_{11} = \alpha_{11} X_6^{g_{11,6}} X_{11}^{g_{11,11}} X_{34}^{g_{11,34}} X_{35}^{g_{11,35}} - \beta_{11} X_{11}^{h_{11,11}} X_{12}^{h_{11,12}} X_{18}^{h_{11,18}} X_{35}^{h_{11,35}} X_{36}^{h_{11,36}} X_{49}^{h_{11,49}}$$

$$\dot{X}_{12} = \alpha_{12} X_{10}^{g_{12,10}} X_{13}^{g_{12,13}} X_{37}^{g_{12,37}} - \beta_{12} X_{11}^{h_{12,11}} X_{12}^{h_{12,12}} X_{18}^{h_{12,18}} X_{36}^{h_{12,36}} X_{49}^{h_{12,49}}$$

$$\dot{X}_{13} = \alpha_{13} X_{11}^{g_{13,11}} X_{12}^{g_{13,12}} X_{18}^{g_{13,18}} X_{36}^{g_{13,36}} X_{49}^{g_{13,49}} - \beta_{13} X_{10}^{h_{13,10}} X_{12}^{h_{13,12}} X_{13}^{h_{13,13}} X_{37}^{h_{13,37}}$$

PC (X_{29}, pyruvate carboxylase); PEPc (X_{30}, cytosolic phosphoenol-pyruvate-carboxikinase); GOT (X_{31}, aspartate aminotransferase); MDHc (X_{32}, cytosolic malate dehydrogenase); TRP3 (X_{33}, mitochondrial malate transport system); TRP2 (X_{34}, mitochondrial pyruvate transport system); ALAt (X_{35}, alanine transaminase); PDH (X_{36}, pyruvate dehydrogenase); CS (X_{37}, citrate synthase); MDHm (X_{38}, mitochondrial malate dehydrogenase); ACN (X_{39}, aconitase); ISCDH (X_{40}, isocitrate dehydrogenase); OXYG (X_{41}, oxygen); RESP (X_{42}, alternative respiratory system); TRP4 (X_{43}, mitochondrial citrate carrier); TRP5 (X_{44}, cytosolic citrate carrier); AK (X_{45}, adenylate kinase); ATPase (X_{46}); NADHase (X_{47}, dehydrogenases).

$$\dot{X}_{14} = \alpha_{14} X_{10}^{g_{14,10}} X_{12}^{g_{14,12}} X_{13}^{g_{14,13}} X_{14}^{g_{14,14}} X_{15}^{g_{14,15}} X_{37}^{g_{14,37}} X_{39}^{g_{14,39}}$$
$$- \beta_{14} X_{14}^{b_{14,14}} X_{15}^{b_{14,15}} X_{39}^{b_{14,39}} X_{43}^{b_{14,43}}$$

$$\dot{X}_{15} = \alpha_{15} X_{14}^{g_{15,14}} X_{15}^{g_{15,15}} X_{39}^{g_{15,39}} - \beta_{15} X_{14}^{b_{15,14}} X_{15}^{b_{15,15}} X_{39}^{b_{15,39}} X_{40}^{b_{15,40}}$$

$$\dot{X}_{16} = \alpha_{16} X_{14}^{g_{16,14}} X_{43}^{g_{16,43}} - \beta_{16} X_{16}^{b_{16,16}} X_{44}^{b_{16,44}}$$

$$\dot{X}_{17} = \alpha_{17} X_5^{g_{17,5}} X_7^{g_{17,7}} X_{17}^{g_{17,17}} X_{18}^{g_{17,18}} X_{28}^{g_{17,28}} X_{30}^{g_{17,30}} X_{45}^{g_{17,45}} X_{48}^{g_{17,48}} X_{49}^{g_{17,49}}$$
$$- \beta_{17} X_1^{b_{17,1}} X_3^{b_{17,3}} X_4^{b_{17,4}} X_5^{b_{17,5}} X_6^{b_{17,6}} X_7^{b_{17,7}} X_{16}^{b_{17,16}} X_{17}^{b_{17,17}} X_{21}^{b_{17,21}} X_{25}^{b_{17,25}}$$
$$\cdot X_{26}^{b_{17,26}} X_{27}^{b_{17,27}} X_{29}^{b_{17,29}} X_{30}^{b_{17,30}} X_{45}^{b_{17,45}} X_{46}^{b_{17,46}} X_{48}^{b_{17,48}}$$

$$\dot{X}_{18} = \alpha_{18} X_5^{g_{18,5}} X_7^{g_{18,7}} X_8^{g_{18,8}} X_9^{g_{18,9}} X_{10}^{g_{18,10}} X_{11}^{g_{18,11}} X_{12}^{g_{18,12}} X_{17}^{g_{18,17}} X_{18}^{g_{18,18}}$$
$$\cdot X_{28}^{g_{18,28}} X_{32}^{g_{18,32}} X_{36}^{g_{18,36}} X_{38}^{g_{18,38}} X_{48}^{g_{18,48}} X_{49}^{g_{18,49}}$$
$$- \beta_{18} X_7^{b_{18,7}} X_8^{b_{18,8}} X_9^{b_{18,9}} X_{10}^{b_{18,10}} X_{18}^{b_{18,18}} X_{32}^{b_{18,32}} X_{38}^{b_{18,38}}$$
$$\cdot X_{41}^{b_{18,41}} X_{42}^{b_{18,42}} X_{47}^{b_{18,47}} X_{49}^{b_{18,49}}. \tag{3.6}$$

The values of all kinetic orders and rate constants were determined with standard methods as shown before. As in the original model, the NADH and NAD$^+$ pools are constrained by a constant total, and the same is true for the AMP, ADP, and ATP pools. As before, the constraints are formulated as power laws and incorporated in the equations (see Appendix). The numerical S-system representation of *A. niger* metabolism under conditions of citric acid production is thus:

$$\dot{X}_1 = 0.1857 \cdot X_{19}^{0.0124} X_{20} - 54.4794 X_1^{0.9915} X_{17}^{0.5798} X_{21}$$

$$\dot{X}_2 = 50.5542 X_1^{0.976} X_2^{-0.026} X_3^{0.04167} X_{17}^{0.5707} X_{21}^{0.9843} X_{23}^{0.0156}$$
$$- 0.0018 X_2^{2.202} X_3^{-1.3389} X_{22}^{0.1968} X_{23}^{0.8031}$$

$$\dot{X}_3 = 0.0019 X_2^{2.2316} X_3^{-1.3956} X_4^{0.1604} X_{23}^{0.8371} X_{24}^{0.1628} - 0.6254 X_2^{-0.0271} X_3^{0.6208}$$
$$\cdot X_4^{0.001} X_{16}^{-0.2462} X_{17}^{0.1212} X_{23}^{0.0162} X_{25}^{0.1628} X_{26}^{0.8208} X_{27}^{0.5335}$$

$$\dot{X}_4 = 0.2462 X_3^{0.96} X_{17}^{0.5833} X_{25} - 0.9767 X_4^{0.9852} X_{24}$$

$$\dot{X}_5 = 1.023 X_3^{0.5004} X_4^{0.0012} X_5^{-0.0016} X_7^{0.026} X_{16}^{-0.2926} X_{17}^{0.0683} X_{26}^{0.9756} X_{27}^{0.6341} X_{30}^{0.0243}$$
$$- 1.6341 \cdot X_5^{0.9107} X_7^{-0.0007} X_{17}^{-3.2606} X_{18}^{-0.0047} X_{28}^{0.9893} X_{30}^{0.0106} X_{48}^{3.7481} X_{49}^{0.1014}$$

$$\dot{X}_6 = 1.6565 X_5^{0.909} X_{17}^{-3.2183} X_{18}^{-0.0047} X_{28} X_{48}^{3.7011} X_{49}^{0.1025}$$
$$- 2.644 X_6^{0.835} X_{17}^{0.1636} X_{29}^{0.5193} X_{34}^{0.4807}$$

$$\dot{X}_7 = 1.3835 X_5^{0.0212} X_6^{0.6733} X_7^{-0.015} X_8^{0.0318} X_{17}^{0.1598} X_{18}^{-0.0152} X_{29}^{0.9619} X_{30}^{0.0199} X_{32}^{0.0181}$$
$$\cdot X_{48}^{0.1618} X_{49}^{0.0196} - 1.7805 \times 10^5 X_5^{-0.0031} X_7^{1.6839} X_8^{-0.6832}$$
$$\cdot X_{17}^{0.0694} X_{18}^{1.618} X_{30}^{0.0456} X_{31}^{0.0507} X_{32}^{0.9035}$$

$$\dot{X}_8 = 3.5235 \times 10^5 \, X_7^{1.7538} \, X_8^{-0.7561} \, X_{18}^{1.7906} \, X_{32} - 0.0033$$
$$\cdot \, X_7^{-0.0151} \, X_8^{0.8278} \, X_{18}^{-0.0168} \, X_{32}^{0.02} \, X_{33}^{0.9799} \, X_{49}^{0.0217}$$

$$\dot{X}_9 = 0.0049 \, X_8^{0.7926} \, X_9^{-0.0007} \, X_{10}^{0.0207} \, X_{18}^{0.0208} \, X_{33}^{0.9799} \, X_{38}^{0.02}$$
$$- 3 \cdot 10^{-5} \, X_9^{1.036} \, X_{10}^{-0.0365} \, X_{18}^{-0.0869} \, X_{38} \, X_{49}^{1.0497}$$

$$\dot{X}_{10} = 3 \cdot 10^{-5} \, X_9^{1.036} \, X_{10}^{-0.0365} \, X_{18}^{-0.0869} \, X_{38} \, X_{49}^{1.0497}$$
$$- 1266.5283 \, X_9^{-0.0007} \, X_{10}^{0.7013} \, X_{12}^{0.9799} \, X_{13}^{0.2449} \, X_{18}^{0.0208} \, X_{37}^{0.9799} \, X_{38}^{0.02}$$

$$\dot{X}_{11} = 0.8236 \cdot X_6^{0.9648} \, X_{11}^{-0.0033} \, X_{34}^{0.9834} \, X_{35}^{0.0165}$$
$$- 0.0214 \cdot X_{11}^{0.089} \, X_{12}^{0.0821} \, X_{18}^{-0.0868} \, X_{35}^{0.022} \, X_{36}^{0.9779} \, X_{49}^{0.0117}$$

$$\dot{X}_{12} = 9.5864 \times 10^2 \, X_{10} \, X_{13}^{0.25} \, X_{37} - 0.0209 \, X_{11}^{0.064} \, X_{12}^{0.0839} \, X_{18}^{-0.0887} \, X_{36} \, X_{49}^{0.012}$$

$$\dot{X}_{13} = 0.0202 \, X_{11}^{0.032} \, X_{12}^{0.0419} \, X_{18}^{-0.0443} \, X_{36}^{0.5} \, X_{49}^{0.006}$$
$$- 1.3359 \, X_{10}^{0.3472} \, X_{12}^{-0.1081} \, X_{13}^{0.125} \, X_{37}^{0.5}$$

$$\dot{X}_{14} = 70.1160 \cdot X_{10}^{0.602} \, X_{12}^{-0.1875} \, X_{13}^{0.2167} \, X_{14}^{-0.1382} \, X_{15}^{0.2604} \, X_{37}^{0.8669} \, X_{39}^{0.133}$$
$$- 0.0902 \cdot X_{14}^{0.5118} \, X_{15}^{-0.1858} \, X_{39}^{0.1774} \, X_{43}^{0.8225}$$

$$\dot{X}_{15} = 5.5 \cdot 10^{-5} \cdot X_{14}^{1.9396} \, X_{15}^{-1.0476} \, X_{39} - 0.4389 \cdot X_{14}^{-0.7789} \, X_{15}^{1.4831} \, X_{39}^{0.75} \, X_{40}^{0.25}$$

$$\dot{X}_{16} = 0.2526 \cdot X_{14}^{0.2038} \, X_{43} - 0.7621 \cdot X_{16}^{0.1135} \, X_{44}$$

$$\dot{X}_{17} = 0.6302 \, X_5^{0.6973} \, X_7^{-0.0002} \, X_{17}^{-4.1888} \, X_{18}^{-0.0036} \, X_{28}^{0.7622} \, X_{30}^{0.0041} \, X_{45}^{0.2336} \, X_{48}^{5.0068}$$
$$\cdot \, X_{49}^{0.0781} - 1.7294 \, X_1^{0.2329} \, X_3^{0.1322} \, X_4^{0.0002} \, X_5^{-0.0006} \, X_6^{0.1385} \, X_7^{0.01} \, X_{16}^{-0.0563}$$
$$\cdot \, X_{17}^{0.5258} \, X_{21}^{0.2349} \, X_{25}^{0.0372} \, X_{26}^{0.1879} \, X_{27}^{0.1221} \, X_{29}^{0.1979} \, X_{30}^{0.0094} \, X_{45}^{0.0778} \, X_{46}^{0.2546} \, X_{48}^{0.1166}$$

$$\dot{X}_{18} = 0.1007 \, X_5^{0.4601} \, X_7^{-0.0037} \, X_8^{0.0086} \, X_9^{0.2558} \, X_{10}^{-0.009} \, X_{11}^{0.0154} \, X_{12}^{0.0203} \, X_{17}^{-1.629}$$
$$\cdot \, X_{18}^{-0.0495} \, X_{28}^{0.5061} \, X_{32}^{0.0049} \, X_{36}^{0.2419} \, X_{38}^{0.2469} \, X_{48}^{1.8734} \, X_{49}^{0.3193}$$
$$- 14.3353 \, X_7^{0.4330} \, X_8^{-0.1866} \, X_9^{-1.8 \cdot 10^{-4}} \, X_{10}^{0.0051} \, X_{18}^{0.9267} \, X_{32}^{0.2469} \, X_{38}^{0.0049} \, X_{41}^{0.0025}$$
$$\cdot \, X_{42}^{0.2495} \, X_{47}^{0.4985} \, X_{49}^{0.2422}. \tag{3.7}$$

Local Stability and Characterization of the Steady State

Local stability is confirmed by examining the eigenvalues, which all have negative real parts (Table 3.6). One pair of complex conjugate eigenvalues indicates the possibility of damped oscillatory responses to perturbations in the dependent variables.

Robustness

Rate Constant Sensitivities

The parameter sensitivities for the revised model are presented in Figures 3.12–3.15. Because the revised model has eighteen equations, there are many more sensitivities

Table 3.6. Eigenvalues of the Revised S-System Model of Citric Acid Metabolism in A. niger

(min^{-1})
-16.4139
-2.2884
-1.7111
-1.0897
-0.7741
-0.4908
-0.3229
-0.0842
-0.0765
$-0.0342 + 0.0337i$
$-0.0342 - 0.0337i$
-0.0280
-0.0021
-0.0016
$-3.8\ 10^{-5}$
$-8.5\ 10^{-5}$
$-7.7\ 10^{-4}$
$-3.7\ 10^{-4}$

and gains than in the original model. They are shown in absolute values, such as $|S(V_i, \alpha_j)|$.

Rate constant sensitivities for fluxes. The absolute rate constant sensitivities with respect to the fluxes of the pathway, $|S(V_i, \alpha_j)|$, have the distribution shown in the three-dimensional plot of Figure 3.12 (panel A). Eighty-six percent of the sensitivities (280 out of 324) are less than 1 in magnitude. This implies that the response to a perturbation in a rate constant will be attenuated in most parts of the system. The largest sensitivities occur with respect to the fluxes through fructose-2,6-bisphosphate (X_4), isocitrate (X_{15}), and to a lesser degree fructose 6-phosphate (X_3), mitochondrial citrate (X_{14}), mitochondrial oxalacetate (X_{10}), and cytosolic citrate (X_{16}). This is most clearly seen in the two-dimensional projections B and C in Figure 3.12, where the magnitudes for a given flux V_i have been summed over all rate constant α_j (panel B) and where the magnitudes for a given rate constant are summed over all fluxes (panel C). In both cases, the solid portion represents the sum of the negative sensitivities, whereas the open portion represents the sum of the positive sensitivities.

Rate constant sensitivities for metabolite concentrations. Figure 3.13A shows the distribution of these sensitivities in absolute values, $|S(X_i, \alpha_j)|$. The majority of the rate constant sensitivities in this model (67.6%) are less than 1 in magnitude. The highest sensitivities occur for the concentrations of mitochondrial pyruvate (X_{11}), cytosolic citrate (X_{16}), Acetyl-CoA (X_{13}), and to a lesser degree, isocitrate (X_{15}), glucose 6-phosphate (X_2), fructose 6-phosphate (X_3), fructose-2,6-bisphosphate (X_4),

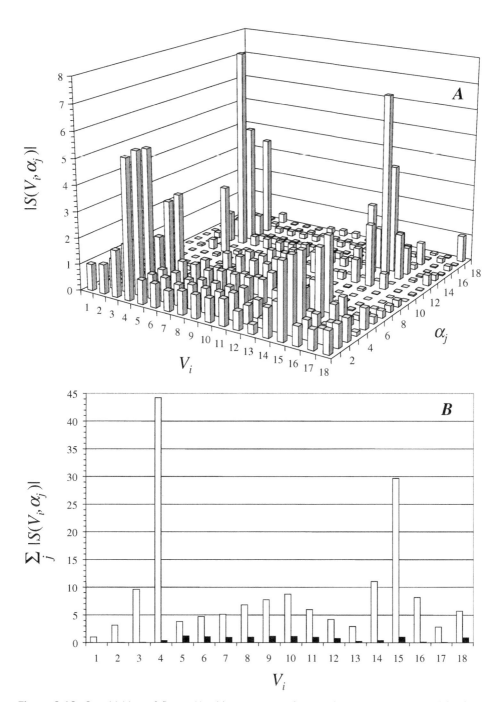

Figure 3.12. Sensitivities of fluxes V_i with respect to changes in rate constants α_j (absolute values). Panel A: The sensitivities indicate small or moderate effects of alterations in most rate constants. Panel B: Magnitudes for a given flux V_i are summed over all rate constants α_j. Panel C: Magnitudes for a given rate constant α_j are summed over all fluxes. The solid bars in each projection represent the sum of the negative sensitivities and the open bars represent the sum of the positive sensitivities. Nomenclature of variables is given in Figure 3.11.

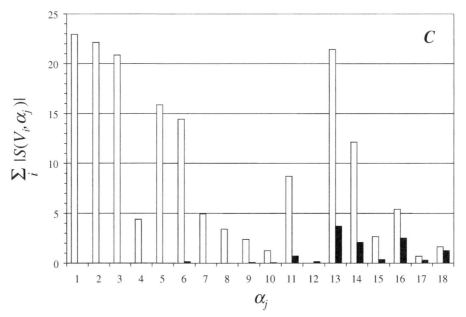

Figure 3.12 *(continued)*

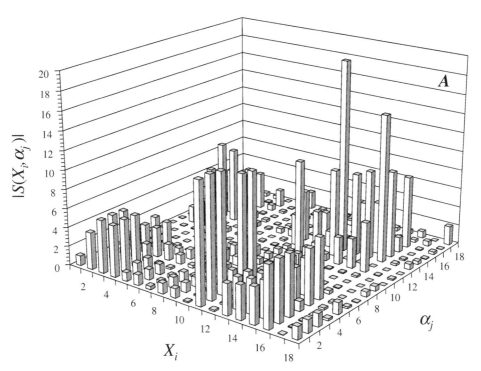

Figure 3.13. Sensitivities of metabolite concentrations with respect to changes in rate constants. Panel A: Absolute values. Panel B: Magnitudes for a given metabolite X_i are summed over all rate constants α_j. Panel C: Magnitudes for a given rate constant α_j are summed over all metabolites. The solid bars in each projection represent the sum of the negative sensitivities and the open bars represent the sum of the positive sensitivities. Nomenclature of variables is given in Figure 3.11.

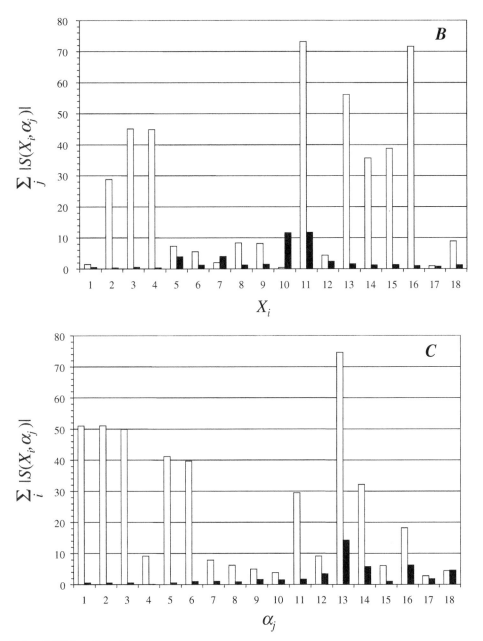

Figure 3.13 *(continued)*

mitochondrial citrate (X_{14}), and isocitrate (X_{15}). Again, this can be seen most clearly in the two-dimensional projections (panels B and C).

Kinetic Order Sensitivities

Kinetic order sensitivities for fluxes. The model contains $104\,g$ and $108\,h$ kinetic orders. Of these, 48 (25 g and 23 h) that exhibit the largest sensitivities are shown in

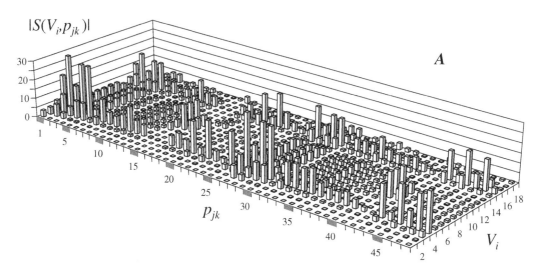

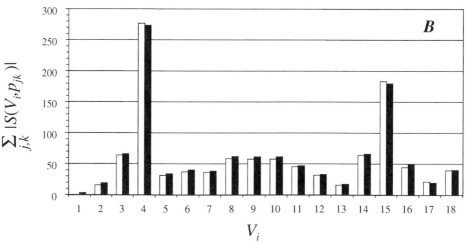

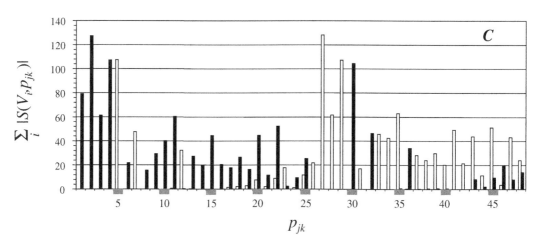

the three-dimensional plot of Figure 3.14. It is interesting to note that their influence is distributed more or less uniformly over all fluxes. Of the full 3,816 kinetic order sensitivities, only 329 are higher than 2, 204 are in the unit range, 815 in the 0.1 range, 900 in the 10^{-2} range, 713 in the 10^{-3} range, and 855 in the 10^{-4} range or less. The kinetic orders showing the most pronounced sensitivities are those associated with the transport of glucose, X_1 ($g_{1,20}$, $h_{1,1}$, and $h_{1,21}$) and the synthesis and transformation of glucose 6-phosphate, X_2 ($g_{2,1}$, $g_{2,21}$, $h_{2,2}$, and $h_{2,3}$) and fructose 6-phosphate, X_3 ($g_{3,2}$ and $g_{3,3}$). The synthesis and degradation of phosphoenolpyruvate, X_5 (represented by $g_{6,5}$ and $h_{5,5}$) are also affected, but to a lesser degree.

Kinetic order sensitivities for metabolite concentrations. The influences of the kinetic-order parameters on the metabolite concentrations are distributed similarly to those of fluxes. The 47 (23 g and 24 h) metabolite sensitivities with the largest values are shown in the three-dimensional plot of Figure 3.15. Again, their influence is roughly distributed uniformly over the pools. Of the total 3, 816 kinetic-order sensitivities, 618 are higher than two, 223 are in the unit range, 982 in the 0.1 range, 918 in the 10^{-2} range, 601 in the 10^{-3} range, and 474 in the 10^{-4} range or less. Most strongly influenced are the cytosolic glucose concentration ($g_{1,20}$ and $g_{2,1}$, $h_{1,1}$ and $h_{1,21}$); glucose 6-phosphate ($g_{3,2}$ and $g_{3,3}$, $h_{2,2}$ and $h_{2,3}$), acetyl-CoA ($h_{13,10}$ and $h_{13,37}$), cytosolic phosphoenlpyruvate ($h_{5,5}$), cytosolic pyruvate ($g_{6,5}$), and phosphoenol pyruvate ($g_{5,5}$).

The results of Figures 3.12–3.15 summarily suggest that the model is much more robust than the initial model. Relatively small perturbations in the parameter values of the model, as they may occur, for instance, in response to temperature changes or to errors in transcription or translation, no longer cause radical changes in concentrations or fluxes.

Logarithmic Gains

Flux logarithmic gains. The flux logarithmic gains (i.e., the influences of the independent variables on the fluxes of the model), are distributed as shown in the

#	1	2	3	4	5	6	7	8	9	10	11	12	13
g	1,20	2,1	2,21	3,2	3,3	3,4	3,23	4,3	5,3	5,26	6,5	6,17	6,28
#	14	15	16	17	18	19	20	21	22	23	24	25	
g	7,29	8,7	8,18	11,6	11,34	13,36	14,10	14,13	14,37	15,14	15,39	16,43	
#	26	27	28	29	30	31	32	33	34	35	36	37	38
h	4,4	1,1	1,21	2,2	2,3	2,22	2,23	3,3	3,26	5,5	5,17	5,28	6,6
#	39	40	41	42	43	44	45	46	47	48			
h	6,29	6,34	7,7	7,18	13,10	13,13	13,37	14,14	14,43	16,44			

Figure 3.14. Largest sensitivities of fluxes with respect to changes in kinetic orders (absolute values). Panel A: The small values indicate that the system is reasonably robust. Panel B: Magnitude for a particular flux V_i summed over all kinetic order parameters p_{jk}. Panel C: Magnitudes for a particular kinetic parameter p_{jk} summed over all the fluxes V_i. The kinetic orders are lexicographically presented according to the assignments in the table. For instance, parameter #38 is $h_{6,6}$.

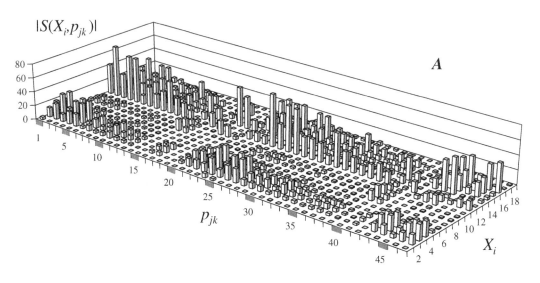

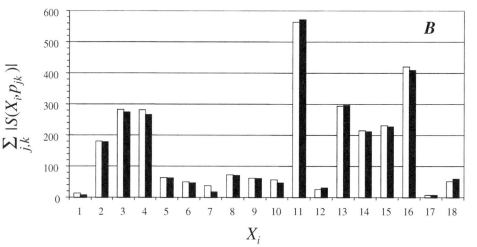

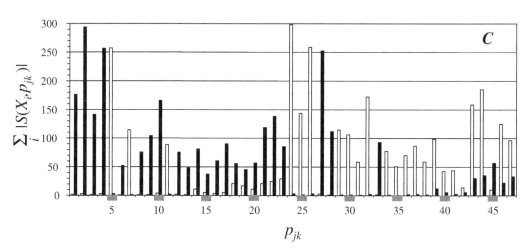

three-dimensional plot of Figure 3.16. Several patterns emerge. First, it is clear that the fluxes most influenced by changes in the independent variables are those through the pools for fructose-2,6-bisphosphate (X_4) and for isocitrate (X_{15}). In the former case, this behavior is not surprising, because fructose-2,6-bisphosphate is involved in a substrate cycle, and it is known that substrate cycles are rather sensitive to small variations in parameters (Newsholme and Start 1973). An indication of the high sensitivities of the fluxes through the isocitrate pool was already found in a preliminary model of this system (Torres 1994b), where aconitase was one of the enzymes with relatively high control over the flux. The remaining fluxes have lower sensitivities with most absolute values below 1 (97.3% or 543 of 558) (see panel B).

The independent variables that have the largest influence on fluxes are clearly those associated with the transport systems. The carrier showing the strongest control is the glucose intake system (TRP1, X_{20}). In previous work, the significance of this process had already become evident (Torres 1994b), and this theoretical insight prompted the experimental characterization of the carrier (Torres et al. 1996). In the present model, the kinetic parameters of this transport system have been included, and the system shows essentially the same behavior. The influence of this transport system is mostly positive, because it primarily affects the fluxes through the initial steps of glycolysis (see Figure 3.16C).

Some transport steps not present in the previous model are among the processes that exert the strongest control over the fluxes in the system. These are the pyruvate and the citrate carriers through the mitochondrial membrane (TRP2, X_{34} and TRP4, X_{43}), as well as the citrate excretion system (TRP5, X_{44}). It is important to note that the influence of the glucose transport system (X_{20}) is almost entirely positive and ubiquitous, affecting essentially all fluxes. This signifies that an increase (decrease) of the carrier activity changes the fluxes throughout the entire system in the same direction. A similar situation, although of lesser magnitude in terms of the logarithmic gains and the number of fluxes affected, is observed for the mitochondrial pyruvate transport system (X_{34}). The profile in the case of the mitochondrial and cytosolic

#	1	2	3	4	5	6	7	8	9	10	11	12	13
g	1,20	2,1	2,21	3,2	3,3	3,4	3,23	5,3	5,26	6,5	6,17	6,28	6,48
#	14	15	16	17	18	19	20	21	22	23			
g	8,7	8,18	11,6	11,34	12,10	12,37	13,36	14,10	14,37	16,43			
#	24	25	26	27	28	29	30	31	32	33	34	35	36
h	1,1	1,21	2,2	2,3	2,23	3,3	3,26	4,4	5,5	5,17	5,28	5,48	6,6
#	37	38	39	40	41	42	43	44	45	46	47		
h	6,29	6,34	7,7	7,18	11,36	12,36	13,10	13,37	14,14	14,43	16,44		

Figure 3.15. Largest sensitivities of metabolites with respect to changes in kinetic orders (absolute values). Panel A: The magnitudes of these sensitivities seem to be reasonable. Panel B: Magnitudes for a particular metabolite concentration X_i summed over all kinetic order parameters p_{jk}. Panel C: Magnitudes for a particular kinetic parameter p_{jk} summed over all the metabolite concentrations X_i. The kinetic orders are lexicographically presented according to the assignments in the table.

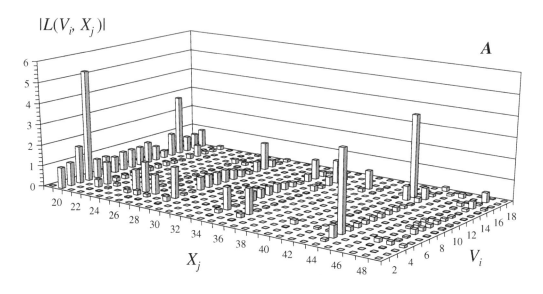

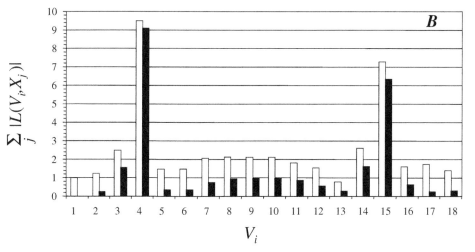

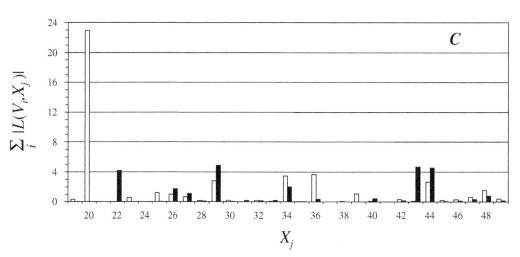

citrate excretion systems (X_{43} and X_{44}, respectively) is different, indicating mostly negative influence on the fluxes.

The influence of the mitochondrial excretion system (X_{43}) is restricted to the fluxes through the mitochondrial citrate and isocitrate pools (X_{13} and X_{14}, respectively). By contrast, the influence of the cytosolic citrate excretion (X_{44}) extends much more widely. In the former case, this means that an increase in the activity of this carrier leaves practically all fluxes unaffected except those listed previously, whereas in the latter case, most fluxes throughout the system are affected. It is interesting to note that the first model (Torres 1994a,b) revealed a similar pattern of control of fluxes by the transport processes, even though this model was different in many respects. The analysis here, combined with earlier findings, also reinforces experimental findings (Kubicek and Röhr 1986) suggesting that transport processes play a leading role in the control of the fluxes involved in the citric acid production. Other processes with substantial influence on fluxes are the steps catalyzed by glucose 6-phosphate dehydrogenase (X_{22}), cytosolic pyruvate carboxylase (X_{29}), and mitochondrial pyruvate dehydrogenase (X_{36}).

Glucose 6-phosphate dehydrogenase (X_{22}) exerts negative influence. This is readily explained because the flux is part of a diverging branch. The negative influence is not so evident in the case of pyruvate carboxylase (X_{29}). In this case, the carboxylase feeds the anaplerotic branch of the citric acid cycle and the recycling of oxalacetate through the phosphoenol pyruvate carboxykinase (X_{30}). The influence of pyruvate dehydrogenase is significant and mostly positive (panel C).

The remaining variables have minimal effect on the fluxes of the model. In some instances, these findings are again in good agreement with previous experimental evidence. Examples include pyruvate kinase X_{28}, which was identified by Meixner-Monori et al. (1983) as not controlling, and the substrate concentration, which is available in the medium at saturating concentrations (Mattey 1992; Wolschek and Kubicek 1998). Less evident is the small negative influence of hexokinase, X_{21}. In our previous study (Torres 1994a), the glucose uptake system and the hexokinase activity were lumped into one reaction, and this combined step showed significant influence on the fluxes. The present study allows us to pinpoint how this influence is shared between the two steps: it is the transport step that is responsible for flux control of the process. At the same time, the system transporting malate (X_{33}) into the mitochondria has no influence and thus acts differently than the other three transport processes. Our earlier model (Torres 1994a,b) showed that citrate synthase (X_{37}) and the respiratory system (X_{42}) have no influence on the fluxes. This conclusion is confirmed by the results here.

One should note the negligible level of control over PFKI (X_{26}). In spite of the many regulatory features of the PFK1 enzyme (Arts et al. 1987) and the importance of

---

Figure 3.16. Logarithmic gains of fluxes with respect to changes in independent variables. Panel A: Fructose-2,6-bisphosphate (X_4) and isocitrate (X_{15}) are affected most strongly. Panel B: Magnitudes for a given flux V_i are summed over all independent variables X_k. Panel C: Magnitudes for a given independent variable X_i are summed over all fluxes.

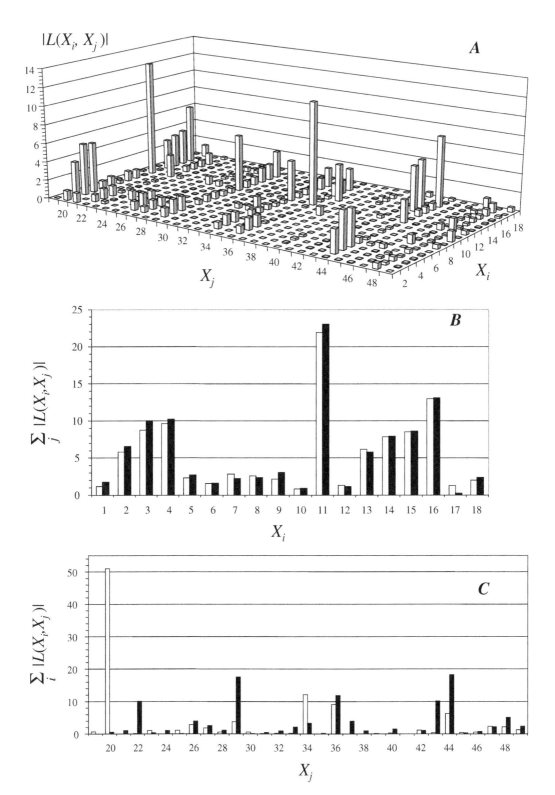

fructose-2,6-biphosphate as an activator of PFK1, it seems that the fluxes at the idiophase steady state are by and large insensitive to changes in enzyme activities and effectors.

Concentration logarithmic gains. The metabolite gains show essentially the same profile as the flux gains (Figure 3.17). One mitochondrial pool, namely, pyruvate (X_{11}), shows the highest sensitivity by far. Also relevant in this regard are the cytosolic citrate (X_{16}), the glycolytic intermediates glucose 6-phosphate (X_2), fructose 6-phosphate (X_3), fructose 2,6-bisphosphate (X_4), and the mitochondrial pools of citrate (X_{14}) and isocitrate (X_{15}). In all these cases, the positive and negative influences are more or less evenly distributed (panel B). This is an interesting result that should be taken into account when one defines the specifics of any optimization of the system, such as the permissible upper and lower limits of metabolite concentrations. This point will be considered in more detail in Chapter 6.

Another observation of interest is that the independent variables with the strongest influence on metabolite concentrations also have the strongest influence on the fluxes. This pattern has been found in studies of several different systems (Shiraishi and Savageau 1992a,b,c; Torres 1994b; Ni and Savageau 1996a,b). It is illustrated in the two-dimensional projection of panel C. The only significant difference appears with respect to citrate synthase (X_{37}) and, to a lesser degree, with respect to X_{21} and X_{24}. In these cases, the independent variables have no influence on the fluxes but negative influence on the pools.

As in the case of fluxes, transport systems play the leading role in controlling the pools. Examples are the substrate carrier (X_{20}, positive influence), the cytosolic citrate carrier (X_{44}, positive and negative influences), the mitochondrial citrate carrier (X_{43}, negative influence), and the mitochondrial pyruvate carrier (X_{34}, mostly positive). Other relevant enzymatic steps are catalyzed by pyruvate carboxylase (X_{29}), pyruvate dehydrogenase (X_{36}), and glucose 6-phosphate dehydrogenase (X_{22}). In the latter case, the situation again resembles the pattern of fluxes; this step has a negative influence on many pools, but especially on the glycolytic pools.

Dynamic Behavior

The first model did not respond adequately to changes in input, and it is interesting to test whether the model revisions have corrected the situation. Figure 3.18 shows the dynamic response, obtained with PLAS, to a twofold increase in the dependent variable, X_1 (cytosolic glucose). The twofold bolus of glucose is converted in glucose 6-phosphate (X_2), which is partially transformed into fructose 6-phosphate (X_3). Both (X_2) and (X_3) overshoot the basal steady-state concentrations by about 1.5–2.0% over a time period of fifteen minutes, before returning to within 0.5%

Figure 3.17. Logarithmic gains of metabolites with respect to changes in independent variables. Panel A: Pyruvate (X_{11}) shows the highest sensitivity by far. Panel B: Positive and negative influences are more or less evenly distributed. Panel C: Magnitudes for a given independent variable X_i are summed over all metabolites.

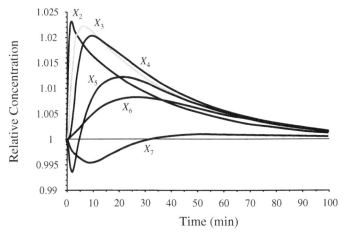

Figure 3.18. Dynamic system response to a twofold increase in cytosolic glucose (X_1). At time zero, the cytosolic glucose pool (X_1) was increased to twice the steady state. The six metabolites represented are glucose 6-phosphate (X_2), fructose 6-phosphate (X_3), fructose-2,6-bisphosphate (X_4), phosphoenol pyruvate (X_5), cytosolic pyruvate (X_6), and cytosolic oxalacetate (X_7). The remaining metabolites exhibit negligible deviations form the basal steady-state values (below 0.5%). Each concentration is normalized with respect to its nominal steady-state value. The response in the revised model is much improved over the initial model (compare y axis with Figure 3.10). (See www.cambridge.org/9780521177481 for color version.)

of the initial steady state. Fructose 6-phosphate is preferentially transformed into fructose-2,6-bisphosphate (X_4), initially (two to four minutes) at the expense of the synthesis of phosphoenol pyruvate (X_5), which exhibits small fluctuations around the steady-state concentration (lower than 1%) before returning to the steady state. From this point on, the perturbation propagates through the anaplerotic oxalacetate synthesis branch and the inner mitochondrial reactions with minimum deviation from the basal steady state (less than 0.5%; results not shown). The transient behavior is characterized by damped oscillations that were expected from the pair of eigenvalues with nonzero imaginary parts. Overall, the time courses show that after twenty to twenty-five minutes, all metabolite concentrations are well within a 1% deviation from the reference steady state, and that only the intermediate concentrations X_2, X_3, X_4, and X_5 deviate by more than 2%. The response is much improved over the previous model (see Figure 3.10) and can be considered biologically reasonable.

QUALITY ASSESSMENT OF THE REVISED MODEL

The results of our stability and sensitivity analyses suggest that the model structure and the parameters corresponding to the experimental conditions considered here provide a reasonably robust model description of the pathway. The logarithmic gain profiles indicate that changes in independent variables are reasonably amplified if these directly affect a particular flux. The steady-state results imply that the transport steps, specifically the substrate uptake system, the mitochondrial carriers of pyruvate and citrate, and the citrate excretion system, are particularly important in controlling the fluxes through the system. These results are in good agreement with previous theoretical and experimental findings (Kubicek and Röhr 1986; Torres 1994a). The comparison of the logarithmic gain profiles obtained from the rather complex, revised model with a previous, simpler one (Torres 1994b) suggests that the increase in complexity has no significant effect on the control structure of this particular system.

Nonetheless, analysis of the sensitivities and logarithmic gains shows that the revised model is far more robust. The absolute values of sensitivities and gains are lower than in the original model, in some instances by an order of magnitude (for instance, compare Figures 3.4 and 3.12). Moreover, the dynamics of the revised model shows that the system is well equipped to handle larger perturbations, whereas the original model showed significant deviations from the basal steady state and exhibited unreasonably long response times.

Overall, we may cautiously conclude that the present model of the pathway accounts appropriately for the metabolic activities and rates of citric acid accumulation in *A. niger*. The good agreement with many experimental findings provides support for the approximations introduced during model design and suggests that the proposed approach may be reliably used for further analysis and for predictions of the regulatory properties of the pathway. Equipped with this tentatively appropriate model, we are now in a position to pursue goals of optimization. The most prominent target will be the improvement of citric acid productivity. In the present case, such an improvement seems to be better accomplished through maximization of the citrate excretion reactions (V_{16}^-), which is explicitly given as the negative term of the sixteenth differential equation in system 3.7), than by an increase in the yield transformation. The details of this optimization will be shown in Chapter 6. We will use of the model developed here and specify a suitable objective function as well as a set of constraints that have to be fulfilled by the system. These constraints define boundaries for cellular conditions that guarantee cell viability. They prevent metabolite and enzyme concentrations from violating the osmolarity requirements and limit the magnitudes of those fluxes that strongly affect other vital parts of cellular metabolism. The optimization will yield optimal profiles of the intracellular enzyme activities, metabolites, and fluxes in cases where all or just a few of the independent variables are accessible to experimental manipulation.

APPENDIX: MATHEMATICAL FORMULATION OF CONSTRAINTS IN THE REVISED MODEL

During the development of the initial model, we discussed how to deal with the total concentrations of NAD^+ and NADH and of ADP and ATP. The revised model requires similar constraints, which, however, involve three components each. We begin with the constancy of the total concentration of the AMP, ADP, and ATP pools, which collectively have the concentration $AMP + ADP + ATP = 0.718$ mM. Retaining ATP as the explicit dependent variable, we need two constraints to eliminate ADP and AMP as dependent variables. The first constraint on the adenosine phosphate pool is now

$$X_{TA} = X_{ATP} + X_{ADP} + X_{AMP}. \tag{A3.1}$$

Under the conditions relevant to the model, we have the following additional relationship among the three pools (Habison et al. 1979):

$$X_{AMP} = \frac{X_{ATP} + X_{ADP}}{4}. \tag{A3.2}$$

The two constraints taken together have the solution ATP $= 0.5$mM, ADP $= 0.075$ mM, AMP $= 0.143$ mM, $X_{TA} = 0.718$ mM.

Substitution of Eq. (A3.2) into (A3.1) yields

$$X_{ADP} = \frac{4}{5}X_{TA} - X_{ATP}, \tag{A3.3}$$

which can be approximated symbolically as

$$X_{ADP} = \gamma_{ADP} X_{TA}^{f_{ADP,TA}} X_{ATP}^{f_{ADP,ATP}}. \tag{A3.4}$$

The kinetic orders in this approximation are

$$f_{ADP,TA} = \frac{\partial X_{ADP}}{\partial X_{TA}} \frac{X_{TA}}{X_{ADP}} = \frac{4}{5} \frac{X_{TA}}{X_{ADP}} \tag{A3.5}$$

and

$$f_{ADP,ATP} = \frac{\partial X_{ADP}}{\partial X_{ATP}} \frac{X_{ATP}}{X_{ADP}} = -\frac{X_{ATP}}{X_{ADP}} \tag{A3.6}$$

and the rate constant is given as

$$\gamma_{ADP} = \frac{X_{ADP}}{X_{TA}^{f_{ADP,TA}} X_{ATP}^{f_{ADP,ATP}}}. \tag{A3.7}$$

In the case of AMP, one obtains similarly

$$X_{AMP} = \frac{1}{5}X_{TA} \tag{A3.8}$$

$$X_{AMP} = \gamma_{AMP} X_{TA}^{f_{AMP,TA}}, \tag{A3.9}$$

where

$$f_{AMP,TA} = \frac{\partial X_{AMP}}{\partial X_{TA}} \frac{X_{TA}}{X_{AMP}} = \frac{1}{5} \frac{X_{TA}}{X_{AMP}} \tag{A3.10}$$

and

$$\gamma_{AMP} = \frac{X_{AMP}}{X_{TA}^{f_{AMP,TA}}}. \tag{A3.11}$$

The revised model distinguishes between NAD$^+$ in cytosol and mitochondria. Thus, the constraint of the original model is replaced with two constraints relating the three nicotine amide pools:

$$X_{TN} = X_{NADH} + X_{NADc^+} + X_{NADm^+} \tag{A3.12}$$

$$X_{NADc^+} = \frac{X_{NADm^+}}{5}. \tag{A3.13}$$

The algebraic solution here is $NADH = 0.035\,\text{mM}$, $NADc^+ = 0.12\,\text{mM}$, $NADm^+ = 0.6\,\text{mM}$, with a total amount of adenine nucleotides of $0.755\,\text{mM}$.

Substituting Eq. (A3.13) into (A3.12), we obtain

$$X_{NADm^+} = \frac{5}{6}(X_{TN} - X_{NADH}) \tag{A3.14}$$

$$X_{NADc^+} = \frac{1}{6}(X_{TN} - X_{NADH}), \tag{A3.15}$$

which yields the power-law representations

$$X_{NADm^+} = \gamma_{NADm^+} X_{TN}^{f_{NADm^+,TN}} X_{NADH}^{f_{NADm^+,NADH}} \tag{A3.16}$$

$$X_{NADc^+} = \gamma_{NADc^+} X_{TN}^{f_{NADc^+,TN}} X_{NADH}^{f_{NADc^+,NADH}}. \tag{A3.17}$$

The kinetic orders in these constraint terms are

$$f_{NADm^+,TN} = \frac{\partial X_{NADm^+}}{\partial X_{TN}} \frac{X_{TN}}{X_{NADm^+}} = \frac{5}{6}\frac{X_{TN}}{X_{NADm^+}} \tag{A3.18}$$

$$f_{NADm^+,NADH} = \frac{\partial X_{NADm^+}}{\partial X_{NADH}} \frac{X_{NADH}}{X_{NADm^+}} = -\frac{5}{6}\frac{X_{NADH}}{X_{NADm^+}} \tag{A3.19}$$

$$f_{NADc^+,TN} = \frac{\partial X_{NADc^+}}{\partial X_{TN}} \frac{X_{TN}}{X_{NADc^+}} = \frac{1}{6}\frac{X_{TN}}{X_{NADc^+}} \tag{A3.20}$$

$$f_{NADc^+,NADH} = \frac{\partial X_{NADc^+}}{\partial X_{NADH}} \frac{X_{NADH}}{X_{NADc^+}} = -\frac{1}{6}\frac{X_{NADH}}{X_{NADc^+}}, \tag{A3.21}$$

and the rate constants are

$$\gamma_{NADm^+} = \frac{X_{NADm^+}}{X_{TN}^{f_{NADm^+,TN}} X_{NADH}^{f_{NADm^+,NADH}}} \tag{A3.22}$$

and

$$\gamma_{NADc^+} = \frac{X_{NADc^+}}{X_{TN}^{f_{NADc^+,TN}} X_{NADH}^{f_{NADc^+,NADH}}}. \tag{A3.23}$$

REFERENCES

Aiba, S., and M. Matsuoka: Identification of metabolic model: Citrate production from glucose by *Candida lipolytica*. *Biotechnol. Bioeng.* 21, 1373–86, 1979.

Alvarez-Vasquez, F.: Modelización, análisis y optimización del metabolismo del hongo *Aspergillus niger* en condiciones de producción de ácido cítrico. PhD dissertation, University of La Laguna, Spain, 2000.

Alvarez-Vasquez, F., C. González-Alcón, and N.V. Torres: Metabolism of citric acid production by *Aspergillus niger*: Model definition, steady state analysis and constrained optimization of the citric acid production rate. *Biotechnol. Bioeng.* 70(1), 82–108, 2000.

Amaranisingham, C.F., and B.D. Davis: Regulation of alpha-ketoglutarate dehydrogenase formation in *Escherichia coli. J. Biol. Chem.* 240, 3664–8, 1965.

Arisan-Atac, I., M. Wolschek, and C.P. Kubicek: Trehalose-6-phosphate synthase A affects citrate accumulation by *Aspergillus niger* under conditions of high glycolytic flux. *FEMS Microbiol. Lett.* 140, 77–83, 1996.

Arts, E., C.P. Kubicek, and M. Röhr: Regulation of phosphofructokinase from *Aspergillus niger*: Effect of fructose 2,6 bisphosphate on the action of citrate, ammonium ions and AMP. *J. Gen. Microbiol.* 133, 1195–9, 1987.

Blázquez, M.A., R. Lagunas, C. Gancedo, and J.M. Gancedo: Trehalose-6-phosphate, a new regulator of yeast glycolysis that inhibits hexokinases. *FEBS Lett.* 329, 51–4, 1993.

Bloom, S., and M. Johnson: The pyruvate carboxylase of *Aspergillus niger. J. Biol. Chem.* 237, 2718–20, 1962.

Boy-Marcotte, E., G. Lagniel, M. Perrot, F. Bussereau, A. Boudsocq, M. Jacquet, and J. Labarre: The heat shock response in yeast: Differential regulations and contributions of the Msn2p/Msn4p and Hsf1p regulons. *Mol. Microbiol.* 33(2), 274–83, 1999.

Briquet, M.: Transport of pyruvate and lactate in yeast mitochondria. *Biochim. Biophys. Acta* 459, 290–9, 1977.

Cleland, W.W., and M.J. Johnson: Tracer experiments on the mechanism of citric acid formation by *Aspergillus niger. J. Biol. Chem.* 208, 679–92, 1954.

Crow, V.L., and C.L. Wittenberger: Separation and properties of NAD^+- and $NADP^+$-dependent glyceraldehyde-3-phosphate dehydrogenases from *Streptococcus mutans. J. Biol. Chem.* 254(4), 1134–42, 1979.

Evans, C.T., A.H. Scragg, and C. Ratledge: Regulation of citrate efflux from mitochondria of oleaginous and non-oleaginous yeast by adenine nucleotides. *Eur. J. Biochem.* 132, 609–15, 1981.

Evans, C.T., A.H. Scragg, and C. Ratledge: A comparative study of citrate efflux from mitochondria of oleaginous and non-oleaginous yeast. *Eur. J. Biochem.* 130, 195–204, 1983.

Feir, H.A., and I. Suzuki: Pyruvate carboxilase of *Aspergillus niger*: Kinetic study of a biotin-containing carboxilase. *Can. J. Biochem.* 47, 697–710, 1969.

Fell, D.A.: Metabolic control analysis – a survey of its theoretical and experimental development. *Biochem. J.* 286, 313–30, 1992.

Ferreira, A.: PLAS©: http://correio.cc.fc.ul.pt/~aenf/plas.html, 2000.

Führer, L., C.P. Kubicek, and M. Röhr: Pyridine nucleotide levels and ratios in *Aspergillus niger. Can. J. Biochem.* 26, 405–8, 1980.

Groen, A.K., R.J.A. Wanders, H.V. Westerhoff, R. van der Meer, and J.M. Tager: Control of metabolic fluxes. In: H. Sies (Ed.), *Metabolic Compartmentation* (pp. 9–37). Academic Press, New York, 1982.

Guebel, D., and N.V. Torres: Optimization of the citric acid production by *A. niger* through a metabolic flux balance model. *Elect. J. Biotechnol.*, in press.

Habison, A., C.P. Kubicek, and M. Röhr: Phosphofructokinase as a regulatory enzyme in citric acid producing *Aspergillus niger. FEMS Microbiol. Lett.* 5, 39–42, 1979.

Habison, A., C.P. Kubicek, and M. Röhr: Partial purification and regulatory properties of phosphofructokinase from *Aspergillus niger. Biochem. J.* 209, 669–76, 1983.

Halestrap, A.P., R.D. Scott, and A.P. Thomas: Mitochondrial pyruvate transport and its hormonal regulation. *Int. J. Biochem.* 11, 97–105, 1980.

Henry, M-F., and E.-J. Nyns: Cyanide-insensitive respiration. An alternative mitochondrial pathway. *Sub-Cell. Biochem.* 4, 1–65, 1975.

Henson, C.P., and W.W. Cleland: Kinetic studies of glutamic-oxalacetate transaminase isoenzymes. *Biochemistry* 3, 338–48, 1964.

Jaklitsch, W., C.P. Kubicek, and M. Scrutton: Intracellular location of enzymes involved in citrate production by *Aspergillus niger*. *Can. J. Microbiol.* 37, 823–7, 1991.

Kirimura, K., Y. Hirowatari, and S. Usami: Alterations of respiratory systems in *Aspergillus niger* under the condition of citric acid fermentation. *Agric. Biol. Chem.* 51, 1299–303, 1987.

Kirimura, K., M. Yoda, H. Shimizu, S. Sugano, M. Mizuno, K. Kino, and S. Usami: Contribution of cyanide-insensitive respiratory pathway, catalyzed by the alternative oxidase to citric acid production in *Aspergillus niger*. *Biosci. Biotechnol. Biochem.* 64(10), 2034–9, 2000.

Kubicek, C.P.: Aspects of control of metabolic fluxes in microorganisms. Dechema-Monogr., *VCH Verlagsgesellschaft* 105, 81–95, 1987.

Kubicek, C.P.: Organic acids. In: C. Ratledge and B. Kristiansen (Eds.), *Basic Biotechnology* (Chapter 14). Cambridge University Press, Cambridge, U.K., 2001.

Kubicek, C.P., and M. Röhr: Influence of manganese on enzyme synthesis and citric acid accumulation in *Aspergillus niger*. *Eur. J. Appl. Microbiol.* 4, 167–173, 1977.

Kubicek, C.P., and M. Röhr: The role of the tricarboxylic acid cycle in citric acid accumulation by *Aspergillus niger*. *Eur. J. Appl. Microbiol. Biotechnol.* 5, 263–71, 1978.

Kubicek, C.P., and M. Röhr: Citric acid fermentation. *CRC Crit. Rev. Biotechnol.* 3(4), 331–73, 1986.

Kubicek, C.P., W. Hampel, and M. Röhr: Manganese deficiency leads to elevated amino acid pools in citric acid accumulating *Aspergillus niger*. *Arch. Microbiol.* 123, 73–9, 1979.

Kubicek, C.P., O. Zehentgruber, H. El-Kalak, and M. Röhr: Regulation of citric acid production by oxygen: Effects of dissolved oxygen tension on adenylate levels and respiration in *Aspergillus niger*. *Eur. J. Appl. Microbiol. Biotechnol.* 9, 101–16, 1980.

Kubicek-Pranz, E.M., M. Mozelt, M. Roehr, and C.P. Kubicek: Changes in the concentration of fructose 2, 6-bisphosphate in *Aspergillus niger* during stimulation of acidogenesis by elevated sucrose concentration. *Biochim. Biophys. Acta* 1033, 250–5, 1990.

Legisa, M., and M. Mattey: Glycerol synthesis by *Aspergillus niger* under citric acid accumulating conditions. *Enzyme Microbiol. Technol.* 8, 607–9, 1988.

Ma, H., C.P. Kubicek, and M. Röhr: Malate dehydrogenase isoenzymes in *Aspergillus niger*. *FEMS Microbiol. Lett.* 12, 147–51, 1981.

Mattey, M.: Citrate regulation of citric acid production by *Aspergillus niger*. *FEMS Microbiol. Lett.* 2, 71–4, 1977.

Mattey, M.: The production of organic acids. *Crit. Rev. Biotechnol.* 12, 87–132, 1992.

Meixner-Monori, B., C.P. Kubicek, and M. Röhr: Pyruvate kinase from *Aspergillus niger*: A regulatory enzyme in glycolysis? *Can. J. Microbiol.* 30, 16–22, 1983.

Meixner-Monori, B., C.P. Kubicek, A. Habison, E.M. Kubicek-Pranz, and M. Röhr: Presence and regulation of the α-ketoglutarate dehydrogenase multienzyme complex in the filamentous fungus *Aspergillus niger*. *J. Bacteriol.* 161, 265–71, 1985.

Meixner-Monori, B., C.P. Kubicek, W. Harrer, G. Schreferl, and M. Röhr: NADP-specific isocitrate dehydrogenase from the citric acid-accumulating fungus *Aspergillus niger*. *Biochem. J.* 236, 549–57, 1986.

Meyrath, J.: Citric acid production. *Proc. Biochem.* 2, 25–7, 1967.

Mischak, H., C.P. Kubicek, and M. Röhr: Citrate inhibition of glucose uptake in *Aspergillus niger*. *Biotechnol. Lett.* 6, 425–30, 1984.

Netik, A., N.V. Torres, J.M. Riol, and C.P. Kubicek: Uptake and export of citric acid by *Aspergillus niger* is reciprocally regulated by manganese ions. *Biochim. Biophys. Acta* 1326, 287–94, 1997.

Newsholme, E.A., and C. Start: *Regulation in Metabolism*. John Wiley & Sons, London, 1973.

Ni, T.-C., and M.A. Savageau: Application of biochemical systems theory to metabolism in human red blood cells. Signal propagation and accuracy of representation. *J. Biol. Chem.* 271(14), 7927–41, 1996a.

Ni, T.-C., and M.A. Savageau: Model assessment and refinement using strategies from biochemical systems theory: Application to metabolism in human red blood cells. *J. Theor. Biol.* 179, 329–68, 1996b.

Osmani, S.A., and M.C. Scrutton: The subcellular localization of pyruvate carboxylase and of some other enzymes in *Aspergillus nidulans*. *Eur. J. Biochem.* 133, 551–60, 1983.

Panneman, H., G.J.G. Ruijter, H.C. Van der Broeck, E.T. Driever, and J. Visser: Cloning and biochemical characterization of an *Aspergillus niger* glucokinase. Evidence for the presence of separate glucokinase and hexokinase enzymes. *Eur. J. Biochem.* 240, 518–25, 1996.

Panneman, H., G.J.G. Ruijter, H.C. Van der Broeck, and J. Visser: Cloning and biochemical characterization of *Aspergillus niger* hexokinase. The enzyme is strongly inhibited by physiological concentrations of trehalose 6-phosphate. *Eur. J. Biochem.* 258, 223–32, 1998.

Pedersen, H., M. Carlsen, and J. Nielsen: Identification of enzymes and quantification of metabolic fluxes in the wild type and in a recombinant *Aspergillus oryzae strain*. *Appl. Environ. Microbiol.* 65(1), 11–19, 1999.

Perkins, M., J.M. Haslam, and A.W. Linnane: Biogenesis of mitochondria. The effects of physiological and genetic manipulation of *Saccharomyces cerevisiae* on the mitochondrial transport sysytem for tricarboxylate-cycle anions. *Biochem. J.* 134, 923–934, 1973.

Prömper, C., R. Schneider, and H. Weiss: The role of the proton-pumping and alternative respiratory chain NADH: Ubiquinone oxidoreductases in overflow catabolism of *Aspergillus niger*. *Eur. J. Biochem.* 216, 223–30, 1993.

Reich, J.G., and E.E. Sel'kov: *Energy Metabolism of the Cell.* Academic Press, London, 1981.

Röhr, M.: A century of citric acid fermentation and research. *Food Technol. Biotechnol.* 36(3), 163–71, 1998.

Röhr, M., and C.P. Kubicek: Regulatory aspects of citric acid fermentation by *Aspergillus niger*. *Proc. Biochem.* 16, 34–7, 1981.

Röhr, M., C.P. Kubicek, O. Zehentgruber, and R. Orthofer: Accumulation and partial reconsumption of polyols during citric acid fermentation by *Aspergillus niger*. *Appl. Microbiol. Biotechnol.* 27, 235–9, 1987.

Sakai, K., K. Hasumi, and A. Endo: Two glyceraldehyde-3-phosphate dehydrogenase isozymes from the koningic acid (heptelidic acid) producer Trichoderma koningii. *Eur. J. Biochem.* 193(1), 195–202, 1990.

Salvador, A.: Synergism analysis of biochemical systems. I. Conceptual framework. *Math. Biosci.* 163(2), 105–29, 2000a.

Salvador, A.: Synergism analysis of biochemical systems. II. Tensor formulation and treatment of stoichiometric constraints. *Math. Biosci.* 163(2), 131–58, 2000b.

Savageau, M.A.: Biochemical systems analysis. II. The steady-state solutions for an n-pool system using a power-law approximation. *J. Theor. Biol.* 25, 370–9, 1969.

Savageau, M.A.: Optimal design of feedback control by inhibition: Dynamic considerations. *J. Mol. Evol.* 5, 199–222, 1975.

Savageau, M.A.: *Biochemical System Analysis: A study of Function and Design in Molecular Biology.* Addison-Wesley, Reading, MA, 1976.

Schmidt, M., J. Wallrath, A. Dörner, and H. Weiss: Disturbed assembly of the respiratory chain NADH: Ubiquinone reductase (complex I) in citric-acid-accumulating *Aspergillus niger* strain B60. *Appl. Microbiol. Biotechnol.* 36, 667–72, 1992.

Schreferl-Kunar, G., M. Grotz, M. Röhr, and C.P. Kubicek: Increased citric acid production by mutants of *Aspergillus niger* with increased glycolytic capacity. *FEMS Microbiol. Lett.* 59, 297–300, 1989.

Shiraishi, F., and M.A. Savageau: The tricarboxylic acid cycle in *Dictyostelium discoideum*.

I. Formulation of alternative kinetic representations. *J. Biol. Chem.* 267(32), 22912–18, 1992a.

Shiraishi, F., and M.A. Savageau: The tricarboxylic acid cycle in *Dictyostelium discoideum*. II. Evaluation of model consistency and robustness. *J. Biol. Chem.* 267(32), 22919–25, 1992b.

Shiraishi, F., and M.A. Savageau: The tricarboxylic acid cycle in *Dictyostelium discoideum*. III. Analysis of steady state and dynamic behavior. *J. Biol. Chem.* 267(32), 22926–33, 1992c.

Sorribas, A., and M.A. Savageau: Strategies for representing metabolic pathways within biochemical systems theory: Reversible pathways. *Math. Biosci.* 94, 239–69, 1989.

Steinbock, F.A.: *Doctoral dissertation*, Technical University of Vienna, Vienna, Austria, 1993.

Steinbock, F.A., I. Held, S. Chojun, H. Harsen, M. Röhr, E.M. Kubicek-Prantz, and C.P. Kubicek: Regulatory aspects of carbohydrate metabolism in relation to citric acid accumulation by *Aspergillus niger*. *Acta Biotechnol.* 116, 571–81, 1991.

Supply, P., A. Wach, and A. Goffeau: Enzymatic properties of the PMA2 plasma membrane-bound H+-ATPase of *Saccharomyces cerevisiae*. *J. Biol. Chem.* 268(25), 19753–9, 1993.

Torres, N.V.: Modeling approach to control of carbohydrate metabolism during citric acid accumulation by *Aspergillus niger*. I. Model definition and stability of the steady state. *Biotechnol. Bioeng.* 44, 104–11, 1994a.

Torres, N.V.: Modeling approach to control of carbohydrate metabolism during citric acid accumulation by *Aspergillus niger*. II. Sensitivity analysis. *Biotechnol. Bioeng.* 44, 112–18, 1994b.

Torres, N.V., J.M. Riol-Cimas, M. Wolschek, and C.P. Kubicek: Glucose transport by *Aspergillus niger*: The low-affinity carrier is only formed during growth on high glucose concentrations. *Appl. Microbiol. Biotechnol.* 44, 790–4, 1996.

Verhoff, F.H., and J.E. Spradlin: Mass and energy balance analysis of metabolic pathways applied to citric acid production by *Aspergillus niger*. *Biotechnol. Bioeng.* 18(3), 425–32, 1976.

Voit, E.O.: *Computational Analysis of Biochemical Systems. A Practical Guide for Biochemists and Molecular Biologists* (xii + 533 pp.). Cambridge University Press, Cambridge, U.K., 2000.

Voit, E.O., and A.E.N. Ferreira: Buffering in models of integrated biochemical systems. *J. Theor. Biol.* 191, 429–38, 1998.

Wallrath, J., J. Schmidt, and H. Weiss: Concomitant loss of respiratory chain NADH: ubiquinone reductase (complex I) and citric acid accumulation in *Aspergillus niger*. *Appl. Microbiol. Biotechnol.* 36, 76–81, 1991.

Wehmer, C.: *Beiträge zur Kenntnis einheimischer Pilze. I. Zwei neue Schimmelpilze als Erreger einer Citronensäure-Gärung.* Halm, Hannover/Leipzig, 1903.

Wolschek, M.F., and C.P. Kubicek: Biochemistry of citric acid accumulation in *Aspergillus niger*. In: B. Kristiansen, M. Mattey, and J. Linden (Eds.), *Citric Acid Biotechnology* (pp. 11–32). Taylor and Francis, London, U.K., 1998.

Woronick, C., and M.J. Johnson: Carbon dioxide fixation by cell-free extracts of *Aspergillus niger*. *J. Biol. Chem.* 235, 9–15, 1960.

Xu, D.-B., C.P. Madrid, M. Röhr, and C.P. Kubicek: The influence of type and concentration of the carbon source on production of citric acid by *Aspergillus niger*. *Appl. Microbiol. Biotechnol.* 30, 553–8, 1989.

Zahorsky, B.: U.S. Patent 1,065,358, 1913.

Zehentgruber, O., C.P. Kubicek, and M. Röhr: Alternative respiration of *Aspergillus niger*. *FEMS Microbiol. Lett.* 8, 71–4, 1980.

## CHAPTER FOUR

Optimization Methods

The goal of optimization is simple: identify the best possible outcome. In mathematical terms, the "outcome" is the value of some function, and "best possible" often is the maximum of the function, which is the point with the highest possible value. The function itself is often called *objective function* and its arguments are called *control variables*. Parameters of the objective function are sometimes called *value coefficients*. If the function describes the yield of a batch process, then the maximum may describe the control settings under which the yield is greatest. In a minimization problem, one might look for the cheapest solution. Searching for the minimum might sound like the opposite problem to maximization, but it is accomplished with the same mathematical concepts. In particular, if one multiplies the objective function with -1, the former maximum becomes the minimum and vice versa. Thus, finding a maximum or a minimum is basically the same, and we talk generically about finding an *extremum* or *optimum*.

Optimization is a huge field, and general methods and specific techniques have been developed not only in mathematics, but also in engineering, operations research, and many branches of quantitative science. To provide some structured overview, one may classify optimization tasks according to a hierarchy of complexity (Vagners 1983). The simplest tasks require the direct algebraic computation of maxima or minima of functions, primarily with tools of differentiation. Slightly more complex is the optimization of multivariate functions, which can sometimes be executed with analytical means, but often requires numerical algorithms that iteratively search for an extremum. This group of problems is relatively simple because it is characterized by one criterion of optimality, one completely defined function describing the optimal solution, and full information about constraints that the optimal solution has to satisfy.

Optimization problems at the next level of complexity involve dynamical systems. There is still one criterion of optimality, and the problem is completely defined, but the state and/or the constraints are defined by ordinary or partial differential equations or have stochastic components. These tasks are more difficult, because they aim at optimizing not only some final point, but also the path toward this point.

This area of *optimal control* leads to *calculus of variations* (Strang 1986) and to criteria of optimality like *Pontryagin's Maximum Principle* (Vagners 1983, p. 1180, footnote).

At the third level are optimization problems where two or more controllers (players) have conflicting goals: each player wants to win. In the simpler of these cases, the players know the payoff for each game strategy, but they do not know their opponents' strategies. The famous zero-sum games fall in this category. In the more complicated tasks of this category, information may be limited or unavailable to the players. Furthermore, the "game" may be dynamic and the task defined by differential equations. The last group of optimization problems has multiple and/or ill-defined criteria and incomplete information. Nonzero-sum games, where several or all players may win or all may lose, belong to this class. These latter types of optimization are addressed by *game theory*, which deals with methods for developing optimal strategies. The exploration of so-called *evolutionary stable strategies*, which became fashionable in the 1970s and 1980s, falls in this area.

In this book, discussions of optimality are limited to the first group of tasks, namely those with a well-defined criterion of optimality and complete information about the state of the process and the constraints that every solution must satisfy. Although biochemical systems are based on differential equations, we will see in the next chapter that important optimization tasks can be reduced to those involving only algebraic equations. This chapter therefore concentrates on optimization methods for algebraic functions.

ANALYTICAL METHODS

Extrema of Functions of One Variable

Many issues concerning optimization can be studied with simple functions of one argument: $y = f(x)$. Although such an analysis is covered by most calculus courses, it is beneficial to review some key aspects, because many of them are similar in higher dimensions but difficult to think about and visualize.

Let us start at the very beginning, namely the linear function $y = a_0 + a_1 x$. There are three possibilities. First, y could be constant. In this case, every value of x yields the same value of y. Obviously, this situation is not very exciting, and whether we call all points maxima, minima, or neither is a matter of semantics. If y is not constant, then it either always grows or it always decreases. It is easy to see that the maximum in these cases is either at $x = +\infty$ or at $x = -\infty$. In reality, x is seldom able to reach $+\infty$ or $-\infty$, but is limited within a range $[a, b]$. For instance, if the yield of a fermentation process is a linear function of batch volume, the optimum would be obtained for a batch of infinite size. If the batch size has an upper limit b, the maximum yield, according to this very simple model, would be obtained for $x = b$ and have the optimal value $y = a_0 + a_1 b$. The batch size also has a lower limit, which is $x = a \geq 0$. Because we are interested in the maximum, this particular limit is irrelevant.

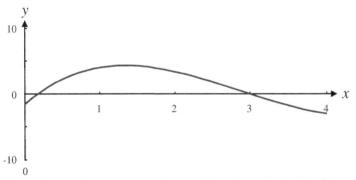

Figure 4.1. Graph of the polynomial $y = -1.5 + 10x - 5x^2 + 0.6x^3$ for $x \in [0, 4]$.

Like this trivial example, almost all optimization problems of practical interest involve constraints. They may be simple, as in the case described, or more complicated and convoluted, as we will see later. Typically, some of the constraints are irrelevant, but others affect the optimal solution very significantly.

Linear functions are easy to analyze, because their entire growth or decline behavior is the same for all arguments and, therefore, can be determined at any value of the independent variable x. If the slope at the particular value x_1 is positive, the slope is positive for all other values of x as well. This is not the case with nonlinear functions and is one of the reasons that nonlinear optimization is challenging. The crucial complication of nonlinear functions is that checking the boundaries of a constraining interval is no longer sufficient. As in the linear case, the maximum may lie at one of the boundaries, but in the nonlinear case it may also lie somewhere in the inside of the interval. Which situation occurs depends on the particularities of the function as well as the constraining interval.

As an example, suppose we are interested in the maximum of the cubic polynomial

$$y = -1.5 + 10x - 5x^2 + 0.6x^3. \tag{4.1}$$

Figure 4.1 shows a graph of the function for x values between 0 and 4. Obviously, the maximum is somewhere between 1 and 2. Several observations about this simple function are typical for many optimization problems.

First, determining the exact location of the maximum requires further work. We can read the location off the graph in Figure 4.1 with some precision, but the exact answer can only be obtained with more math.

Second, the graph in Figure 4.1 is limited to the interval $[0, 4]$ of the independent variable x. Is it possible that there is another maximum or maybe even a higher value outside this interval? Indeed, if we extend the search for the maximum to the enlarged interval $[0, 6]$, the value at the upper end of the interval has a higher value than the former maximum (Figure 4.2). Going to the extreme, if x goes to infinity, y eventually grows without bounds also. Thus, the obvious maximum between 1

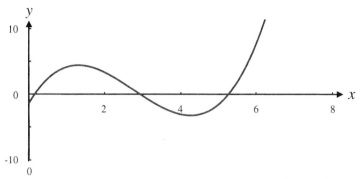

Figure 4.2. Graph of the polynomial $y = -1.5 + 10x - 5x^2 + 0.6x^3$ for $x \in [0, 6]$.

and 2 is not the absolute, *global* maximum, but only a *local* maximum, that is, a maximum within a certain range. This observation implies a very important aspect of essentially all nonlinear optimization problems: Relevant maxima and minima are critically affected by constraints on the system.

In this example, limiting the search to the interval $[0, 4]$ as opposed to $(-\infty, +\infty)$, leads to a maximum between 1 and 2, but excludes the true maximum at infinity or at the border of a bigger interval like $[-10, 10]$.

Third, there is information in the fashion with which the function *approaches* the maximum. If the maximum is inside the relevant interval, one sees that the slope of the function becomes smaller and smaller in magnitude, when the x-values approach the maximum from either below or above (Figure 4.3). At the maximum itself, the slope is zero. If it were not zero, one could move up or down the slope to some neighboring points, some of which would have higher values and some of which would have lower values, and the point could not be an extremum. Searching for an extremum of a nonlinear function can thus be restricted to points where the slope is zero. We will see later that the slope actually has to change its sign from positive to negative or vice versa to make a true extremum possible.

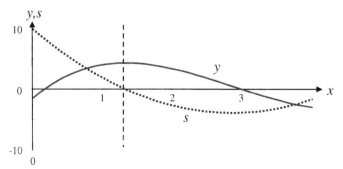

Figure 4.3. Graph of the polynomial $y = -1.5 + 10x - 5x^2 + 0.6x^3$ and its slope s for $x \in [0, 4]$. The dashed vertical line indicates the local maximum, which coincides with s crossing zero.

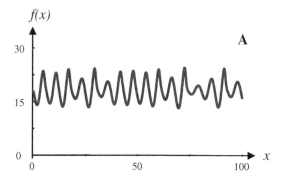

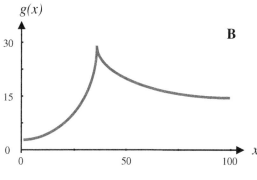

Figure 4.4. Differentiation requires smoothness. Panel A: The function $f(x)$ is complicated, yet smooth. Panels B: The function $g(x)$ is smooth and differentiable except at the cusp located at $x = 37.5$. Panel C: The function $h(x)$ is generally linear and smooth, but assumes values at $x = 25$, 50, 75, and 100 that do not lie on the linear graph. At these points, the function is not continuous and not differentiable.

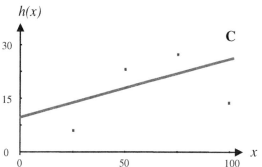

Analytical Computation of Extrema of Nonlinear Functions in One Variable

It is useful to study in more detail the slope of a one-variable function at an extremum. Although the extremum could be a maximum or minimum, it is sufficient to concentrate on the maximum, because the analogous considerations apply to the minimum of a function, if the function is flipped over (e.g., multiplied with -1). For this section, we assume that the function is smooth in the entire interval of interest. In other words, it has no gaps and no sharp corners. In more formal terms, it can be differentiated and has a unique slope for every value of x. Figure 4.4 shows a smooth function, along with examples of nonsmooth functions.

One needs to distinguish two cases. In the first case, the maximum is at the border of the range under consideration. This border could be some real number a or b or it could be $-\infty$ or $+\infty$. In most applications, this case can be analyzed by inspection, that is, by comparing the value of the function at the border with candidates for maxima inside the range or by computing limits.

In the second case, the maximum is truly inside the range. An example was the maximum between 1 and 2 in the polynomial (Eq. 4.1). This case is more interesting in practical applications. As we observed, the slope at the maximum has to be zero. The slope is mathematically represented by the derivative, and the problem of finding a maximum is thus reformulated into one of determining where the derivative is zero. Using the same example as given previously, one computes the derivative as

$$\text{slope}(y) = \frac{dy}{dx} = y' = 10 - 10x + 1.8x^2. \tag{4.2}$$

The expression in Eq. (4.2) is a function that represents all slopes of the polynomial (4.1). Thus, finding extrema requires setting the function in Eq. (4.2) equal to zero. The result in this particular case is a quadratic equation that has two solutions, which are called *roots*. They are $x_1 = 1.3079$ and $x_2 = 4.2476$. Comparing these solutions with the graph in Figure 4.2, one sees that the first value corresponds to the maximum of the function and the second to the minimum.

For more complicated nonlinear functions, there may be just one solution, there may be many, there may be infinitely many, or there may be none at all.

Examples

1. The function $y(x) = \exp(x) - x$ has the derivative $y'(x) = \exp(x) - 1$. Setting the derivative equal to zero yields the unique solution $x = 0$.
2. The function $y(x) = \exp(x) + x$ has the derivative $y'(x) = \exp(x) + 1$. Setting the derivative equal to zero shows that there is no solution, because there is no real value x for which the exponential function has the value -1.
3. The function $y(x) = \sin(x)$ has the derivative $y'(x) = \cos(x)$. Setting the derivative equal to zero leads to infinitely many isolated solutions $x = (2k + 1)\pi/2$, where k is any integer.
4. The function $y(x) = 3$ has the derivative $y'(x) = 0$. The problem has infinitely many solutions (for every real x) that form a continuum.

Setting the derivative equal to zero and solving the equation yields *candidates* for extrema. To determine which among these candidates are actually maxima, as opposed to minima or no extrema at all, one needs to do some additional evaluation. Intuitively, the simplest method is checking values of the function close to the candidate in the graph or analytically. If neighboring points on both sides of the candidate are smaller, the candidate is indeed a maximum. If the neighboring points are larger, the candidate must be a minimum. However, even if the derivative is zero, it is still possible that neighbors on one side are smaller and neighbors on the other side are larger. The standard example is

$$y = x^3. \tag{4.3}$$

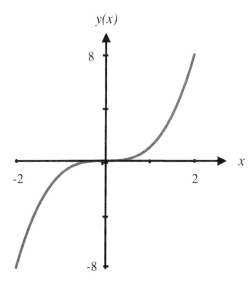

Figure 4.5. Graph of the function $y = x^3$.

The function is graphed in Figure 4.5. Clearly, the derivative

$$y' = 3x^2 \tag{4.4}$$

is zero for $x = 0$. Thus, $x = 0$ is a candidate for a maximum. Nonetheless, y is negative for all negative x and positive for all positive x, and $x = 0$ is neither a maximum nor a minimum. Another indication of the problem is that the derivative at this point decreases toward zero but never becomes negative. This suggests (and one can actually prove it) that a necessary and sufficient condition for an extremum is that the derivative changes its sign, either from positive to negative or from negative to positive.

Instead of an inspection of the function or its derivative, one commonly uses a test that characterizes the *curvature* of the function at the candidate point. The curvature is mathematically determined by the second derivative. The rule is as follows: if the second derivative at the candidate point is negative, the function is called *concave downward* and has a maximum. Conversely, if the second derivative is positive, the function is called *convex* or *concave upward* and has a minimum (Saaty and Bram 1964; Stewart 1999).

If the second derivative is zero, one has to study even higher derivatives or investigate points close to the candidate point. If the first k derivatives are zero, and the $(k+1)$st derivative is not, then the answer depends on whether k is even or odd. If k is even, the function does not have an extremum at this point. For odd k, the function has a maximum or minimum, if this derivative is negative or positive, respectively.

Examples
1. The second derivative of the polynomial $y = -1.5 + 10x - 5x^2 + 0.6x^3$ is $y'' = 3.6x - 10$. For the first solution, $x_1 = 1.3079$, which is derived from setting $y' = 0$,

the second derivative is negative, indicating a maximum, whereas for the second solution, $x_2 = 4.2476$, the second derivative is positive, indicating a minimum. These numerical solutions confirm results from an intuitive inspection of Figures 4.1–4.3.

2. The function $y = x^3$ has the first derivative $y' = 3x^2$, which identifies $x = 0$ as the only candidate for an extremum. The second derivative is $y'' = 6x$, which is zero at $x = 0$. Because this is neither positive nor negative, one computes the next derivative, which is $y''' = 6$. Because $k = 2$ is even, the function does not have an extremum at $x = 0$, which, of course, was also easy to find out from the graph in Figure 4.5.

3. The derivatives of $y = -x^4$ are $y' = -4x^3$, $y'' = -12x^2$, $y''' = -24x$, and $y'''' = -24$. The only candidate is again $x = 0$. The last zero-valued derivative is number 3, which is odd. There is an extremum, and because the fourth derivative is negative, the function has a maximum.

Analytical Computation of Extrema of Nonlinear Functions in Several Variables

The optimization of nonlinear functions of several variables can be a very difficult task. In fact, if the functions or constraints are sufficiently complicated, analytical solutions can no longer be obtained, and one has to resort to computer algorithms that *search* for the optimal solution. Before we go into computerized searches, though, it is useful to discuss the principles and limitations of analytically optimizing multivariate nonlinear functions. We will try to keep the math as simple as possible, but need to introduce some concepts and terms that are not always taught in elementary calculus. If necessary, this section may be skipped, as its material is not a prerequisite for the rest of the book. However, reading this part will provide the fundamental understanding of how search algorithms work and why they are necessary in the first place.

Many of the principles for optimizing functions of a single variable apply analogously to higher-dimensional functions. To describe them efficiently, we need additional terminology. A critical concept is the *partial derivative*. It represents the slope of a multivariate function in one direction. To illustrate the partial derivative, consider the function $z = f(x, y) = 3 - 2 \cdot \sqrt{1 + 6.25 \cdot (x^2 + y^2)}$. Its graph is a curved hyperboloid surface (von Seggern 1990, p. 220), which is shown in Figure 4.6. If one picks some values for the independent variables, say x_1 and y_1, the corresponding z value z_1 lies on the hyperboloid surface. Imagine cutting the surface with a vertical plane that is parallel to the x axis. The intersection between this vertical plane and the curved surface is some curved line, sometimes called a *projection*. In fact, we can say specifically which function describes this line. It is z, if its second argument y is fixed to be y_1; thus $z(x, y_1)$. For instance, if $y_1 = 2$, the line is given by the function $z(x, 2) = 3 - 2 \cdot \sqrt{1 + 6.25 \cdot (x^2 + 4)}$. Indeed, this is a function of just one independent variable, namely x. As such, it has a slope at the point x_1, which is given by its regular derivative. The same arguments hold if we cut the curved surface

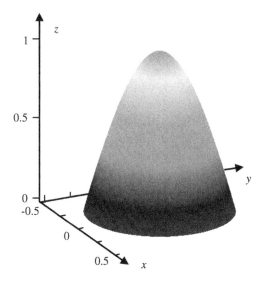

Figure 4.6. The graph of the two-variable function $z = 3 - 2 \cdot \sqrt{1 + 6.25 \cdot (x^2 + y^2)}$ is a smooth surface.
(See www.cambridge.org/9780521177481 for color version.)

of $z = f(x, y)$ at x_1 with a vertical plane parallel to the y axis. Now the result is a function just of y. An example is shown in Figure 4.7, where $x_1 = 0.22$.

Equipped with this intuition in two independent variables, x and y, the concept of a partial derivative, which we already used in previous chapters to describe sensitivities, is not so difficult to understand. The partial derivative of $f(x_1, x_2, \ldots, x_n)$ in x_j-direction is the usual derivative of the function f, with respect to x_j, when all other variables $x_1, \ldots x_{j-1}, x_{j+1}, \ldots x_n$ are held constant. To signify that a derivative is a partial derivative, it is customary to use the "curly d," ∂. Thus, the (partial) derivative of f with respect to x_4 is $\frac{\partial f}{\partial x_4}$. An equivalent notation is f_{x_4}.

The *overall change in f* is described by the *total* or *exact differential*, which is a sum of the slopes in each direction, each multiplied with unit increases in this

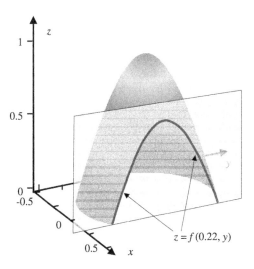

Figure 4.7. The intersection of the graph in Figure 4.6 and the plane $x = 0.22$ shows the projection of $z = f(0.22, y)$, which now is a univariate function of y.
(See www.cambridge.org/9780521177481 for color version.)

$z = f(0.22, y)$

direction, which are written as $dx_1, dx_2, \ldots, dx_n$. The total differential of f is thus

$$df = \frac{\partial f}{\partial x_1} dx_1 + \frac{\partial f}{\partial x_2} dx_2 + \cdots \frac{\partial f}{\partial x_n} dx_n = f_{x_1} dx_1 + f_{x_2} dx_2 + \cdots f_{x_n} dx_n. \quad (4.5)$$

To obtain the derivative in the direction of one coordinate, such as x_3, the appropriate increase dx_3 is set equal to 1 and all others $(dx_1, dx_2, dx_4, dx_5, \ldots, dx_n)$ equal to zero.

The concept of a total differential may at first be confusing, because there seems to be a mixture of all kinds of derivatives. One way to obtain an intuitive impression may be the comparison with a simple univariate linear function $y = mx + b$ with slope m and intercept b, which passes through the two points (x_1, y_1) and (x_2, y_2). It is well known that the slope m may be written as

$$m = \frac{y_2 - y_1}{x_2 - x_1} = \frac{\text{change in } y}{\text{change in } x} = \frac{dy}{dx}. \quad (4.6)$$

Rearranging terms, we immediately obtain

$$dy = m \cdot dx, \quad (4.7)$$

which has the form of the total differential in Eq. (4.5) with the partial derivative f_x interpreted as the slope m.

It is useful to represent the partial derivatives $\frac{\partial f}{\partial x_i}$ collectively as a vector, which is called the *gradient* of f and abbreviated as *grad f*. An alternate notation is an upside-down Greek delta, called *Nabla*, and the function f:

$$grad\ f = \nabla f = \left(\frac{\partial f}{\partial x_1}, \frac{\partial f}{\partial x_2}, \frac{\partial f}{\partial x_3} \right) = (f_{x_1}, f_{x_2}, f_{x_3}). \quad (4.8)$$

Instead of using this form, it is sometimes more convenient to write the gradient in terms of unit vectors, which have the length of one unit and point in the directions of the coordinates. In three dimensions, the unit vectors are $\mathbf{i} = (1, 0, 0), \mathbf{j} = (0, 1, 0)$, and $\mathbf{k} = (0, 0, 1)$. In terms of these vectors, the gradient is given as

$$grad\ f = \nabla f = \frac{\partial f}{\partial x_1} \mathbf{i} + \frac{\partial f}{\partial x_2} \mathbf{j} + \frac{\partial f}{\partial x_3} \mathbf{k}. \quad (4.9)$$

One readily confirms that the two definitions are really the same. The right-hand side is a sum of three products between a scalar (namely the partial derivative, which is here treated like a number) and a vector. Executing the multiplication, according to the rule

$$a \cdot (v_1, v_2, v_3) = (a \cdot v_1, a \cdot v_2, a \cdot v_3) \quad (4.10)$$

yields

$$\nabla f = \frac{\partial f}{\partial x_1} \mathbf{i} + \frac{\partial f}{\partial x_2} \mathbf{j} + \frac{\partial f}{\partial x_3} \mathbf{k} = \frac{\partial f}{\partial x_1}(1, 0, 0) + \frac{\partial f}{\partial x_2}(0, 1, 0) + \frac{\partial f}{\partial x_3}(0, 0, 1)$$

$$= \left(\frac{\partial f}{\partial x_1}, 0, 0 \right) + \left(0, \frac{\partial f}{\partial x_2}, 0 \right) + \left(0, 0, \frac{\partial f}{\partial x_3} \right) = \left(\frac{\partial f}{\partial x_1}, \frac{\partial f}{\partial x_2}, \frac{\partial f}{\partial x_3} \right). \quad (4.11)$$

At this point, it is sufficient to know that the gradient provides a measure for the slopes of a multivariable function in different directions. We will discuss more properties of the gradient in the section on search algorithms.

Examples
1. The partial derivatives of $f(x, y) = x \exp(-2y)$ are $f_x = \exp(-2y)$ and $f_y = -2x \exp(-2y)$. The gradient is $\nabla f = (\exp(-2y), -2x \exp(-2y))$.
2. The Euclidean distance between a point (x_1, x_2, x_3, x_4) and the origin of the co-ordinate system is defined as $f(x_1, x_2, x_3, x_4) = \sqrt{x_1^2 + x_2^2 + x_3^2 + x_4^2}$. Its partial derivatives, which are the components of the gradient vector ∇f, are

$$\frac{\partial f}{\partial x_i} = \frac{x_i}{\sqrt{x_1^2 + x_2^2 + x_3^2 + x_4^2}} \qquad i = 1, 2, 3, 4. \tag{4.12}$$

In analogy to the one-variable case, all candidates for extrema have to have zero slopes in all directions; otherwise, some neighboring points would have higher, and some lower values, and the point would not be an extremum. Thus, a necessary condition to be satisfied at an extremum is that the gradient consists entirely of zeros:

$$\nabla f = (0, 0, \ldots, 0). \tag{4.13}$$

A (candidate) point $\mathbf{X}_0$ with coordinates $(x_{10}, x_{20}, \ldots, x_{n0})$ in n-dimensional space that satisfies this condition is called a *stationary point*. A stationary point may be a maximum or minimum, but it may not be either. However, maxima and minima always are stationary points. If the graph of the function is a curved surface in three-dimensional space, as in the previous example (Fig. 4.6) then a marble at any stationary point would not roll in any direction. As a different way of describing the situation, the *tangent plane*, which touches the graph at $\mathbf{X}_0$, is exactly horizontal, because all slopes are zero. In the example of Figure 4.6, the only stationary point is exactly at the top of the hyperboloid, and this point is obviously a maximum.

Summarizing the results so far, the search of a maximum (or minimum) can be restricted to points:

- where the tangent plane is horizontal;
- where the gradient is a vector with all zeros;
- where all partial derivatives are zero;
- where the total differential is zero; or
- where the slopes in all directions are zero.

Each of these statements can be used to identify candidates for extrema.

As in the simpler case of one variable, these conditions are necessary but not sufficient. That is, there are stationary points with horizontal tangent planes that are not extrema. An illustrative example is that of a saddle, as it is shown in Figure 4.8. It has the functional form of a *hyperbolic paraboloid*: $z(x, y) = (x/0.5)^2 - (y/0.3)^2$ (von Seggern 1990, p. 216). At the center of the saddle, the tangent plane is

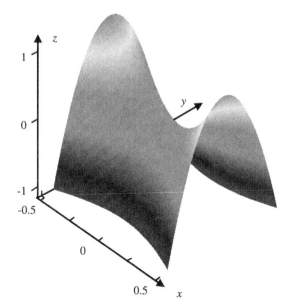

Figure 4.8. The center of the saddle is characterized by a zero gradient, yet this center is not a maximum or minimum. It is minimal in x-direction and maximal in y-direction.
(See www.cambridge.org/9780521177481 for color version.)

horizontal. In x-direction, the function has a minimum, but in y-direction, it has a maximum.

Sufficient conditions for an extremum are derived in analogy with the one-variable case (Saaty and Bram 1964; Courant 1972). Specifically, one needs to study higher derivatives of the function at the stationary point. As a representative example, suppose we are interested in a function f of two variables, x, and y. For an extremum, the derivatives f_x and f_y must be zero at the stationary point (x_0, y_0). Furthermore, for a maximum, all points close to the mountain peak are lower, and for a minimum, all points close the bottom of the valley are higher. Formulated mathematically, the expression

$$f(x_0 + h, y_0 + k) - f(x_0, y_0) \tag{4.14}$$

must have the same sign for all sufficiently small values of h and k. The expression is most efficiently analyzed with its Taylor approximation for small h and k. Taylor showed in the early eighteenth century that a smooth function of two variables x and y can be formulated as a two-variable polynomial, expressed at some point of interest, such as (x_0, y_0). The polynomial has the form

$$f(x_0 + h, y_0 + k) = f(x_0, y_0) + \sum_{i=1}^{n} \frac{1}{i!} \left(h\frac{\partial}{\partial x} + k\frac{\partial}{\partial y} \right)^i f(x_0, y_0) + R_n \tag{4.15}$$

(Saaty and Bram 1964), where n must approach infinity for the polynomial to be an exact representation of f for all values of h and k and thus for all values of x and y. The terms in the sum represent the ith partial derivatives of f with respect to x and y, which are to be evaluated at (x_0, y_0). For finite values of n, the function $f(x_0 + h, y_0 + k)$ and the polynomial differ by a remainder R_n, which usually becomes smaller with increasing n. Of interest here is that even a small n leads to an approximation of the function f that is very accurate near the point (x_0, y_0). Because we are indeed interested in small h and k, we may choose $n = 2$ and approximate

$$f(x_0 + h, y_0 + k) - f(x_0, y_0) = \tfrac{1}{2}(h^2 f_{xx} + 2hk f_{xy} + k^2 f_{yy}) + R_2. \qquad (4.16)$$

Note that the first derivatives do not appear in this equation. They are zero at (x_0, y_0), because this point is a candidate for an extremum. The quantities f_{xx}, f_{xy}, and f_{yy} are second partial derivatives. For instance, f_{xy} signifies that one first computes the partial derivative of f with respect to x and then differentiates the result with respect to y. The remainder R_2 is small if h and k are small. Specifically, it is of the order $h^2 + k^2$ and goes to zero if h and k go to zero (Courant 1972). Except for R_2, the right-hand side is the *second differential* of the function f. Further analysis (Saaty and Bram 1964) shows that f has an extremum if

$$f_{xx} f_{yy} - f_{xy}^2 > 0. \qquad (4.17)$$

For this to be possible at all, f_{xx} and f_{yy} must have the same sign. Otherwise, their product, diminished by the positive term f_{xy}^2 could not exceed 0. Thus, if in addition to Eq. (4.17), $f_{xx} < 0$ and $f_{yy} < 0$, the stationary point is a maximum If $f_{xx} > 0$ and $f_{yy} > 0$, the point is a minimum. If

$$f_{xx} f_{yy} - f_{xy}^2 < 0, \qquad (4.18)$$

the function neither has a maximum nor minimum. If the expression is equal to zero, one has to study even higher derivatives. Thus, the results are very much parallel to those for functions of only one variable. Functions of more than two variables are treated in a similar manner.

Examples
1. Consider the paraboloid $f(x, y) = x^2 + y^2 + 1$. It has the form of a bowl, centered around the z axis, and has its lowest point for $x = y = 0$, for which $f = 1$. The partial derivatives are $f_x = 2x$ and $f_y = 2y$. They both disappear at the stationary point $(x, y) = (0, 0)$, indicating that the function possibly has an extremum at the origin of the x-y plane. The second derivatives are $f_{xx} = f_{yy} = 2$, $f_{xy} = f_{yx} = 0$. Because $f_{xx} f_{yy} - f_{xy}^2 > 0$ and $f_{xx} = f_{yy} > 0$, the extremum is a minimum.
2. The hyperbolic paraboloid $f(x, y) = x^2 - y^2$, which is similar to the function in Figure 4.8, forms a saddle. Looking at the projection on the vertical plane $y = 0$, the function reduces to the standard parabola. Looking at the projection on the vertical plane $x = 0$, the parabola in y-direction is upside down. The partial derivatives are $f_x = 2x$ and $f_y = -2y$. They both disappear at $(x, y) = (0, 0)$, indicating that the function possibly has an extremum at the origin of the x-y plane. The second derivatives are $f_{xx} = 2$, $f_{yy} = -2$, $f_{xy} = f_{yx} = 0$. Because $f_{xx} f_{yy} - f_{xy}^2 = -4 < 0$, the stationary point is not an extremum.

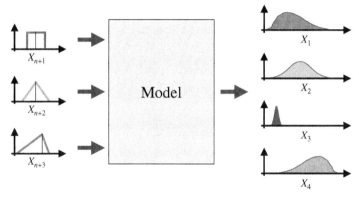

Figure 2.6. Schematic illustration of a Monte-Carlo simulation. Three independent (input) variables X_{n+1}, X_{n+2}, and X_{n+3}, are distributed as shown on the left. Their distributions lead to output, such as X_1, X_2, X_3, and X_4, that is also distributed. The output distributions may have a variety of shapes that depend on the model and are often difficult to predict without an actual execution of the simulation.

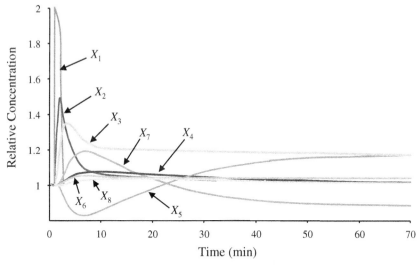

Figure 3.10. Dynamic system response to a twofold increase in glucose 6-phosphate (X_1). At time 1, the glucose 6-phosphate pool (X_1) was increased to twice the steady state. The eight metabolites represented are glucose 6-phosphate (X_1), fructose 6-phosphate (X_2), phosphoenol pyruvate (X_3), pyruvate (X_4), oxalacetate (X_5), malate (X_6), NADH (X_7), and ATP (X_8). Each concentration is normalized with respect to its nominal steady-state value. The long time scale of the response is cause for concern.

The following plates are available for download in colour from www.cambridge.org/9780521177481

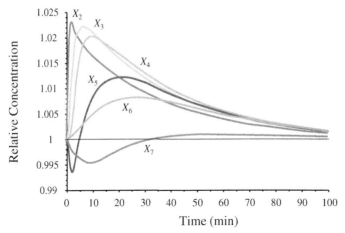

Figure 3.18. Dynamic system response to a twofold increase in cytosolic glucose (X_1). At time zero, the cytosolic glucose pool (X_1) was increased to twice the steady state. The six metabolites represented are glucose 6-phosphate (X_2), fructose 6-phosphate (X_3), fructose-2,6-bisphosphate (X_4), phosphoenol pyruvate (X_5), cytosolic pyruvate (X_6), and cytosolic oxalacetate (X_7). The remaining metabolites exhibit negligible deviations form the basal steady-state values (below 0.5%). Each concentration is normalized with respect to its nominal steady-state value. The response in the revised model is much improved over the initial model (compare y axis with Figure 3.10).

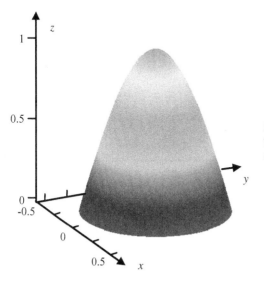

Figure 4.6. The graph of the two-variable function $z = 3 - 2 \cdot \sqrt{1 + 6.25 \cdot (x^2 + y^2)}$ is a smooth surface.

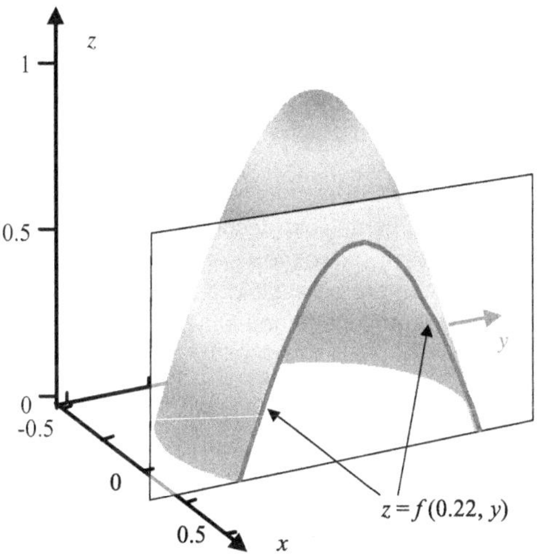

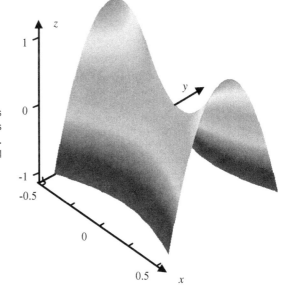

Figure 4.7. The intersection of the graph in Figure 4.6 and the plane $x = 0.22$ shows the projection of $z = f(0.22, y)$, which now is a univariate function of y.

Figure 4.8. The center of the saddle is characterized by a zero gradient, yet this center is not a maximum or minimum. It is minimal in x-direction and maximal in y-direction.

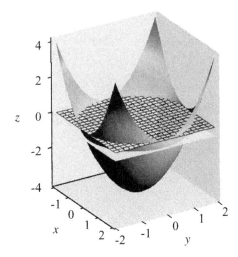

Figure 4.9. Paraboloid described in Eq. (4.47). The wire mesh shows the x-y plane. The intersection between the paraboloid and the solid plane is the circle of constrained minima.

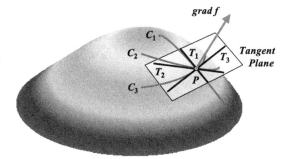

Figure 4.12. Three paths, C_1, C_2, and C_3, run along a three-dimensional surface, defined by the function f. They intersect at point P, where their tangents, T_1, T_2, and T_3, along with the tangents of all other paths intersecting at P, form the tangent plane at P. The gradient is perpendicular to the tangent plane at P.

Optimization of Nonlinear Functions with Constraints

We have seen in the one-dimensional case that the involvement of constraints may drastically affect the location of an extremum. For linear functions, the extrema simply moved to the boundaries of the constraint interval. In the nonlinear case with one variable, by contrast, a local extremum could become the global extremum if the domain for the search was constrained. For nonlinear functions of several variables, constraints can alter the set of candidates for an extremum quite dramatically.

In the single-variable case, only inequality constraints made real sense. After all, searching for the maximum of a function and simultaneously requiring that this maximum be located exactly at one fixed point, resembles Henry Ford's famous saying "You may have a car in any color as long as it is black." With a constraint like that, there is not much optimizing left to do.

The situation is different in higher-dimensional optimization tasks, where it may make sense to have a constraint given as a nonlinear equation, rather than an inequality. The search for extrema is then limited to candidates that exactly satisfy this equation. To get an intuitive feel for this type of problem, suppose we are looking for the point on some surface $\varphi(x, y, z) = 0$ in three-dimensional space that has the shortest distance to the origin of the coordinate system, $(0, 0, 0)$. This distance is given as

$$f(x, y, z) = \sqrt{x^2 + y^2 + z^2}. \tag{4.19}$$

The optimization problem is, therefore, minimize the distance f under the constraint that the solution point (x_0, y_0, z_0) lies on the surface given by $\varphi = 0$.

If one can solve $\varphi(x, y, z) = 0$ in such a way that one of the variables, say z, is expressed in terms of the other two, x and y, then one could simply substitute the expression for z in Eq. (4.19). This would reduce the constrained problem to an unconstrained problem. However, this is not always possible and, besides, it would give z a special role over x and y, which is not always desirable.

For a more general approach (Courant 1972), suppose we are interested in an extremum of a function of four variables, $f(x, y, z, t)$. Such a function is difficult to visualize, because its graph is an object in five-dimensional space. Nonetheless, one can readily find real-world examples for such functions. For instance, the temperature at all points in a room over some time period has four independent variables, namely, the three space coordinates and time. The fifth dimension shows the dependent variable, namely temperature.

Suppose the extremum must satisfy two constraints, namely

$$\varphi(x, y, z, t) = 0 \quad \text{and} \quad \psi(x, y, z, t) = 0. \tag{4.20}$$

Let us assume initially that two of the variables can be expressed in terms of the other two, at least in the close vicinity of the extremum. We will see that this assumption can ultimately be relaxed. But for now suppose

$$z = g(x, y) \quad \text{and} \quad t = h(x, y). \tag{4.21}$$

The functions g and h may be substituted for z and t in the objective function f.

As an example, consider

$$\varphi(x, y, z, t) = x \cdot y \cdot z \cdot t - 1 = 0 \tag{4.22}$$

and

$$\psi(x, y, z, t) = \sqrt{x^2 + y^2 + z^2} - 2 = 0. \tag{4.23}$$

The first equation can be reformulated such that t is expressed in terms of x, y, and z:

$$t = \frac{1}{x \cdot y \cdot z}. \tag{4.24}$$

The second equation yields

$$z = \sqrt{4 - x^2 - y^2}, \tag{4.25}$$

which is already a function of x and y only. Substitution of this relationship in the equation for t produces the functional dependence of t on x and y:

$$t = \frac{1}{x \cdot y \cdot \sqrt{4 - x^2 - y^2}}. \tag{4.26}$$

Thus, t and z can be replaced with functions of x and y.

For f to have an extremum, its total differential must be zero:

$$df = f_x \, dx + f_y \, dy + f_z \, dz + f_t \, dt = 0. \tag{4.27}$$

Because of the constraints, z and t are actually functions of x and y. Thus, the evaluation of the total differential requires application of the chain rule. Specifically, we can write the differentials

$$dz = \frac{\partial z}{\partial x} \, dx + \frac{\partial z}{\partial y} \, dy \tag{4.28}$$

and

$$dt = \frac{\partial t}{\partial x} \, dx + \frac{\partial t}{\partial y} \, dy. \tag{4.29}$$

These expressions may be substituted in Eq. (4.27), effectively making f a function of the two variables x and y with the total differential

$$df = \left[f_x + f_z \cdot \frac{\partial z}{\partial x} + f_t \cdot \frac{\partial t}{\partial x} \right] dx + \left[f_y + f_z \cdot \frac{\partial z}{\partial y} + f_t \cdot \frac{\partial t}{\partial y} \right] dy = 0. \tag{4.30}$$

To compute the derivatives of f in x-direction or y-direction, we evaluate this expression for $dx = 1$ and $dy = 0$, or for $dx = 0$ and $dy = 1$, respectively. This is so, because, for instance, setting $dx = 1$ and $dy = 0$ means stepping forward one unit in x-direction, while y remains unchanged. Focusing on derivatives in the principal directions produces necessary conditions that any extremum must satisfy, namely

$$f_x + f_z \cdot \frac{\partial z}{\partial x} + f_t \cdot \frac{\partial t}{\partial x} = 0, \tag{4.31}$$

$$f_y + f_z \cdot \frac{\partial z}{\partial y} + f_t \cdot \frac{\partial t}{\partial y} = 0. \tag{4.32}$$

The relationships between the differentials dz and dt on one hand and dx and dy on the other are determined by the constraints φ and ψ. They are most elegantly obtained from the total differentials of φ and ψ. For instance, differentiation of $\varphi(x, y, z, t) = 0$ yields

$$d\varphi = \frac{\partial \varphi}{\partial x} dx + \frac{\partial \varphi}{\partial y} dy + \frac{\partial \varphi}{\partial z} dz + \frac{\partial \varphi}{\partial t} dt = 0. \tag{4.33}$$

As shown previously, z and t are actually functions of x and y, which allows us to substitute dz and dt with x- and y-derivatives. This leads to

$$\frac{\partial \varphi}{\partial x} dx + \frac{\partial \varphi}{\partial y} dy + \frac{\partial \varphi}{\partial z} \frac{\partial z}{\partial x} dx + \frac{\partial \varphi}{\partial t} \frac{\partial t}{\partial x} dx + \frac{\partial \varphi}{\partial z} \frac{\partial z}{\partial y} dy + \frac{\partial \varphi}{\partial t} \frac{\partial t}{\partial y} dy = 0. \tag{4.34}$$

To obtain the derivative in x-direction, we set again $dx = 1$ and $dy = 0$, with the result

$$\frac{\partial \varphi}{\partial x} + \frac{\partial \varphi}{\partial z} \frac{\partial z}{\partial x} + \frac{\partial \varphi}{\partial t} \frac{\partial t}{\partial x} = \varphi_x + \varphi_z \frac{\partial z}{\partial x} + \varphi_t \frac{\partial t}{\partial x} = 0. \tag{4.35}$$

The same operations are executed for ψ and with respect to y, and the result is

$$\varphi_y + \varphi_z \frac{\partial z}{\partial y} + \varphi_t \frac{\partial t}{\partial y} = 0. \tag{4.36}$$

Equations (4.35) and (4.36) remain valid if we multiply them with some real numbers λ and μ. These numbers, one per constraint, are called *Lagrange multipliers*. They are defined such that, at the extremum, the two equations

$$f_z + \lambda \varphi_z + \mu \psi_z = 0 \tag{4.37}$$

and

$$f_t + \lambda \varphi_t + \mu \psi_t = 0 \tag{4.38}$$

are satisfied. This is always possible, except if $\varphi_z \psi_t - \varphi_t \psi_z = 0$. If this expression actually is zero, one needs to sort the variables differently and, for instance, express x and t as functions of y and z. If none of these combinations leads to a value different from zero, one is out of luck and needs to characterize the extremum in an entirely different fashion. This, however, is rare.

What the values of these multipliers actually are is not so important at this point, as long as they exist. Upon multiplication of Eq. (4.35) with λ and of the corresponding equation for y with μ, the resulting equations

$$\lambda \cdot \left(\varphi_x + \varphi_z \frac{\partial z}{\partial x} + \varphi_t \frac{\partial t}{\partial x} \right) = 0 \tag{4.39}$$

and

$$\mu \cdot \left(\psi_x + \psi_z \frac{\partial z}{\partial x} + \psi_t \frac{\partial t}{\partial x} \right) = 0 \tag{4.40}$$

are added to Eq. (4.31). This yields

$$f_x + \lambda \varphi_x + \mu \psi_x + (f_z + \lambda \varphi_x + \mu \psi_x) \cdot \frac{\partial z}{\partial x} + (f_t + \lambda \varphi_t + \mu \psi_t) \cdot \frac{\partial t}{\partial x} = 0. \tag{4.41}$$

Because of the definitions of the Lagrange multipliers in Eqs. (4.37) and (4.38), the terms in parentheses drop out, and the result is

$$f_x + \lambda \varphi_x + \mu \psi_x = 0. \tag{4.42}$$

The same arguments lead to an analogous equation in which the derivatives are computed with respect to y.

Thus, taken together with Eqs. (4.37) and (4.38), any extremum is characterized by the six equations

$$\begin{aligned}
\varphi(x, y, z, t) &= 0 \\
\psi(x, y, z, t) &= 0 \\
f_x + \lambda \varphi_x + \mu \psi_x &= 0 \\
f_y + \lambda \varphi_y + \mu \psi_y &= 0 \\
f_z + \lambda \varphi_z + \mu \psi_z &= 0 \\
f_t + \lambda \varphi_t + \mu \psi_t &= 0.
\end{aligned} \tag{4.43}$$

These equations have six unknowns, namely, the four coordinates of the extremum plus λ and μ. Note that these equations are entirely "symmetric" in the variables. Even though z and t were originally singled out, because they were expressed in terms of x and y, their special role is no longer visible in the result. Thus, as long as *any* two variables can be expressed in terms of the other two, the procedure leads to the same final result.

The result can be generalized to more variables and further streamlined as follows: Any extremum of a function f of n variables $x_1, \ldots, x_n$ and m constraints $(m < n)$ satisfies the constraint equations

$$\varphi_i(x_1, x_2, \ldots, x_n) = 0 \qquad (i = 1, \ldots, m) \tag{4.44}$$

and the equations

$$\frac{\partial F}{\partial x_j} = 0 \qquad (j = 1, \ldots, n), \tag{4.45}$$

where

$$F = f + \lambda_1\varphi_1 + \lambda_2\varphi_2 + \cdots + \lambda_m\varphi_m. \tag{4.46}$$

This is always true, except for the unusual occurrence that all Jacobian determinants [which contain the derivatives of the φs with respect to the xs (Kolman 1997)] disappear. In this case, as illustrated by $\varphi_z\psi_t - \varphi_t\psi_z = 0$ in the previous example, it is not possible to use the set of constraints to express any m variables in terms of the remaining $n - m$, and other types of analysis are needed.

The Lagrange multipliers came into the picture for mathematical convenience. Nevertheless, they have an interpretation. They are the sensitivities of the extremum with respect to changes in the constraints (Vagners 1983, p. 1154). In other words, if the value of λ is large, a slight change in the constraint $\varphi(x, y, z, t) = 0$ has a strong effect on the location of the extremum.

If the function f to be optimized is replaced by a dynamical system, described by a system of differential equations, and including time-dependent control functions, the corresponding function F, consisting of f and the Lagrange terms, is the *Hamiltonian* of the system. The optimization task in this case leads to the calculus of variations and Pontryagin's Maximum Principle (Vagners 1983, p. 1183).

Example
Find the point of the paraboloid

$$\varphi(x, y, z) = x^2 + y^2 - z - 4 = 0 \tag{4.47}$$

that is closest to $(0, 0, 0)$. The paraboloid is centered around the z axis (Figure 4.9), and touches it at -4, so it is easy to rationalize that $(0, 0, -4)$ is its unconstrained minimum. The question is: Is it also the *constrained* minimum? Let us try out the theory.

This distance between $(0, 0, 0)$ and any (x, y, z) is given as

$$f(x, y, z) = \sqrt{x^2 + y^2 + z^2}. \tag{4.48}$$

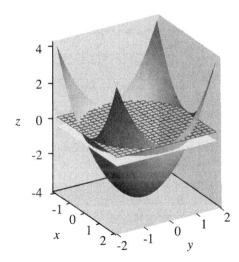

Figure 4.9. Paraboloid described in Eq. (4.47). The wire mesh shows the x-y plane. The intersection between the paraboloid and the solid plane is the circle of constrained minima.
(See www.cambridge.org/9780521177481 for color version.)

This function needs to be minimized. Without the requirement $\varphi = 0$, the solution, of course, would be $(0, 0, 0)$, but this point does not lie on the paraboloid.

According to the general result shown previously, we form the function

$$F = f + \lambda\varphi = \sqrt{x^2 + y^2 + z^2} + \lambda \cdot (x^2 + y^2 - z - 4) \tag{4.49}$$

with one Lagrange multiplier λ and compute its derivatives. They are

$$\frac{\partial F}{\partial x} = \frac{x}{\sqrt{x^2 + y^2 + z^2}} + 2\lambda x = 0$$

$$\frac{\partial F}{\partial y} = \frac{y}{\sqrt{x^2 + y^2 + z^2}} + 2\lambda y = 0 \tag{4.50}$$

$$\frac{\partial F}{\partial z} = \frac{z}{\sqrt{x^2 + y^2 + z^2}} - \lambda = 0.$$

The first two equations are solved either for $x = 0$ and $y = 0$ or for

$$\lambda = -\frac{1}{2}\frac{1}{\sqrt{x^2 + y^2 + z^2}}. \tag{4.51}$$

According to the third equation,

$$\lambda = z \cdot \frac{1}{\sqrt{x^2 + y^2 + z^2}}. \tag{4.52}$$

Substituting the first solution, $x = y = 0$, in the equation of the paraboloid, $\varphi = 0$, we obtain $z = -4$. At this point, the distance is $f = 4$. So, is it true that the unconstrained minimum is the global minimum? We could compute second derivatives, but it is easier here to check close-by points. For instance, suppose $x = y = \varepsilon$ is a small number. The corresponding z-coordinate on the paraboloid is

$$z = 2\varepsilon^2 - 4. \tag{4.53}$$

Substituting these quantities for x, y, and z in the distance function yields

$$f = \sqrt{\varepsilon^2 + \varepsilon^2 + (2\varepsilon^2 - 4)^2}$$
$$= \sqrt{2\varepsilon^2 + 4\varepsilon^4 - 16\varepsilon^2 + 16}. \tag{4.54}$$

Because ε is small, ε^4 is smaller than ε^2. Therefore, all ε-terms taken together are negative, and f is smaller than 4. This means that $(0, 0, -4)$ is actually a local *maximum*! It is not a global maximum, because the distance grows without bounds if x and y go to infinity.

If x or y or both are not zero, equating (4.51) and (4.52) yield $z = -1/2$. For any point with a z-coordinate of $-1/2$, x and y have to satisfy the equation $\varphi = 0$. Thus, the requirement is

$$x^2 + y^2 = 3.5. \tag{4.55}$$

This equation describes a circle with radius $\sqrt{3.5}$, centered at zero. In other words, there is a circular line on the paraboloid at height $-1/2$ whose points have a distance from zero that could be minimal (see Fig. 4.9). Checking close-by values confirms that all these points are minimal in a sense that no other points on the paraboloid are closer to zero. Because all points on this circle have the same (minimal) distance, there is an infinite number of minima.

What happens for an elliptical paraboloid, such as $\varphi(x, y, z) = 2x^2 + y^2 - z - 4 = 0$? What happens if the intersection point with the z axis is changed from -4 to $-1, -1/2$, or -0.1? How would you compute how far a big ball would fall into a paraboloid bowl before it gets stuck?

For high-dimensional optimization tasks, the problem with the Lagrange multiplier method is the need to solve a system of nonlinear equations. In stark contrast to linear systems, where linear algebra has produced very powerful tools even for big systems, systems of nonlinear equations can be extremely complicated. Even seemingly simple equations, such as $\cos(x) = x$ cannot be solved analytically. In higher-dimensional cases, the problems can become overwhelming and require numerical methods as they are described next.

SEARCH ALGORITHMS

Functions of One Variable

In the examples so far, it was easy to compute the derivatives, set them to zero, and thus to determine the candidates for maxima and minima. If the functions are more complicated, the same procedures could often be applied in principle, but they tend to fail in practical applications. A very simple example is the function $y(x) = 0.5 \exp(x) - x^2$, which has the derivative $y'(x) = 0.5 \exp(x) - 2x$. Although simple-looking, there is no algebraic solution to the equation $\exp(x) = 4x$, even though a solution clearly exists: one only needs to graph $\exp(x)$ and $4x$ and check where the two lines intersect. In fact, there are two solutions, one between $x = 0.35$ and 0.36 and one between $x = 2.15$ and 2.16. If problems can emerge with simple examples like this, it is easy to imagine that they are much more severe in more complicated optimization situations. And obviously, graphical solutions are not very helpful in higher dimensions either.

The complications with the analytical methods render it evident that an entirely different approach is needed. Such an approach is an *iterative search* for the optimal solution with some computer algorithm. A good introduction to the topic is Glantz and Slinker (1990, Chapter 10). *Iterative search* basically means the following: Begin with a guess of what the maximum might be. Unless one is a wizard, this guess is probably wrong. Nonetheless, starting with this point (sometimes called a *guestimate*), check whether close-by points have smaller or higher values. Improve the guestimate by moving in the direction of higher values, because the maximum is more likely there than in other direction. The extent of the move should depend

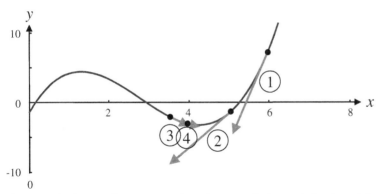

Figure 4.10. Schematic representation of a numerical search for the minimum of a univariate function, the polynomial analyzed before (see Figure 4.2). Steps are shown with numbers in circles. Dots represent the first guestimate and subsequent updates. See text for details.

on how steeply the function is increasing. Thus, move for some distance and check again whether the points in the neighborhood have higher or smaller values. Move again toward higher values. The algorithm needs some termination rule that tells it when to stop searching. A typical rule is something like: If the new estimate (it is not so much guessing anymore) is within 10^{-8} of the previous estimate, stop the routine.

Actually focusing on the slope of the function, instead of the function itself, the algorithm attempts to minimize the absolute value of the slope. The absolute value is always positive, so the best possible outcome is a slope of zero, which indicates a stationary point and thus a candidate (i.e., a necessary condition) for a maximum or minimum, as we discussed previously.

Even though we successfully computed the (internal) maximum and minimum in our earlier example of a polynomial with analytical means, it is useful to explore how a typical search algorithm would perform. It must be emphasized that modern search algorithms are much more sophisticated than the search that follows. However, the principles are the same, and simplicity is of the essence for illustrative purposes.

Suppose we are looking for the minimum inside the search interval $[0, +8]$ and our initial guestimate is $x = 6$. Keep in mind that in an actual situation no graph like Figure 4.10 is available that would give us better intuition. As the first step, the computer algorithm computes the slope of the function at $x = 6$. If the computation is too complicated (as may be the case in optimization problems with many variables) the slope is estimated by computing values of the function close to the guestimate. In this particular case, the slope is easy to compute. According to our previous calculation, it is $y'(x) = 10 - 10x + 1.8x^2$, which has a value of 14.8 for $x = 6$. It is positive, which implies that the function has larger y-values for larger x-values. Because we are looking for a minimum, the next estimate should be a smaller x-value. The algorithm also determines how far the search should move to the left. In practice, this determination can be quite complicated and specific to the particular algorithm. For the present illustration, suppose the search is moved to $x = 5$. The slope is computed again, and because it is still positive, the next estimate is moved further to the left. If the movement goes in the same direction several times in a row, typical algorithms increase the distance. For instance, the next estimate might be $x = 3$. Now the slope is negative, the direction is reversed, and the distance is decreased. Suppose the next estimate is 4; the slope is negative. The next estimate may be 4.5, the direction of movement is reversed, the distance is decreased. This process should really be iterated until the slope is exactly zero. However, it is actually unlikely that the bouncing back and forth eventually hits *exactly* the point of the extremum. More likely, the algorithm would go back and forth until the slopes have values at the level of the computational precision. At this point, the computed slope would still not be zero, and the algorithm would possibly enter a futile search without end. Because of this problem, algorithms are either limited by a predetermined level of precision or by the number of iterations. In the first case, the algorithm stops (*converges*) if the absolute value of the slope is less than a *tolerance* like 10^{-8}. In the second case, the algorithm simply stops after 500, 1,000, and 10,000 iterations and presents the relatively best result.

These types of iterative searches have significant advantages and disadvantages. The main advantage is that the method very often works. In fact, it often converges much faster than one might expect. Besides, in complicated situations, an iterative method is one of only a very few choices. The drawbacks can be manifold. They fall into two categories. First, the algorithm may not converge. It may bounce back and forth between two estimates, neither of which is a maximum or minimum. It may explore the functions for x-values in the gazillions, not "noticing" that these values are irrelevant. It may not find a solution within the preset number of iterations, even if it is large. Many of these algorithmic problems are common enough to have led to safeguards in commercial software. Still, in complex optimization problems, there is no guarantee that an algorithm ultimately yields useful results.

The second category of problems is of a mathematical nature. In many practical applications, the function to be optimized has several local maxima and minima. A simple example is shown in Figure 4.11. Imagine the first guestimate is $x = 8$. (Keep

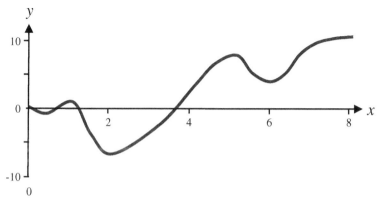

Figure 4.11. The function has local minima at 0.4, 2, and 6. If a search algorithm starts at $x = 8$, it is likely to get trapped in the local minimum at 6, rather than to converge to the minimum at 2.

in mind that, in reality, one does not know the graph of the function!) It does not require much intuition to see that even if the algorithm works perfectly, it is likely to converge to the wrong point, namely $x = 6$, instead of wandering over to the true minimum at $x = 2$. In the jargon of the field, the *algorithm gets trapped in a local minimum* (of the slope function). Avoiding these traps is in general difficult, if not impossible. Even worse, if the algorithm converges, one often does not know whether the identified maximum or minimum is really the true, overall maximum or minimum, or if it is just a local extremum. In practical applications, one way of improving the chances of finding the global optimum is to start the algorithm several times with different guestimates. If the algorithm always (or mostly) converges to the same extremum, this extremum has a reasonable chance of being global. Still, there is no guarantee.

Searching Nonlinear Functions of Several Variables

Most nonlinear searches deal with several variables, and although the principles are similar to those in the one-variable case, numerous techniques and tricks have been developed to deal with special applications. A large collection of methods and applications from various fields can be found in Vagners (1983) and Bertsekas (1998), and in references therein.

As in the one-variable case, multivariate searches make use of the slope of the function, which, at the extremum, must be zero. As a multivariate analogue of the slope, we introduced earlier the gradient, which is a vector of partial derivatives. For three independent variables, it was defined as

$$\nabla f = \frac{\partial f}{\partial x_1}\mathbf{i} + \frac{\partial f}{\partial x_2}\mathbf{j} + \frac{\partial f}{\partial x_3}\mathbf{k} = \frac{\partial f}{\partial x_1}(1, 0, 0) + \frac{\partial f}{\partial x_2}(0, 1, 0) + \frac{\partial f}{\partial x_3}(0, 0, 1)$$

$$= \left(\frac{\partial f}{\partial x_1}, \frac{\partial f}{\partial x_2}, \frac{\partial f}{\partial x_3}\right). \tag{4.56}$$

It is helpful to develop an intuitive feel for the gradient. Imagine some surface in three-dimensional space, such as the surface of a mountain, and consider a point traveling along a smooth path on this surface. Its coordinates are x, y, and z, and because they obviously depend on time, they are in fact $x(t)$, $y(t)$, and $z(t)$. Each point on a path of this type can be described as a vector of the form

$$\mathbf{p}(t) = x(t)\mathbf{i} + y(t)\mathbf{j} + z(t)\mathbf{k}, \tag{4.57}$$

where $\mathbf{i}$, $\mathbf{j}$, and $\mathbf{k}$ are again unit vectors. Interpreted in a slightly different way, the definition says: to find the location of the point at time t, start at the origin of the coordinate system and go $x(t)$ units far in the first direction, $y(t)$ units far in the second direction, and $z(t)$ units far in the third direction. The tangent vector of the path is computed by differentiation with respect to time, t. It is

$$\mathbf{p}'(t) = x'(t)\mathbf{i} + y'(t)\mathbf{j} + z'(t)\mathbf{k}. \tag{4.58}$$

If we imagine many paths, all on the same surface and all going through the same

Figure 4.12. Three paths, C_1, C_2, and C_3, run along a three-dimensional surface, defined by the function f. They intersect at point P, where their tangents, T_1, T_2, and T_3, along with the tangents of all other paths intersecting at P, form the tangent plane at P. The gradient is perpendicular to the tangent plane at P. (See www.cambridge.org/9780521177481 for color version.)

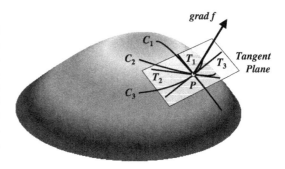

point P, then all tangent vectors form a plane that touches the surface at P. It is called the *tangent plane* (see Figure 4.12).

Now let us go back to the gradient. Suppose the surface is mathematically characterized by a function $f(x, y, z) = c$. As an example, consider the function $f(x, y, z) = x^2 + y^2 + z^2 = 1$, which represents all points on the surface of a sphere that is located at the center of the three-dimensional coordinate system and has a radius of one unit. If the path runs along this surface, its coordinates must satisfy the surface equation. Thus,

$$f(x(t), y(t), z(t)) = c \qquad \text{for all points of the path } \mathbf{p}(t). \tag{4.59}$$

If we differentiate this equation with respect to t, we obtain

$$\frac{\partial f}{\partial x} \cdot x' + \frac{\partial f}{\partial y} \cdot y' + \frac{\partial f}{\partial z} \cdot z' = (\nabla f) \bullet \mathbf{p}' = 0. \tag{4.60}$$

The center part of this equation represents a product of two vectors, namely, the gradient and the tangent vector. The theory of vector geometry says that, if the product of two vectors is zero, as it is here, the two vectors are orthogonal to each other. Thus, the gradient forms a right angle with all tangent vectors at a given point P. In other words, the gradient is perpendicular to the tangent plane (see Figure 4.12).

In the case of maximizing a function of a single variable, the search algorithm had to move in the direction of positive slope. In the case of a surface in three-dimensional space, one must determine which direction on the tangent plane has the steepest slope. Studying again a smooth function $f(x, y, z)$, one can compute its slope at a given point P in any direction. Suppose the direction is given by some vector $\mathbf{b}$ of length one. This vector may be defined by its coordinates or by its angles with the x, y, and z axes. All points in direction b lie on a ray that is given by

$$\mathbf{r}(s) = \mathbf{p}_0 + s\,\mathbf{b} = x(s)\mathbf{i} + y(s)\mathbf{j} + z(s)\mathbf{k}, \tag{4.61}$$

where $\mathbf{p}_0$ is the location of P, and s is the distance between P and some other point Q in direction $\mathbf{b}$ (see Figure 4.13).

The rate of change of f at point P in direction $\mathbf{b}$ is called the *directional derivative* and denoted by $D_b f$ or by df/ds (Kreyszig 1993). Analogous to the calculus of

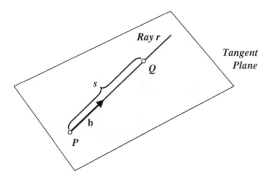

Figure 4.13. The slope of the tangent plane is steepest in direction b, and the search moves from P to along ray r toward point Q, a distance of s units (adapted from Kreyszig 1993).

univariable functions, it is obtained as

$$D_b f = \frac{df}{ds} = \lim_{s \to 0} \frac{f(Q) - f(P)}{s}. \tag{4.62}$$

Application of the chain rule yields

$$D_b f = \frac{df}{ds} = \frac{\partial f}{\partial x} \cdot \frac{dx}{ds} + \frac{\partial f}{\partial y} \cdot \frac{dy}{ds} + \frac{\partial f}{\partial z} \cdot \frac{dz}{ds}. \tag{4.63}$$

The derivatives of x, y, and z with respect to s can be expressed in terms of **b**, by differentiation of Eq. (4.61):

$$\frac{d\mathbf{r}}{ds} = \mathbf{b} = \frac{dx}{ds}\mathbf{i} + \frac{dy}{ds}\mathbf{j} + \frac{dz}{ds}\mathbf{k}. \tag{4.64}$$

It is now not difficult to see that the directional derivative is just the vector product of the direction vector **b** and the gradient:

$$D_b f = \mathbf{b} \bullet \nabla f. \tag{4.65}$$

This result was the result of some strenuous math, but it constitutes an important insight for our purposes of searching for an optimum. One can see that the directional derivative is directly proportional to the gradient. Thus, the larger the gradient, the steeper the slope in direction **b**. In other words, the gradient indicates the direction in which the function increases or decreases most. Furthermore, the length of the gradient vector is a measure for the steepness. Finally, the same arguments hold for dimensions higher than 3.

Over 150 years ago, the French mathematician Cauchy used these insights to propose a *gradient method* or *method of steepest descent* (Vagners 1983, p. 1200; Kreyszig 1993, p. 1086). This method is the prototype for a variety of algorithms in modern-day use. Because it is typically described for finding the minimum of a function (rather than the maximum), we follow this tradition. The principle is the following. Start at some guestimate point x_0. The direction of the steepest descent is given by $-\nabla f$. Compute the minimum of f in this direction along a straight line. Take this minimum as the next estimate and iterate the process until an acceptable solution is found.

The computation of the minimum occurs as follows. Determine the value of s and the point

$$\mathbf{x}_1(s) = \mathbf{x}_0 - s\nabla f(\mathbf{x}_0) \tag{4.66}$$

for which the function

$$F(s) = f(\mathbf{x}_1(s)) \tag{4.67}$$

has a minimum.

As Strang (1986) noted, the method of steepest descent ideally would work like water running down a mountain. At each point, the water would follow the steepest slope. However, because the algorithm does not update direction continuously, the descent is more like that of a skier who cannot turn while in motion. He slides in a straight line until the terrain starts to climb again and then selects the new direction according to the steepest slope.

It is well known that the simple method of steepest descent can run into severe convergence problems, especially if the graph of the objective function forms a deep "banana-shaped" valley whose center contains the minimum. In this situation, the algorithm often bounces between the steep walls of the valley, using very small steps, and sometimes stops far away from the true minimum. As a favorite solution, some algorithms do not just use the gradient, which essentially is a linear approximation, but approximate the objective function locally with a quadratic model (Vagners 1983, p. 1200). This model requires second derivatives, which are collectively represented by the Jacobian matrix. This matrix may be provided by the user or is approximated numerically by the algorithm.

Genetic Algorithms

Search algorithms of the gradient type are not the only possibility for optimizing nonlinear functions. Genetic algorithms have a totally different philosophy and approach. From a methodological point of view most interesting may be the fact that in spite of their success, apparently nobody *truly* understands why these algorithms work.

The key concept behind genetic algorithms is the coding of the numerical coordinates of a point as a "gene." This is typically achieved in the form of binary strings of zeros and ones. Simplistically, the first piece of the gene represents the x_1-coordinate, the second piece represents the x_2-coordinate, and so fourth. Each gene is thus the blueprint of a point, and the idea is to recombine genes in such a fashion that the corresponding point is optimal. The process starts with a population of genes (points that might be extrema), which are either randomly determined or designed with some additional knowledge of the optimization problem. The population is evaluated according to preset criteria. Foremost is the value of the objective function, but it is also tested whether the genes (points) satisfy all constraints. The "fittest" fifty or so among all genes, which correspond to the highest values of the objective function, while satisfying all constraints, are used to "produce offspring." The best five or ten genes are kept unchanged in the next generation. The remaining offspring consists of new genes that derive from recombining appropriate portions from pairs of parent genes. The new genes again represent points, which are evaluated for fitness and selection for the next generation. Good introductions to genetic algorithms were presented by Goldberg (1989), Ribeiro Filho, Treleaven, and Alippi (1994), Srinivas and Patnaik (1994), and Yao and Sethares (1994).

The great advantage of genetic algorithms is their almost unlimited flexibility. Any number of constraints can be included, and these may be linear or nonlinear functions or even differential or integral equations. As long as the objective function and the constraints can be evaluated at a given point, fitness can be determined, and the fitness measure can be used to select genes for the next generation.

S-system models are particularly suited for analyses with genetic algorithms because of their regular structure, which suggests straightforward coding rules. Although the combination of S-systems and genetic algorithms is appealing, only two articles have been published so far (Zhang, Voit, and Schwacke 1996; Okamoto et al. 1997; Sakamoto and Iba 2001; Maki et al. 2001). In Zhang's analysis, yield data of oil palms in Malaysia were used to parameterize a growth model with two variables. The limited experience from this and some other examples suggests that genetic algorithms may be successful for fast, preliminary estimations, but that they may not be so well suited for producing highly accurate solutions.

REFERENCES

Bertsekas, D.P.: *Network Optimization: Continuous and Discrete Models*. Athena Scientific, Belmont, MA, 1998.

Courant, R.: *Vorlesungen über Differential – und Integralrechnung*. Zweiter Band. Springer-Verlag, Berlin, 1972.

Glantz, S.A., and B.K. Slinker: *Primer of Applied Regression and Analysis of Variance*. McGraw-Hill, New York, 1990.

Goldberg, D.E.: *Genetic Algorithms in Search, Optimization, and Machine Learning*. Addison-Wesley, Reading, MA, 1989.

Kolman, B. *Introductory Linear Algebra with Applications* (6th ed.). Prentice Hall, Upper Saddle River, NJ, 1997.

Kreyszig, E.: *Advanced Engineering Mathematics* (7th ed.). John Wiley & Sons, New York, 1993.

Maki, Y., D. Tominaga, M. Okamoto, S. Watanabe, and Y. Eguchi: Development of a system for the inference of large scale genetic networks. *Pacific Symp. Biocomputing*, 446–58, 2001.

Okamoto, M., Y. Morita, D. Tominaga, K. Tanaka, N. Kinoshita, J.-I. Ueno, Y. Miura, Y. Maki, and Y. Eguchi: Design of virtual-labo-system for metabolic engineering: Development of biochemical engineering system analyzing tool-kit (BEST KIT). *Comp. Chem. Eng.* 21(Suppl.), S745–50, 1997.

Ribeiro Filho, J.L., P.C. Treleaven, and C. Alippi: Genetic-algorithm programming environments. *IEEE Comp.*, pp. 28–43, June 1994.

Saaty, T.L., and J. Bram: *Nonlinear Mathematics*. Dover, New York, 1964.

Sakamoto, E., and H. Iba: Inferring a system of differential equations for a gene regulatory network by using genetic programming. *Proc. of the 2001 Congress on Evolutionary Computation CEC 2001*. Seoul, Korea, pp. 720–6, 2001.

Srinivas, M., and L. M. Patnaik: Genetic algorithms: A survey. IEEE Comp., pp. 17–26, June 1994.

Stewart, J.: *Single Variable Calculus. Early Transcendentals* (4th ed.). Brooks/Cole, Pacific Grove, 1999.

Strang, G.: *Introduction to Applied Mathematics*. Wellesley-Cambridge Press, Wellesley, MA, 1986.

Vagners, J.: Optimization techniques. In: C.E. Pearson (Ed.), *Handbook of Applied Mathematics. Selected Results and Methods* (2nd ed., pp. 1140–216). 1983.

von Seggern, D.H.: CRC *Handbook of Mathematical Curves and Surfaces*. CRC Press, Boca Raton, FL, 1990.

Yao, L., and W.A. Sethares: Nonlinear parameter estimation via the genetic algorithm. *IEEE Trans. Signal Proc.* 42(4), 927–35, 1994.

Zhang, Z., Voit, E.O., and L.H. Schwacke: Parameter estimation and sensitivity analysis of S-systems using a genetic algorithm. In: T. Yamakawa and Matsumoto, G. (Eds.), *Methodologies for the Conception, Design, and Application of Intelligent Systems*. World Scientific, Singapore, 1996.

CHAPTER FIVE

Optimization of Biochemical Systems

INTRODUCTION

The optimization of metabolic pathways has its roots in biochemistry and chemical engineering. Biochemistry has contributed with knowledge about enzyme kinetics and catalytic processes in vitro and in vivo, whereas chemical engineering has developed strategies of optimizing processes at the bench and large-scale industrial levels. Optimization of biochemical and biotechnological systems is a crucial component of the young field of metabolic engineering (e.g., Bailey 1991, 1998; Simpson, Colon, and Stephanopoulos 1995; Simpson, Follstad, and Stephanopoulos 1999; Stephanopoulos 1998; Stephanopoulos, Aristidou, and Nielsen 1998; Koffas et al. 1999).

In broad strokes, one may classify metabolic optimization tasks in five categories (Heinrich and Schuster 1996, Chapter 6). The first category deals with the maximization of reaction rates and steady-state fluxes. The second addresses the minimization of transition time, which yields higher productivity per time unit; we will discuss in Chapter 7 that this task is actually closely related to the optimization of concentrations and fluxes. The third is the minimization of intermediate metabolite concentrations. The fourth task is the optimization of thermodynamic efficiency, and the fifth task addresses optimal stoichiometry and network design.

Savageau (1976) established criteria specifically for the optimal functioning of metabolic pathways. He suggested the following seven: (1) minimization of changes in the level of end product, as the system shifts from one steady state to another in response to a change in demand; (2) responsiveness to changes in the availability of initial substrate; (3) insensitivity to perturbations in the structure of the systems itself; (4) limited accumulation of metabolic intermediates; (5) reduction of the synthesis of intermediate metabolites when end product is supplied exogenously; (6) stability; and (7) temporal responsiveness to change.

Questions of optimality in stoichiometry and network design have been approached from two viewpoints. First, attempts were made to explain why evolution has led to particular pathways that seem to be rather efficient, if not optimal, for the environmental surroundings of the organism. Savageau and collaborators (Savageau 1974a,b, 1975, 1976, 1985, 1989; Irvine and Savageau 1985a,b; Irvine 1991; Hlavacek and Savageau 1995, 1996, 1997; Alves and Savageau 2000a,b,c,d) and Meléndez-Hevia and co-workers (Meléndez-Hevia and Isidoro 1985; Meléndez-Hevia and Torres 1988; Meléndez-Hevia 1990; Meléndez-Hevia, Waddell, and Montero 1994; see also Mittenthal et al. 1998, 2001; Ebenhöh and Heinrich 2001) addressed this issue with methods of kinetic and stoichiometric systems analysis.

Second, optimal pathway and network design can be a task for biotechnology. The key question in this context is: Is it possible to manipulate a biochemical system in such a fashion that it is optimal with respect to some desired goal of biotechnological relevance? The goal in this case is not survival of the organism, but an anthropogenic criterion such as optimal yield. Hatzimanikatis, Floudas, and Bailey (1996a,b) addressed this question in the recent past. In addition to these optimization tasks that address systems at steady state, Conejeros and Vassiliadis (1998a,b, 2000) proposed optimization strategies for biochemical systems under dynamic conditions.

The early phases of pathway optimization followed two venues. On the biochemical side, organisms were altered through random mutagenesis, and subsequent selection led toward increased synthesis of a desired product (for recent introductions, see Archer, MacKenzie, and Jeenes 2001; Harwood and Wipat 2001). This strategy has been very successful, in some cases creating organisms that produce over a hundred times more yield than the parent strain.

Chemical engineers meanwhile developed tools for optimal online and offline process control. Online control is particularly useful for complicated chemical and biochemical processes. Especially when microorganisms are involved, comprehensive models are not often available for predicting the dynamics of such processes, perturbations are difficult to incorporate in offline models, and efficient real time control with offline models is limited; in such cases, online optimization offers a feasible alternative to offline process control (Arkun and Stephanopoulos 1980; Garcia and Morari 1981; Lübbert and Simutis 2001).

A good example is Petersen and Whyatt's (1990) analysis of a continuous bioreactor. An accurate model of the microbial system was not available, and thus the first step of their procedure was the formulation of the system as a time-series model. This model automatically accounted for disturbances and process variations, and it was not necessary to wait for the system to reach a new steady state before further control actions were taken (see also Rolf and Lim 1985; Semones and Lim 1989). In the next step, the steady-state productivity of the system was expressed in terms of the control variable of interest, namely the dilution rate. Finally, the process was maximized by studying the first and second derivatives of the steady-state productivity function. As a direct extension of this optimization strategy, online control can be made adaptive (Rolf and Lim 1985; Harmon et al. 1987; Semones and Lim 1989). In some cases, adaptive online control may even simplify the mathematical analysis, because processes that are actually nonlinear may be viewed as linear processes with time-varying parameters (Harmon, Svoronos, and Lyberatos 1987).

For targeted manipulations at the kinetic or pathway levels, these more global approaches are unsuitable, and more detailed knowledge is needed. This knowledge is to be integrated in a mathematical model, which henceforth is the subject of optimization. This optimization strategy requires more effort but promises new insights and specific prescriptions of which catalytic or transport steps should be amplified or attenuated for improved yield and to what degree. Thus, the key element of this type of optimization is the search for appropriate mathematical models. As we shall see, there are several general options that all offer advantages in particular situations. The overriding criteria that such models have to satisfy are that they are biochemically valid and at the same time permit efficient mathematical analysis and, particularly, optimization.

One major challenge posed by biological processes in comparison to purely chemical processes is the fact that organisms exhibit dynamical features that can seldom be captured with algebraic functions, which would allow optimization with some of the analytical methods described in the previous chapter (Lübbert and Simutis 2001). Furthermore, organisms usually respond to normal and sometimes abnormal input in a reasonable fashion, adapt to a new environmental milieu, buffer excesses in substrates, and have the ability to counteract unforeseen shortages and demands. At a fundamental level, biotechnology is therefore a fight against these natural dynamic responses and adaptations. Biotechnological optimization attempts to "motivate" (i.e., force) a microorganism to synthesize a product for which it itself has no use, at least not in quantities that are of biotechnological relevance. Because the organism has evolved to resist conditions that are not in its own best interest, it is likely to fight back.

The adaptive measures of the organism in response to variations in the experimental milieu create a complex system that metabolic engineers hope to capture with a mathematical model, usually in the form of a system of differential equations. Even under the audacious assumption that one succeeds with this crucial step of model design, that illusionary "perfect model" alone is only the beginning. In addition, methods are needed for manipulating the model in a way that the output is optimized. The implicit, underlying hope is that the actual organism would behave like the model and therefore synthesize the desired output in the optimal fashion predicted by the model.

What one ideally would like to accomplish is an optimization of the entire dynamics of the system. Beginning with a culture of the organism at a (natural) baseline steady state, the dynamical model would correctly predict the organism's responses to any sets or series of artificial inputs. Equipped with this rich body of information, one would design an optimization strategy that would take these predicted responses and prescribe a dynamic path of time-dependent artificial input conditions that would in optimal time lead to the best possible output. As a decision-maker in multiobjective linear programming might say, this is the *bliss point*: the ideal situation, which satisfies all wishes but is seldom within reach (see later in this chapter).

From a mathematical point of view, this *dynamic optimization* of biochemical or biotechnological systems falls within the realm of *optimal control theory* (Vagners

1983), which is an order of magnitude more difficult than steady-state optimization. The objective function and constraints of an optimal control problem are functions of time, and the time dependence is typically represented by a system of nonlinear differential equations, which may be augmented with some nonlinear algebraic and integral equations. Because differential equations are involved, dynamic optimization becomes very complicated very fast, often requiring calculus of variations. The difficulties of this approach must be overcome if the biotechnological system of interest does not operate at a steady state.

As a simple example of dynamic optimization in biotechnology, consider a scenario where a single bolus of substrate is being supplied at time zero and the fermentation is allowed to run until the substrate is used up, both for intrinsic metabolic processes and for the synthesis of the desired product. Obviously, at time zero, the system is in a different state than at all further time points, and the steady state is only reached once the cells stop metabolizing or are dead.

Another example is the realistic and complicated situation in which the kinetic properties of the organism are not constant throughout the fermentation process. For instance, before the substrate bolus, the (starving) microorganism may be dormant. As soon as substrate becomes available, genes that transport substrate into the cell are induced. Once substrate is amply available, these genes may be turned off and others, responsible for metabolizing the substrate, are expressed. As substrate becomes limiting toward the end of the fermentation process, yet another set of genes may assume control of the dynamics.

In an optimal control problem, one may ask a question such as: how can one obtain the maximum yield at a given time, if the fermentation process starts at time zero? The problem can be approached either by an exclusive focus on achieving a desired output at a given time point (no matter what is happening in between) or by addressing the transients of the system between some perturbation and some final time. The final time of the process may not even be fixed, and the optimization task may be the generation of as much product as possible, irrespective of duration. If the optimization addresses transient responses, the difference between some desired target response and the actual response under a particular setting of control variables is to be minimized. If the desired outcome is fixed, the classical literature calls the optimization task a *regulator problem*; if it changes due to its own dynamics, it is called a *servomechanism problem*. The reader interested in these types of optimal control problems is referred to Oldenburger (1966) for details.

Conejeros and Vassiliadis (1998a,b, 2000) are apparently among a very few who have attacked dynamic optimization in the context of biochemical pathways. Faced with the enormous complexity of the situation, they somewhat simplified the problem with the following steps. First, instead of a variable final time, they investigated optimal production at a series of fixed end points. Secondly, based on sensitivities of the time-dependent objective function with respect to magnitudes of fluxes, they ranked the importance of all relevant reaction steps at different times of the experiment. This ranking identified the bottlenecks of the biochemical system throughout the fermentation process. As indicated previously, these bottlenecks are not necessarily constant, but reflect the genetic and physiological responses of the organisms to the changing external milieu. The ranking was followed by a decision on how many steps could experimentally be amplified or attenuated, and the most sensitive steps were proposed for modification. Further tests showed the consequences of modifying larger or smaller numbers of reaction steps. In addition to modifying enzyme activities (reaction steps), the authors subsequently allowed for alterations in some of the metabolite concentrations and their regulatory effects. They used for this purpose modification factors in the form of products of power-law functions, as Hatzimanikatis et al. (1996a,b) had suggested for optimization tasks under steady-state conditions.

Conejeros and Vassiliadis applied their methods to the production of penicillin V in a fed-batch culture of *Pencillium chrysogenum* (for a description of the biochemical system, see Pissara, Nielsen, and Bazin 1996). In this culture, the changes in external medium affect the substrate intake characteristics of the organism. This physiological adaptation alters the dynamics of the fermentation process and renders a steady-state approach inappropriate. Although not a perfect application system, the investigation suggested modifications of the metabolic pathway at two levels: amplification/attenuation of enzyme activities and regulation by metabolites. Theory indicated that the most promising alteration would be the modification of active sites, which however presented practical problems. The results also showed how changes in regulation would affect productivity. Based on these results, the authors concluded that the metabolites – rather than the enzyme concentrations, as previously thought – might be the truly controlling factors in the enhancement of the production flux.

Except for these recent studies, metabolic optimization has mostly dealt with systems under steady-state conditions, and we will do the same here. The reasons for focusing on this situation are obvious. At steady state, the describing differential equations are set equal to zero, which reduces the optimization task to one involving only algebraic equations. Furthermore, steady-state processes are not just simpler to analyze, they are also advantageous from an operational point of view, because they are easier to control, do not require vessels to accumulate materials between process units, and yield products with uniform quality (Duncan and Reimer 1998).

The prime example is maximization of yield in a fermentation setting, in which a batch culture is continuously fed and none of the metabolite concentrations change over time. Of course, yield is the output of an often large and highly nonlinear system, and having encountered all the challenges and pitfalls of nonlinear optimization in the previous chapter, one might expect that such an optimization task would be rather complicated. However, we will see that there are ways of working around some of the problems of nonlinear optimization.

ARE NONLINEARITIES REALLY NECESSARY?

Of course this is a loaded question. It is well known that even small biochemical systems can exhibit responses that are without doubt not linear. Essentially all biological processes are saturated, which is an immediate problem for linear representations, and some biological systems show dynamics like stable (limit cycle) oscillations or even chaos, which simply cannot be captured with linear models. Thus, the *prima facie* answer is *yes*: nonlinearities are really necessary.

It is an answer we really do not like. Recalling the previous chapter, there are just a lot of problems and challenges in dealing with nonlinear systems. Even for the simplest of enzymatic pathways, an unbranched chain of reactions, the nonlinear optimization (at the kinetic level) is very involved (for instance, see Heinrich and Schuster 1996, pp. 309–15). It is hard to imagine executing that type of analysis for a pathway of relevant, moderate size. By contrast, there are very many advantages of linear optimization. As a consequence, a lot of effort has been expended to change the answer to the question in the heading to *no*. Yes, biochemical systems have nonlinear features, but one would greatly benefit if one could restrict the nonlinear optimization problem to one amenable to linear methods. As we will see, it is indeed possible in many cases of relevance to limit the analysis to aspects that can be validly linearized.

A direct linearization of the model equations at some operating point, such as the baseline steady state, would be an option. However, the accuracy of the result is rather limited, and because we are interested in moving the system to a new steady state whose declared purpose it is to be substantially different from baseline, direct linearization is not such a good idea.

Three avenues have been taken to allow the use of linear methods for nonlinear biochemical system models. The first focuses on the network of fluxes. Flux networks are traditionally represented by matrix equations, whose central component is the stoichiometric matrix (see Chapter 2). Because of the linear structure of this approach, it is independent of the scale of the system and can be applied successfully even to very large networks. Flux networks ignore information at the kinetic level, but maintain the integrated nature of the biochemical system. We will discuss some aspects of this stoichiometric network approach in the next section.

The second approach is a restriction of optimization tasks to steady states and the use of S-system models for the underlying pathway description. The S-system model itself is nonlinear and rich enough to capture relevant responses, but its steady-state equations are linear in logarithmic coordinates. The key argument now is that the logarithmic transformation does not change the locations of maxima and minima. In other words, the two functions $f(x)$ and $f(\ln(x))$, for positive x-values, have their maxima and minima for the same values of x, and this is also true for functions of many variables. The transformation between $f(x)$ and $f(\ln(x))$ just stretches the x axis in a smooth (logarithmic) fashion, and the maxima and minima follow this stretching (Figure 5.1). Imagine drawing the graph of the function on a sheet of rubber. If the rubber is stretched, the axis and the graph stretch at the same rate, and the maxima and minima still correspond to the same tick marks on the axis. Some other types of stretching or shrinking, such as by an exponential function, also lead to a valid linearization, but the logarithmic stretching is for our purposes the most relevant.

The fact that the logarithmic transformation does not affect the position of the extrema allows us to make use of the rich repertoire of methods from linear algebra and to employ straightforward methods of linear programming for the optimization

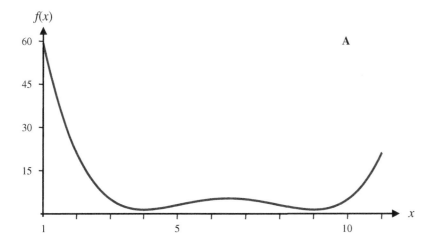

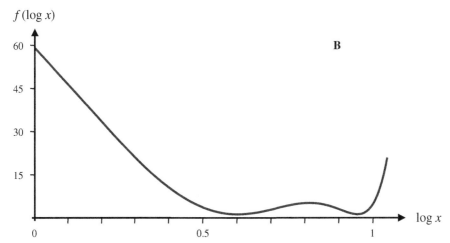

Figure 5.1. The graph of the function $f(x)$ in panel A becomes nonlinearly distorted, when the independent variable x is logarithmically transformed (panel B). Nonetheless, the locations of maxima and minima follow the same distortion and may be obtained from either representation. For instance, the local maximum at $x = 6.5$ in panel A moves in panel B to about 0.813, which corresponds to log (6.5).

of arbitrarily complex biochemical systems under steady-state conditions, as long as they are formulated as S-systems.

The philosophy of this optimization approach is as follows. Suppose we are interested in maximizing the production of a particular metabolite. The tools we have available are (limited) controls of several input variables, which act either as precursors, exogenous products or intermediates, or as modulators of some of the involved processes. Of course, there are constraints that must not be violated. Input variables can only be varied within a certain physiological range, internal concentrations must not exceed thresholds of toxicity, and the fluxes cannot be increased indefinitely. Furthermore, for continual production, the optimized system must be at a steady state.

We will see that all these constraints, as well as typical objective functions, are readily translated into linear functions and inequalities, and that the optimization task within the S-system methodology becomes a standard linear program. This approach will be the subject of most of this chapter.

A third approach was proposed by Hatzimanikatis and Bailey (1996). Instead of making use of the linearity of steady states in S-systems or of linearizing a traditional biochemical model directly at some operating point, they suggested linearizing *logarithmic deviations* from baseline. Thus, rather than using the original variables X_i, they used new variables $z_i = \ln(X_i / X_{i0})$, where X_{i0} is the baseline concentration of metabolite X_i. They tested the accuracy of this "(log)linear kinetic model" and deemed it sufficient for optimization purposes. This approach has the advantage that it uses parameters, such as elasticities, which are traditionally measured in metabolic control analysis. Incidentally, this system of linear equations in logarithmic deviations is identical to the linearization used for stability analyses of S-systems (Savageau 1975, 1976).

SEARCHING FOR EXTREMA OF LINEAR FUNCTIONS WITH MANY VARIABLES

The possibility of valid linearization of biochemical systems suggests that it is worth our while to study optimization methods that are specially tailored for linear systems, because one should expect them to be simpler than those for nonlinear systems with all their problems discussed in the previous chapter. Indeed, constrained linear optimization problems permit specialized, very effective approaches that are not directly applicable to nonlinear tasks. Some of these are discussed in this chapter.

A linear function of one variable and without constraints has its maximum at $+\infty$ or $-\infty$, unless it happens to be constant, which is not very interesting. For functions of many variables, the situation is similar. Because more than three dimensions are hard to intuit, let us look at a linear function of two variables, x and y. Its graph is a plane in three-dimensional space. Observations from the single-variable case and some hand waving are sufficient to imagine what happens with linear functions with more than two variables. So the "2-d case" is a good illustration. For the following, we exclude the boring case of a constant function, which has the same value, no matter what the arguments x and y are. Because the function is by decree not constant, it has a non-zero slope in x-direction, in y-direction, or both. Playing around with a flat piece of cardboard reveals these scenarios. As in the 1-d case, one is easily convinced that the global maximum is located at one of the four combinations $(+\infty, +\infty)$, $(+\infty, -\infty)$, $(-\infty, +\infty)$, or $(-\infty, -\infty)$, where the first component of each pair refers to x and the second to y. As in the 1-d case, the situation only becomes interesting if constraints are involved. Parallel to the 1-d case, one could confine x and y to fall within some interval each. This would correspond to a rectangular domain for the function and, again, one easily finds the maximum in one of the four corners (Figure 5.2). However, there are more interesting options that derive from other types of constraints.

The simplest case in this category is the definition of general linear constraints. These have the form of straight-line relationships between y and x. For instance,

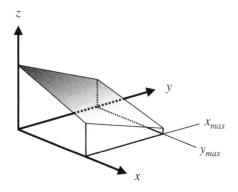

Figure 5.2. The maximum of the linear objective function $z(x, y)$ within the admissible domain $[0, x_{max}] \times [0, y_{max}]$ is at $(0, 0)$. The minimum occurs at (x_{max}, y_{max}).

suppose a batch process can be run under two different conditions, and the objective is to produce as much yield as possible. Under condition A, the production yield per batch is P_a, and under B it is P_b. Thus, for any combination of A-type and B-type batches, the total production is

Total production P

$$= P_a \times \text{number of } A\text{-type batches} + P_b \times \text{number of } B\text{-type batches.} \quad (5.1)$$

Coding the number of batches of A- and B-type as x and y, respectively, the objective can be formulated as

$$\text{maximize } P = P_a\, x + P_b\, y. \quad (5.2)$$

Because of physical limitations of the lab, there is a fixed maximum number N of processes that can be run in a given period of time, which immediately suggests the constraint $x + y \leq N$. Suppose further that the cost of an A-type batch is C_a and the cost for a B-type is C_b. In reality, one would probably try to minimize cost, but for simplicity of illustration, let us instead assume that there is just a limited budget M that must not be exceeded. This leads directly to a second constraint, namely $C_a\, x + C_b\, y \leq M$. One could imagine that, because of other resource limitations, a maximum of R_a A-type experiments and R_b B-type experiments can be run. This information is readily converted into the constraints $0 \leq x \leq R_a$ and $0 \leq y \leq R_b$, which are reminiscent of the interval constraints in the one-dimensional case. For our illustration, this might be enough in terms of constraints, so let us summarize the optimization task, which is also called a *linear program*:

> *maximize* the objective function $P = P_a\, x + P_b\, y$
> *subject to* the constraints
>
> | | |
> |---|---|
> | $x + y \leq N$ | (fixed maximum number of batches) |
> | $xC_a + yC_b \leq M$ | (fixed maximum budget) |
> | $0 \leq x \leq R_a$ | (resources for A-type experiments) |
> | $0 \leq y \leq R_b$ | (resources for B-type experiments). |

$$(5.3)$$

Linear programs have received enormous attention over the past fifty years (for reviews, see, e.g., Charnes and Cooper 1961; Dantzig 1963; Luenberger 1984; Strang 1986; Kreyszig 1993; Kolman 1997; Bertsekas 1998).

It is noted that the constraints might be redundant or irrelevant. For instance, if $C_a = C_b$, then the second constraint becomes $x + y \leq M/C_a$, and depending on whether M/C_a exceeds N or vice versa, one constraint is not limiting the admissible domain. In the former case, the second constraint is immaterial, and in the latter, the first constraint is immaterial. If $M/C_a = N$, the first and second constraints are equivalent, and one of them could be dropped without consequence. Except for some loss in efficiency, redundancies among constraints do no harm, and unless one deals with a huge system and many redundancies, it is not necessary to expend a lot of effort trying to determine the minimum set of constraints.

For the following, we simply ignore the fact that x and y in our example must actually be *integer* numbers and will be satisfied for this illustration even if the optimal solution calls for something like 10.3 A-type and 15.7 B-type experiments. The subject of *integer programming* adds more complication than is desirable right now. The interested reader is referred to Garfinkel and Nemhauser (1972).

It is informative to study a graphical representation of the situation. For this purpose, it is beneficial to replace the parameters with actual numbers. Let us suppose the following:

$$P_a = 10, \; P_b = 25,$$
$$N = 18, \tag{5.4}$$
$$C_a = 50, \; C_b = 30, \; M = 600,$$
$$R_a = 10, \; R_b = 10.$$

The objective function (Eq. 5.2) and the constraints in (5.3) are defined through functions of x and y that can be drawn in the x-y-plane. Actually, the constraints are borders of domains, because of the $<$ signs. Let us initially ignore the $<$ signs for a moment and pretend the constraints were regular equalities. If so, the last two constraints are simply intervals, which appear in the plane as vertical and horizontal lines, respectively (Figure 5.3). The remaining constraints and the objective function become more obviously relationships between x and y upon separating y and moving it to the left-hand side. For instance, the second constraint (without the $<$ sign) becomes

$$y = M/C_b - x \, C_a/C_b. \tag{5.5}$$

This constraint equation says *spend exactly $M on the project* and has a simple interpretation. It tells us how many B-type experiments one can execute, given one executes a given number of A-type experiments and spends M dollars. As an example, suppose we decide on nine A-type experiments. The number of B-type experiments then is

$$y = 600/30 - 9 \times 50/30 = 5. \tag{5.6}$$

Nine A-type batches yield $9 \times P_a = 90$ units of product, and five B-type batches yield $5 \times P_b = 75$ units, for a total of $P = 165$ units in fourteen batches.

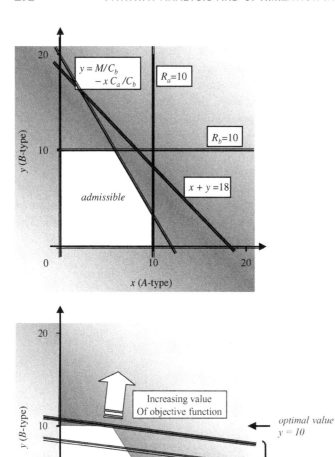

Figure 5.3. Graphical determination of optimal solution. Top panel: Constraints limit the admissible domain. Bottom panel: The objective function is shifted parallel toward the optimal vertex of the simplex ($x = 6$, $y = 10$).

As an extreme strategy, suppose we do not run any B-type batches (i.e., $y = 0$). According to Eq. (5.5), $x = 12$ and $P = 12 \times 10 = 120$. Thus, this strategy yields less product. Furthermore, we need to consider the other constraints. In particular, the constraint on x requires $0 \leq x \leq R_a = 10$, which is not satisfied. In the jargon of the field, this solution is *not admissible*.

The solution in the scenario with no B-type batches is inconsistent with the combination of Eq. (5.5) and the constraint $0 \leq x \leq R_a$. This changes, as soon as we switch Eq. (5.5) back from an equation to the actual inequality $x\,C_a + y\,C_b \leq M$. Specifically, if we reduce the number of A-type batches to the maximum of ten, both conditions, $0 \leq x \leq R_a$ and $x \, C_a + y \, C_b \leq M$ are satisfied. However, this solution produces limited yield, namely, $P = 10 \times P_a + 0 \times P_b = 100$ units.

To check possible solutions in a more straightforward fashion, it is useful to transform the inequality $x \, C_a + y \, C_b \leq M$ into

$$y \leq M/C_b - x \, C_a/C_b. \tag{5.7}$$

Its interpretation is that y has to be below (or equal) the expression on the right-hand side, which corresponds to Eq. (5.5). Thus, y has to be below (or at most on) this straight line. The analogous is true for all other linear constraints. The admissible domain for the optimization problem is shown as the clear area in Figure 5.3.

Let us illustrate this with a scenario of *no A-batches*. Now $x = 0$, and the corresponding y, according to Eq. (5.5), is $y = M/C_b = 20$, and the total product is $P = 20 \times 25 = 500$. That would be a much better solution, but we are only allowed a maximum of ten B-type batches, and these would only produce $P = 10 \times 25 = 250$. Still, this is better than former results.

Graphing all constraints and shading the nonadmissible side shows the clear area of admissible points (Figure 5.3, top panel). The optimization task now can be imagined as follows. Remember that the graph of the objective function $P(x, y)$ for all values of P is a plane in three-dimensional space. This plane is not horizontal but slanted; in our example, it increases with both x and y. Imagine constructing walls along the straight lines that enclose the admissible area. These walls intersect with the plane of $P(x, y)$ and constitute the boundaries within which we have to search for the maximum. It is not difficult to see that the maximum must be in one of the corners, which in linear optimization are called *vertices*. Because $P(x, y)$ increases with x and y, the highest point of the plane within the admissible area is to be found for large x and y values. To find out which vertex is optimal, we express the objective function in the form $y = P/P_b - x \cdot P_a/P_b$. For any given value P, this straight line shows the yield for any combination of A-type and B-type batches. If we change the value of P, the line shifts in a parallel fashion up or down. The graphically obtained optimal solution consists of moving this line "up the plane" until it cuts the admissible domain exclusively at the most desired vertex. In this case, this happens for six A-type experiments and $600/30 - 6 \times 50/30 =$ ten B-type experiments. The overall yield in this situation is $P = 6 \times P_a + 10 \times P_b = 270$ in fourteen batches.

It is theoretically possible that all values of the objective function have the same maximum value along one of the walls. This can happen if the objective function is parallel to one of the constraints. Even in this case, we would find one maximum (though not the only one) at one of the vertices of the admissible domain.

Optimization of higher-dimensional problems follows exactly the same principles. The constraints become *hyperplanes*, which are the analogues of lines in two dimensions and planes in three dimensions, and the admissible domain consists of some convex body that is limited by these hyperplanes. In this context, *convex* means that if two points belong to the body, then all points of the straight-line connection between the points belong to the body. Such a domain is called a *simplex*, and the standard method for linear optimization is called a *simplex algorithm*

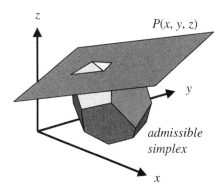

Figure 5.4. In three-dimensional space, linear constraints are planes that limit the admissible domain to a so-called *simplex*. The objective function $P(x, y, z)$ has the geometric structure of a plane. It is drawn close to the optimal admissible point.

(Dantzig 1963). In three dimensions, the shape of a simplex looks like an asymmetrically cut diamond in which the faces comprise the constraints; an illustration is shown in Figure 5.4. The objective function (which in three dimensions is a plane) moves through this simplex until it hits the vertex with the highest (optimal) value.

The fact that the optimum (or at least *one* optimum) is always found at a vertex of the admissible simplex is crucially important and generally true only for *linear* optimization tasks. It reduces the search for a maximum (or minimum) from an infinite number of possible points within the admissible simplex to the comparatively much smaller number of corners. Even for very large numbers of variables and constraints, this linear search can be made relatively efficient.

For large simplexes with millions of vertices, an exhaustive search among all vertices would be wasteful. Modern implementations of simplex algorithms are therefore designed such that they do not have to check every single vertex. There are strategies that avoid the most irrelevant parts of the simplex and climb toward the maximum, more or less without wasting search effort. Unless there is a whole line or (hyper-)plane of maxima along one of the edges or (hyper-)planes of the simplex, the maximum is unique. In particular, there is no possibility that it is only local and that the true global maximum would lie somewhere in the inner part of the simplex. This is very different from nonlinear optimization and a crucial advantage.

It is noted that the simplex search is not always the most effective method and that other strategies, such as Karmarkar's algorithm (Strang 1986), have been proposed. Because of the widespread importance of large linear optimization tasks, new algorithms often receive a lot of attention.

The details of simplex algorithms or alternate methods are certainly important. However, if one has some faith in commercially available software, the conceptual ideas about linear optimization are probably sufficient for our purposes. There are numerous books detailing linear optimization or *linear programming*, as it is often called. Some of them are written by mathematicians, some by engineers, and some by business experts in operations research. Some established standards of persisting value are Hadley (1962), Dantzig (1963), Luenberger (1984), Gass (1985), Groetsch and King (1988), and Bertsekas (1998). Brief introductions are found in Strang (1986) and Kreyzsig (1993).

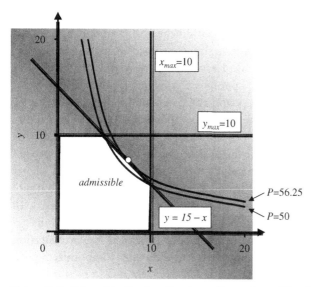

Figure 5.5. If the objective function is nonlinear (here $P(x, y) = x \cdot y$), the optimum may not be located at a vertex. In this case, $P(7.5, 7.5) = 56.25$ is superior to the values at the candidate vertices, which both have a value of $P(5, 10) = P(10, 5) = 50$.

It is important to realize that nonlinearities in either the objective function $f(x, y)$ or in the constraints (or, of course, both) can immediately annihilate the simplicity of the linear case and, in particular, the feature that the optimal solution lies at a vertex. The former case is illustrated in Figure 5.5. In the latter case, return to the previous linear batch process example, but "bulge out" the constraint $y \leq M/C_b - x\, C_a/C_b$. It is not difficult to imagine that the maximum may now be located at the bulge and no longer at one of the vertices. The same is true for higher dimensions, where the overall optimization problem may become very complex.

OPTIMIZATION OF STOICHIOMETRIC NETWORKS

In Chapter 2, we introduced the concept of a stoichiometric network that describes the topology of fluxes in a biochemical system. As discussed there, the key formulation of the system is that of a matrix equation of the type

$$\frac{d\mathbf{S}}{dt} = \mathbf{N} \cdot \mathbf{v}, \tag{5.8}$$

where $\mathbf{S}$ is the vector of metabolite concentrations, $\mathbf{N}$ denotes the stoichiometric matrix, and $\mathbf{v}$ the vector of reaction rates (Gavalas 1968; Glansdorff and Prigogine 1971; Horn and Jackson 1972; Heinrich, Rapoport, and Rapoport 1977). At steady state, the differential equations are set equal to zero, and the result is a system of linear algebraic equations

$$\mathbf{N} \cdot \mathbf{v} = \vec{0}, \tag{5.9}$$

where $\vec{0}$ is a vector of 0's.

Of greatest interest here is the fact that stoichiometric network models permit constrained optimization with standard methods of linear programming, as outlined previously. The objective function often reflects some monetary costs, shadow prices (Savinell 1991), or investment costs (Górak, Krasławski, and Vogelpohl 1987), but it can also include quality of the product, "paracosts" such as stability, flexibility, or controllability, or other costs such as safety and environmental impact (Górak et al. 1987). The constraints are dictated by the system components and by technical and economical factors. They can be linear or nonlinear. In particular, they include the linear equations in Eq. (5.9). Examples for this stoichiometric optimization approach are found in a series of papers by Papoutsakis (1984) and Papoutsakis and Meyer (1985ab) who studied the saccharolytic production of oxychemicals like butanol and acetone in butyric-acid and propionic-acid bacteria. The analysis began with the formulation of fermentation equations, which were stoichiometric equations accounting for the biochemistry and the bioenergetics that governed the fermentation process. The equations were solved and validated, and used later to optimize a yield of interest.

Majewski and Domach (1990) pursued a similar strategy for an analysis of the Krebs cycle in *E. coli*. Fell and Small (1986) used stoichiometric optimization to analyze lipogenesis in adipose tissue, and Savinell (1991) and Savinell and Palsson (1992a–d) optimized metabolic flux patterns in *Hybridoma* cells for biotechnological purposes. Also using a stoichiometric approach, but analyzing it in a different way, Kacser and Acerenza (1993) proposed a method of genetically manipulating organisms for optimized yield. Seressiotis and Bailey (1988) and Mavrovouniotis, Stephanopoulos, and Stephanopoulos (1990) combined a stoichiometric approach with methods of artificial intelligence to determine possible metabolic pathways leading from some substrate to a product of interest. Heinrich and Schuster (1996) discussed a variety of optimization tasks in biochemistry, cell regulation, and evolution.

An interesting extension of these methods arises when hierarchical objective functions need to be optimized, that is, when the task is the optimization of a function $F(x, y)$ whose argument y is the solution of an optimization problem. The strategy for this type of problem is multilevel, or multi-objective, programming (Candler and Norton 1977; Candler and Townsley 1982; for a review in the context of chemical process design, see Clark 1990; Clark and Westerberg 1990). We will discuss this type of optimization later in this chapter and in Chapter 7.

Stoichiometric optimization can be employed for analyzing structural or topological questions about metabolic networks. For instance, Majewski and Domach (1990) showed the importance of flux reversal in the optimal production of ATP in *E. coli*. Optimization of stoichiometric equations was also used to demonstrate that some central metabolic pathways optimally solve the problem of rearranging carbons in the intermediates. Examples include the pentose cycle (Meléndez-Hevia and Isidoro 1985; Meléndez Hevia and Torres 1988), the Krebs cycle (Meléndez-Hevia, Waddell, and Cascante 1996) and the glycolytic pathway (Meléndez-Hevia et al. 1997). Papoutsakis and Meyer (1985a) suggested stoichiometric equations as a means for classifying microorganisms.

More recently, Stephanopoulos and Simpson (1997) combined a stoichiometric pathway representation with the so-called *top-down approach* developed within the framework of metabolic control analysis (Brown, Hafner, and Brand 1990; Fell 1992; Liao and Delgado 1993; Quant 1993). This approach focuses on groups of reactions, rather than on all reactions individually. Given the oftentimes large size of biotechnologically relevant pathways, this grouping is quite intuitive and maybe even unavoidable. The goal of Stephanopoulos and Simpson's analysis was the identification of a small number of reaction steps within a larger pathway (in their case, the aromatic amino-acid biosynthetic pathway) that would be suitable for biotechnological manipulation. The grouping of steps led to a subset of the stoichiometric matrix, which the authors called the *steady-state internal metabolite stoichiometry* (SIMS) *matrix*. Sets of independent pathways of the system were shown to correspond to vectors of the kernel (null space) of this SIMS matrix (Groetsch and King 1988; Heinrich and Schuster 1996). Because the null-space matrix is not unique, several groupings were explored. For each grouping, the authors computed group control coefficients (similar to logarithmic gains), which identified those parts of the pathway that exhibited the strongest kinetic control and were thus deemed to have the highest potential for improved yield. The methods also prescribed which groups of steps would have to be modulated in a coordinated fashion to increase pathway flux while maintaining internal metabolites at physiological levels.

In many cases, the stoichiometric matrix may be all that is definitively known about a metabolic system, and optimization of flux patterns is the best solution. However, the biochemical knowledge base is rapidly increasing, and the number of pathways with detailed kinetic information is growing. This information should be used in the control and optimization of integrated systems. A convenient way of accomplishing this is the formulation of integrated biochemical systems with S-system differential equations.

OPTIMIZATION OF S-SYSTEMS

The great advantage of representing a pathway with an S-system is the fact that its steady state can be expressed explicitly in the form of a system of linear equations (Savageau 1969). For a system that includes both dependent and independent variables, the steady-state solution is characterized by the equation

$$\vec{y}_D = \mathbf{M} \cdot \vec{b} + \mathbf{L} \cdot \vec{y}_I \tag{5.10}$$

where the matrix $\mathbf{M}$ contains as elements the sensitivities of the system with respect to changes in rate constants, and the matrix $\mathbf{L}$ contains as elements the log gains, which quantify changes in the steady-state values that result from changes in independent variables [see Chapter 2, Eq. (2.26), and Savageau 1971a,b, 1972, 1976; Voit 1991, 2000].

Log gains are of prominent importance for our purposes here, because our strategy of optimization calls for targeted manipulation of some or all of the independent "control" variables, which are typically inputs, enzyme activities, or the magnitudes of transport steps, but could also include factors like temperature or pH. Because the goal of the optimization is to move the system to a new steady state that is sufficiently far from baseline, altering an independent variable with high gains has a better chance of success than altering a variable that is not influential (i.e., has a low gain). Many optimization studies, whether under steady-state or dynamical conditions (Torres, Voit, and Alcón 1996, 1997, 1998; Conejeros and Vassiliadis 1998a,b, 2000; Rodríguez-Acosta, Regalado, and Torres 1999; Marín-Sanguino and Torres 2000), therefore begin with the development of a gain profile.

High log gains are not a pure blessing though. Because independent variables with high gains strongly affect the system, any experimental inaccuracies during the implementation of the optimal profile of altered enzymes and transport steps are amplified, and the theoretical optimum may not be achievable in practice (Voit and Del Signore 2001; see also Chapter 6).

The explicit, linear steady-state equations of the S-system form are of enormous benefit to optimization tasks. In any other nonlinear system, which does not reduce to linearity at steady state, the right-hand sides of the differential equations constitute a system of nonlinear equations, and even though these are algebraic equations, some or all of the problems discussed in the previous chapter become relevant. In contrast, linearity throughout the optimization task allows us to use a straightforward simplex algorithm that efficiently reduces the search for extrema to vertices of the admissible simplex.

In addition to the steady-state constraints, the optimization task with S-systems includes other components. It typically reads

linear program:

(1) maximize ln(flux)

subject to

(2)   steady-state equations, expressed in logarithms of variables
(3)   ln(dependent or independent variable) ≤ constant
(4)   ln(dependent or independent variable) ≥ constant
(5)   ln(dependent or independent variable) = constant
(6)   ln(dependent or independent variable) unrestricted
(7)   ln(flux) ≤ constant
(8)   ln(flux) ≥ constant
(9)   ln(flux) unrestricted
(10) ln(flux1/flux2) ≤ constant   (5.11)

(Voit 1992; Torres et al. 1996, 1997, 1998). A linear program (optimization task) for biochemical networks in S-system form may contain some or all of the previous components. It always includes an objective function like (1). In Eq. (5.11), a flux is to be optimized. A flux in an S-system model is given as a product of power-law functions, such as

$$V_i^- = \beta_i \prod_{j=1}^{n+m} X_j^{h_{ij}}. \tag{5.12}$$

Taking the logarithm and expressing the result in terms of y-variables transforms this flux into the linear form

$$\ln(V_i^-) = \ln(\beta_i) + \sum_{j=1}^{n+m} h_{ij} \cdot y_j. \tag{5.13}$$

For optimization purposes (such as maximizing this expression), the constant $\ln(\beta_i)$ is immaterial and often omitted. The typical objective function is thus

$$\text{maximize} \sum_{j=1}^{n+m} f_{ij} \cdot y_j, \tag{5.14}$$

where f_{ij} stands for either g_{ij} or h_{ij} and may be positive, negative, or zero.

As a variation on the theme, the task could be to maximize the efficiency of the system, which could be defined as the output flux in relation to the input flux. For instance, if the input and output fluxes are given as

$$V_1^+ = \alpha_1 \prod_{j=1}^{n+m} X_j^{g_{1j}} \tag{5.15}$$

and

$$V_n^- = \beta_n \prod_{j=1}^{n+m} X_j^{h_{nj}} \tag{5.16}$$

respectively, the objective function (with the omission of the unalterable rate constants) would read

$$\text{maximize} \sum_{j=1}^{n+m} (h_{nj} - g_{1j}) \cdot y_j. \tag{5.17}$$

Instead of a flux or a flux ratio, the objective function may simply consist of the concentration of a metabolite X_j. To retain y-variables throughout the linear program, such an objective function reads equivalently

$$\text{maximize } y_j. \tag{5.18}$$

Similar to the output/input flux ratio, the objective may also be to maximize the yield of the system, which one could define as the ratio of output over input. With an input X_k and an output X_j, the objective function in y-variables reads

$$\text{maximize } y_j - y_k. \tag{5.19}$$

Constraint (2) in the linear program (5.11) ensures that the optimized system is in a steady state, no matter what the altered independent (control) variables are. As discussed in detail (see also Chapter 2), these equations are linear in y-variables.

Constraints (3) and (4) force variables to stay within certain limits. For consistency, the original X-variables are again replaced with y-variables, which does not change the validity of the inequalities. In some cases, a variable should not be altered at all. This situation is captured by constraint (5). At the other extreme, a variable may assume any positive value. Constraint (6) is a translation of this *carte blanche* into the corresponding logarithm.

Constraints (7)–(9) are the analogous constraints on fluxes, which in S-system formulation also adhere to the linearity of the optimization task. Finally, constraint (10) forces the ratio of the two fluxes V_i and V_j to remain below a certain limit. Due to the product nature of the fluxes, their ratio again becomes linear in logarithmic space. The overall result is that the entire constrained optimization task becomes one of straightforward linear programming (Voit 1992; Regan, Bogle, and Dunhill 1993). Numerous software packages are available for linear programming tasks. They usually are well behaved, cannot get trapped in local minima (because there are none!), and are able to deal with thousands of variables and constraints.

OPTIMIZATION OF GMA SYSTEMS

GMA systems and S-systems derive from the same type of power-law approximation of the underlying processes in a metabolic pathway. The only difference is that the approximation leading to GMA systems occurs at the level of individual fluxes, so that every diverging or converging process at a branch point is represented by a separate product of power-law functions. By contrast, the S-system form derives from representing all entering fluxes collectively as one product of power-law functions and, analogously, from representing all departing fluxes collectively as one product of power-law functions. This seemingly banal difference has far-reaching consequences. Most important in the present context is the fact that the GMA form does not in general permit us to compute steady states with analytical means. In the S-system form, we simply moved the beta terms of the steady-state equations

$$0 = \alpha_i \prod_{j=1}^{n+m} X_j^{g_{ij}} - \beta_i \prod_{j=1}^{n+m} X_j^{h_{ij}} \qquad i = 1, \dots, n \tag{5.20}$$

to the left-hand side and took logarithms, which yielded linear equations in $y_i = \ln(X_i)$. This procedure is not effective in the case of GMA systems, whose steady-state equations are

$$0 = \gamma_{i1} \prod_{j=1}^{n+m} X_j^{f_{ij1}} \pm \gamma_{i2} \prod_{j=1}^{n+m} X_j^{f_{ij2}} \pm \cdots \pm \gamma_{ik} \prod_{j=1}^{n+m} X_j^{f_{ijk}} \pm \dots. \qquad i = 1, \dots, n. \tag{5.21}$$

Of course, one could move the terms with negative rate constants to the left-hand side, but the logarithmic transformation would not lead to linearity.

To some degree, optimization of GMA equations may still be advantageous over the optimization of differential equations that could have any mathematical form. A trick to simplify the representation is the definition of new variables z for every product of power-law functions in the system. For instance, we could define

$$z_{ik} = \gamma_{ik} \prod_{j=1}^{n+m} X_j^{f_{ijk}}. \tag{5.22}$$

The steady-state equations are now linear in the new variables. Furthermore, taking the logarithm of each definition leads to a second set of linear equations. The connection between these two sets is a set of simple logarithmic constraints. For streamlined notation, we may define $w_{ik} = \ln(z_{ik})$ and $\zeta_{ik} = \ln(\gamma_{ik})$ and let p_i denote the number of products of power-law functions in equation i. Furthermore, $\varphi_{ik}, \psi_{ik},$ $u_{ik}, U_{ik}, l_{ik}, L_{ik}, u_k,$ and l_k are real-valued coefficients and boundaries, whose meaning is evident from the equations and constraints that follow. With these settings, the optimization task for a GMA system may be formulated as

maximize objective

$$\sum \varphi_{ik} z_{ik} + \sum \psi_k y_k \tag{5.23}$$

subject to

steady-state equations

$$\sum_{k=1}^{p_i} z_{ik} = 0 \tag{5.24}$$

$$\zeta_{ik} + \sum_{j=1}^{n+m} f_{ijk} y_j = w_{ik} \tag{5.25}$$

flux constraints of the type

$$\sum_{k=1}^{p_i} u_{ik} z_{ik} \leq U_{ik} \tag{5.26}$$

$$\sum_{k=1}^{p_i} l_{ik} z_{ik} \geq L_{ik} \tag{5.27}$$

metabolite constraints of the type

$$\sum_{k=1}^{n+m} u_k y_k \leq U_k \tag{5.28}$$

$$\sum_{k=1}^{n+m} l_k y_k \geq U_k \tag{5.29}$$

logarithmic constraints of the type

$$w_{ik} = \ln(z_{ik}). \tag{5.30}$$

The objective function allows for the optimization of an arbitrary sum of fluxes and variables. The steady-state constraints fall into two classes. The first corresponds directly to the GMA system at steady state, and the second is a reformulation of Eq. (5.22) in logarithmic coordinates. The flux and metabolite constraints are self-explanatory. Finally, the logarithmic constraint connects the two types of variables in the optimization program, namely the metabolites, expressed in logarithmic form, and the flux terms z_{ik}.

The interesting aspect of this formulation is that the objective function and all equations and constraints in this optimization task are linear, with the notable exception of the logarithmic constraints. Because of the monotonicity and simple structure of the logarithmic function, the task is considerably less complex than a general optimization problem.

The so-called *indirect optimization method* (IOM; e.g., Torres et al. 1997; Marín-Sanguino and Torres 2000) pursues a different approach to optimizing biochemical pathways that are represented as GMA systems. In this approach, the GMA system is approximated by the corresponding S-system. This S-system model is optimized, and the optimal solution is tested in various ways with the original GMA system. The indirect optimization method will be discussed in detail in Chapter 7.

MULTIPLE CRITERIA OPTIMIZATION

So far, we have discussed a situation where the optimum is determined by a single goal, such as *maximize yield*. In many real-world situations, such single-mindedness is unrealistic, and any optimal solution must strike a balance between different goals. An example may be the maximization of yield, while simultaneously minimizing cost, minimizing time, or maximizing long-term return on investment. "Optimal" in this case is measured against several criteria. If one is very lucky, a solution optimizes all criteria at once. However, that is very rare. The following describes some approaches to optimizing systems with multiple criteria. An application to biochemical systems in S-system form is found in Chapter 7.

The formal study of *multiple criteria optimization* (MCO) began over a hundred years ago with Pareto (1896), who explicitly recognized the difficulties of reducing realistic decision problems to single objective optimization tasks. Not much was done about these problems in the first half of the twentieth century, but interest was renewed in the 1970s, when J. Cochrane and M. Zeleny organized a conference dealing specifically with multiple criteria decision making (Cochrane and Zeleny 1973; see also French et al. 1983). Many of the techniques for this type of optimization, as well as for linear programming, were developed and applied in the context of decision making in business and economy, which has influenced the terminology (Charnes and Cooper 1961). For instance, the person performing the optimization is often called *decision maker* or, simply, DM.

It is not the purpose of this book to discuss theory and techniques for dealing with multiple criteria optimization. Nonetheless, because the topic is of some importance in our context, we will review general strategies that have a chance of producing feasible solutions. In contrast to linear programming, multiple criteria optimization is not straightforward. It requires intuition, experience, judgment, and some willingness to trade one objective for another, at least to some degree.

For our review, we follow very closely the structure of the excellent book by Steuer (1986). Not only does this book proceed at a reasonable pace, it also supports almost every detail with a two- or three-dimensional example, which is illustrated in intuitive, graphical form. Although the book is not the newest on the market, it provides a very good introduction.

In its general form, a multiple criteria optimization problem may be written as

$$\text{maximize } \{ f_1(\mathbf{x}) = \mathbf{z}_1 \}$$
$$\text{maximize } \{ f_2(\mathbf{x}) = \mathbf{z}_2 \}$$
$$\vdots \qquad\qquad\qquad\qquad (5.31)$$
$$\text{maximize } \{ f_k(\mathbf{x}) = \mathbf{z}_k \}$$
$$\text{subject to } \mathbf{x} \in S.$$

The vector $\mathbf{x}$ contains the (real number) quantities $x_1, \ldots, x_n \in \Re$, which can be varied to yield optimal results. There are k *objective functions* of $\mathbf{x}$, which are called f_i; they may be linear or nonlinear. The value of the ith objective function for a given vector $\mathbf{x}$ is called the *criterion value* or *objective value*; it is denoted as z_i. S is the *feasible region*, which is the set of all points $\mathbf{x}$ that are *acceptable* or *admissible*, according to subject area specific constraints. As in the case of single-objective optimization, S summarizes the conditions under which the optimization occurs. For instance, it limits cost or sets a maximum for the number of particular items produced. The feasible region may be formulated explicitly in terms of (linear or nonlinear) constraints $s_1, \ldots, s_m$ as

$$s_1(\mathbf{x}) \leq b_1$$
$$s_2(\mathbf{x}) \leq b_2$$
$$\vdots \qquad\qquad\qquad\qquad (5.32)$$
$$s_m(\mathbf{x}) \leq b_m.$$

The desired, best-of-all-worlds outcome of the optimization has been called the *bliss point* (Rustem 1983). This point maximizes all objective functions and satisfies all constraints. If such a point exists, it means bliss for the decision maker: the work is done; the result cannot be improved. In the vast majority of cases, this point does not exist. It may not be possible simultaneously to maximize all objectives, or, if that is indeed possible, the solution may violate one or more constraints. If the bliss point is not feasible, the objective becomes to find the *best feasible alternative*. Such an alternative is called an *efficient solution*. There may be several such solutions, and these form an *efficient set*. Each of these efficient points (solutions) is admissible (i.e., it satisfies the constraints) and *nondominated* in the following sense: A solution $\bar{\mathbf{x}} \in S$ is called *efficient* or *Pareto optimal*, if and only if no other $\mathbf{x} \in S$ leads to a better outcome. In other words, $\bar{\mathbf{x}}$ is one of the relatively best results.

Before we go into any detail, it is useful to contrast some of the peculiarities accompanying multiple criteria optimization with single-objective optimization problems, as we discussed them previously, and, in particular, with the standard linear program (LP) that has a single linear objective and exclusively linear constraints.

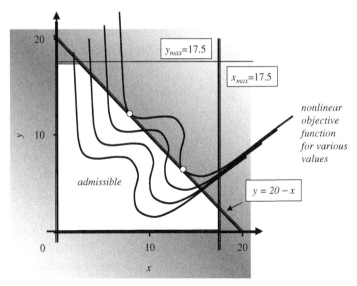

Figure 5.6. A multiple criteria optimization may have distinct optimal solutions that are not connected by other optimal solutions. In this example, there are two optimal solutions, which are indicated by white circles.

1. *An optimal solution may be achieved by distinct criterion vectors.* The optimal solution of an LP is uniquely characterized by one criterion vector. In MCO, distinct criterion vectors may yield equally desirable solutions.
2. *The efficient set may be nonconvex.* In an LP, the set of efficient points is convex, which means that all straight-line connections between efficient points are also efficient points. That is not necessarily the case in MCO.
3. *The optimal point may occur at a nonextreme point.* In two dimensions, for instance, it can occur anywhere within a line segment, but not at the ends of the segment. In three dimensions, the optimal point may lie in the interior of some *face*, which is a part of a plane defined by the constraints.
4. *There may be local optima that are not global.* We found out earlier that single-objective nonlinear optimization tasks may exhibit this feature. In MCO, this may happen even if all objectives and constraints are linear.
5. *The optimal solution set may be finite and disconnected.* In an LP, the optimal solution is either a single point (vertex) or a connected part of a whole line, plane, or hyperplane. Figure 5.6 shows the case of an MCO in two dimensions where two isolated points are equally desirable, whereas all points "in between" are not. In the special case where all constraints and objectives are linear, the efficient set is connected.

A typical case of multiple criteria optimization is the *multiple objective linear program* (MOLP). "Typical" in this context refers to cases solved rather than to cases that could come up in actuality. "Typical" real-world examples more often than not have some nonlinear aspects, but nonlinearities anywhere in the task often create such severe difficulties that few general solution strategies exist. Thus, it is useful to spend time on the more amenable MOLP. It has the form

$$\text{maximize } \{c^1x = z_1\}$$
$$\text{maximize } \{c^2x = z_2\}$$
$$\vdots$$
$$\vdots$$
$$\text{maximize } \{c^kx = z_k\}$$
$$\text{subject to } x \in S.$$

(5.33)

The form of MOLP is a direct extension of the usual LP that we discussed previously, at least, if the constraint equations are linear, that is, if the feasible region S is bounded by straight lines, planes, or hyperplanes. The vector x again contains the quantities to be optimized, and z_i is the ith criterion value. Each vector $c^i (i = 1, \ldots, k)$ is the gradient of the ith objective function. If $k = 1$, MOLP reduces to LP. In streamlined notation, MOLP can be written as

$$\text{"max" } \{Cx = z \,|\, x \in S\},$$

(5.34)

where C is the $k \times n$ *criterion matrix*, which is the matrix of objective function coefficients, whose rows are the gradients c^i. z is the *criterion vector* or *objective function vector*. The word "max" is set in quotation marks, indicating simultaneous maximization of all objectives, which usually is not possible. In reality, maximization here means *maximization of each objective to the greatest extent possible*.

Of course, this is a rather vague goal, which is hard to put into rigorous mathematical terms. Somehow the (relative) importance of each goal must be quantified. This is often – explicitly or implicitly – accomplished with a *utility function* $U(z)$ that specifies how better performance in one task weighs against inferior performance in another competing task. Mathematically speaking, the utility function maps all criterion vectors onto one real number in such a way that a more desirable combination of accomplishments toward the various tasks receives a higher value than a less desirable solution. Although appealing in principle, the utility function idea is hampered by the fact that its functional form is usually nonlinear and seldom known in detail. If it were known, the multiple objective optimization task would read in streamlined form

$$\max\{U(z) \,|\, z = Cx, x \in S\}.$$

(5.35)

One difficulty in determining a utility function is the challenge of comparing apples and oranges. Is a 0.5% increase in profit worth laying off 200 workers? Should investments go toward product improvement or marketing? Is the risk of one fatality in two hundred years acceptable if a new power plant provides electricity for a whole city?

A special case that is much easier to handle than the general utility function concept is the *weighted-sums* approach, which assigns a weight λ_i to each objective. Typically, these weights are chosen between zero and one such that they sum to one: $\sum \lambda_i = 1$. If the weights are collected in a column vector λ, their inclusion in MOLP is accomplished by multiplying the objectives in matrix form with the row vector λ^T. The result is a weighted-sums MOLP of the form

$$\max\{\lambda^T \mathbf{Cx} \mid \mathbf{x} \in S\}. \tag{5.36}$$

Because of the linear nature of including weights, the weighted-sums method deals with strictly convex combinations of objectives.

It will not come as a surprise that the choice of weights very heavily influences which point $\mathbf{x}$ is deemed optimal. Not much more can be said in general about utility functions that can assume any nonlinear form. Replacing the utility function with a linear combination of objectives reduces MOLP to a single-criterion LP, which can be solved with standard methods, as discussed previously. The only requirement is a good weight vector, which, however, is usually difficult to determine. A slight inconvenience of this formulation is also that the term $\lambda^T \mathbf{Cx}$ has no simple interpretation. The corresponding terms $\mathbf{c}^i \mathbf{x}$ in the LP were readily interpreted in terms of the relative contribution of a component x_j on the objective value. By contrast, $\lambda^T \mathbf{Cx}$ is some composite of all criterion functions, which Steuer calls the *figure of merit*. The higher this figure of merit, the more desirable is the solution.

One may ask why is it so difficult to estimate the optimal weighting vector. Problems include the following.

1. The optimum depends on the decision-maker's preference, which again reflects the issue of having to compare actually incomparable goals. The decision maker just may not be willing to weigh two criteria against each other, like profit and employment level.
2. The optimum depends on the relative lengths of the gradients of the objective functions and on the geometry of the feasible region.
3. Correlations between objectives may lead to strange results. For instance, a very important criterion may be assigned a very low weight, if the criterion is correlated with another criterion that may not be all that important, but in the solution happens to have a large weight.
4. The optimal point may not be extreme, but lies on a boundary. The LP does not consider this point as a candidate and is therefore not able to find it.

Some strategies have been developed for dealing with some of these difficulties. Not much can be done with objectivity when two criteria are truly incomparable. By contrast, the second issue can often be ameliorated with appropriate scaling of the gradients. For instance, if one objective is $1,200\,x_1 + 35,000\,x_2 - 6,400\,x_3$ and the other objective is $0.02\,x_1 - 0.003\,x_2 + 0.1\,x_3$, one may divide the first objective function by 1,000 and multiply the second function with one hundred. The optimal points for these functions per se are not affected, but the relative weights given to the two functions are better balanced.

Rustem (1983) suggests an iterative approach involving interactions between the decision maker and a computer algorithm. Rustem's strategy is based on a *weighting matrix* that characterizes the distance between some current solution $\mathbf{x}$ and the bliss point. The weighting matrix specifies the relative importance of each component of x in approaching its desired value. The diagonal elements penalize the current deviation from the desired value, whereas the remaining elements characterize the relative importance of the deviation in one variable versus the deviation in another variable. At each iteration, the weighting matrix is changed, and the effects on the current solution and its distance from the bliss point are recorded. The decision maker obtains an optimal solution update as well as the changes in shadow prices that occurred since the last iteration. If these changes are unacceptable – for whatever reason – the decision maker may intervene by moderating some of the requirements on the solution.

Another strategy is *trapping the solution*. This method is somewhat ad hoc and requires insights in the subject area. It also is not quite straightforward but requires an interactive mode of operation, in which one explores different possibilities to find a series of candidate solutions, which serve as a starting point for refined searches. The trapping is accomplished in the following way.

One picks one objective function (that is of particular interest) and replaces all others with lower bounds. For instance, instead of trying to maximize simultaneously profit, market share, and employment under a set of constraints, one selects profit as the overriding objective and specifies minimum levels for market share and employment that must be achieved and should preferably be exceeded. The method is very intuitive but requires a lot of interaction between the decision maker and the algorithm. By identifying different objectives as prime tasks and by using different sets of lower bounds, the decision maker develops a series of candidate solutions, which either provide an acceptable solution or suggest further optimization.

Vector Optimization

A MOLP can be considered a vector-maximization problem in the sense that the desired solution "maximizes" the criterion vector. In the usual case where the bliss point is not feasible, the goal of the optimization is the characterization of the set of all efficient points. Steuer formulates a MOLP in vector-maximum format as

$$\text{eff}\{Cx = z \mid x \in S\}. \tag{5.37}$$

The terminology eff indicates that the set of all solutions to the problem is the efficient set.

Algorithms for vector optimization may be classified in two ways (Hartley 1983). The first classification consists of two classes. The first of these contains algorithms that attempt to reduce the vector problem to a parametric family of scalar-value problems, whereas the second class contains approaches that generalize scalar algorithms to vector algorithms. More specifically, the algorithms in the first class use a characterization of efficiency to transform the vector problem into a family of single-objective problems, which are specified in terms of a number of parameters. The algorithms then couple single objective techniques with parametric methods to find all solutions of this family. A standard method for the second class of algorithms is dynamic programming.

The second classification distinguishes algorithms by their characterization of efficient sets. In algorithms of the first class, each iteration generates a subset of the efficient set. The union of these subsets is the efficient set. In the second class, each iteration generates a set that includes the efficient set as a subset. The intersection of these sets is the efficient set. A third class contains algorithms that combine features from the first two or use other ad hoc methods.

Even for the (relatively) simplest case of a MOLP, vector maximization is computationally very expensive. For realistically large systems, the number of efficient extreme points, which are candidates for the best solution, may be so large that it is impossible for the decision maker to evaluate the results at each iteration with reasonable effort. To some degree, this problem can be alleviated with filtering methods that select efficient points according to specified criteria. A further complication is that the decision maker's preferred solution may lie somewhere in the interior of a large efficient facet. Because many standard methods focus exclusively on vertices, additional methods are needed to deal with this situation. Such methods tend to be fairly involved.

Goal Programming

Instead of optimizing simultaneous objectives, goal programming (GP) establishes a *goal level* g_i of achievement for each objective. These goal levels, also called *targets* or *thresholds*, are "soft" constraints that are included in the definition of the feasible region. Some differences between GP and a linear program (LP) are the following:

1. The objectives are conceptualized as goals that are to be achieved as closely as possible.
2. Variables are defined that measure the deviation from the goal level in terms of overachievement (d_i^+) or underachievement (d_i^-).
3. Each goal is assigned a priority in the form of a *penalty weight* w_i^+ or w_i^- for nonachievement. The idea here is that one might rather accept overachievement by 10% than underachievement by 10%.
4. The penalty-weighted sum of deviations from the goal levels is minimized to find the solution that most closely satisfies the goals.

Two strategies of goal programming are the *Archimedian Model* and the *Preemptive Model*. The difference between the two is the way that deviations from the goal level are evaluated. In both models, solutions are generated whose criterion vectors are closest to the "utopian" set, which, analogous to the bliss point, satisfies all goals. In the Archimedian Model, closeness is measured in terms of a weighted L_1 metric, which is a simple (weighted) sum of absolute differences. This model can be analyzed with standard LP methods. In the Preemptive Model, goals are grouped according to priorities and required to be satisfied sequentially from the top to the bottom of the list. This *lexicographic* measure of closeness ensures that the most important goals are satisfied, no matter what happens to the remaining goals. Sometimes, the most efficient GP strategy is an interactive combination of the two models.

An example of an Archimedian goal program might read

minimize$\{w_1^+ d_1^+ + w_1^- d_1^- + w_2^+ d_2^+ + w_3^- d_3^- + w_4^+ d_4^+ + w_4^- d_4^-\}$
subject to

$$
\begin{array}{lll}
\mathbf{c}^1 \mathbf{x} - d_1^+ + d_1^- & = g_1 \\
\mathbf{c}^2 \mathbf{x} \qquad\qquad -d_2^+ & \geq g_2 \\
\mathbf{c}^3 \mathbf{x} \qquad\qquad +d_3^- & \leq g_3 \\
\mathbf{c}^4 \mathbf{x} \qquad\qquad\qquad +d_4^- & \geq g_4^L & \qquad (5.38) \\
\mathbf{c}^4 \mathbf{x} \qquad\qquad\qquad\qquad -d_4^- & \leq g_4^U \\
\qquad\qquad d_i^+, d_i^- & \geq 0 \\
\end{array}
$$
$$\mathbf{x} \in S$$

The deviations for overachievement and underachievement, d_i^+ and d_i^-, as well as the symbols $=$, $\geq$, and $\leq$ indicate the specifics of each goal. Goal 1 should be matched as closely as possible; each over- or underachievement is being penalized. Goal level 2 should be reached, but may also be exceeded. Goal level 3 constitutes a maximum not to be exceeded, but closely approached. Finally, goal 4 should be attained within a certain range, which is defined by the lower and upper limits g_4^L and g_4^U, respectively. Should a deviation from a goal level be immaterial, the corresponding penalty weights w_i^+ and w_i^- are zero, and no deviation variable needs to be formulated.

OPTIMIZATION OF STRUCTURE

As mentioned in the Introduction, numerous articles have addressed explanations of optimal designs found in nature. We will here briefly review a related, yet different aspect, namely the question of whether the structure of a network can be optimized for a particular purpose. Again, the S-system form is particularly well suited for such analyses. The reason is simple. The structure of a model is essentially mapped into differential equations in a one-to-one fashion. Every parameter has a unique specific meaning and, vice versa, every feature of a system is represented essentially by one parameter in the S-system model. For instance, if a variable X_k affects the production of X_i, the production term of X_i includes a power-law factor $X_k^{g_{ik}}$. The kinetic order g_{ik} characterizes the strength of the effect, and if this effect vanishes, so does g_{ik}. No other parameters, with the potential exception of the rate constant α_i and of some functional constraints between parameters, are affected by changes in the effect of X_k on the production of X_i. The one-to-one relationship between system features and model parameters facilitates the comparative analysis of models with and without the particular feature of interest.

In the context of explaining observed natural designs, the unique meaning of parameters has led to the *method of controlled mathematical comparisons* (MCMC; e.g., Savageau 1985; Irvine 1991). A typical example is the assessment of feedback inhibition by product X_n on the synthesis of the lead intermediate X_1. Suppose the synthesis of X_1 is represented by the power-law term $\alpha_1 X_0^{g_{10}} X_j^{g_{1j}} X_k^{g_{1k}} X_n^{g_{1n}}$ where X_0 is the substrate for the process, and X_j and X_k are two modulators. The inhibition is represented by the kinetic order g_{1n}, which is a real number with negative value, indicating inhibition rather than activation. According to MCMC, a system A with inhibition is compared to an otherwise equivalent system B without inhibition in the following manner. First, the latter system B is characterized by $g_{1n} = 0$. Because $X_n^{g_{1n}} = X_n^0 = 1$, this setting reflects that X_n does not affect the synthesis of X_1. Second, the two systems should be equivalent to the outside observer. For instance, their steady states, which are typically observed before the internal structure of the system is known, should be the same and of the value experimentally determined. To enforce this "external equivalence," the term for synthesis of X_1 in systems A and B should have the same numerical value. The difference in g_{1n} is accounted for by adjusting α_1 in system B, such that $\alpha_{1A} X_0^{g_{10}} X_j^{g_{1j}} X_k^{g_{1k}} X_n^{g_{1n}} = \alpha_{1B} X_0^{g_{10}} X_j^{g_{1j}} X_k^{g_{1k}}$. With the adjustments in g_{1n} and α_1, the systems behave exactly the same, as long as X_n is not involved. At the same time, a comparison of their stability, sensitivities, robustness, and dynamic responses reveals the specific and unique effect that the inhibition by X_n has on the pathway. Savageau and collaborators have used this type of analysis extensively for explaining features of metabolic pathways (Savageau 1974a,b, 1975, 1976; Alves and Savageau 2000a,b,c,d), regulation of the immune response (Irvine and Savageau 1985a,b; Irvine 1991), and the regulation of gene expression (Savageau 1976, 1985, 1989; Hlavacek and Savageau 1995, 1996, 1997).

Hatzimanikatis et al. (1996a,b) used the regular structure of S-systems and their unique relationships between phenomenological features and parameters for structure optimization. By keeping key parameters numerically unspecified, they created a *regulatory superstructure* that contained all relevant alternative regulatory structures as special cases. Given this superstructure, they investigated a *synthesis problem* dealing with the following question: "What kind of regulation (activation or inhibition, by which metabolite, and of what strength) should be assigned to each enzyme in the network, and what associated changes should be made in the manipulated parameters (such as enzyme expression levels, environmental conditions, and effectors external to the system), to optimize the performance of the metabolic network?"

In some sense, this type of synthesis problem involves discrete decisions, such as the presence or absence of some modulation. However, in the S-system representation presence and absence are simply coded on a continuous scale for one of the kinetic orders, and because this continuum contains zero, seemingly discrete scenarios of absence of the modulation or a switch from inhibition to activation are naturally embedded in the regulatory superstructure. This enabled Hatzimanikatis et al. (1996a,b) to formulate the synthesis problem as a mixed-integer linear optimization (MILP) task. The objective function consisted of a sum of metabolic inputs or combinations of inputs. Four types of constraints were defined: mass balance equations for all metabolites; bounds on metabolites, manipulated variables, rates, and metabolic outputs; constraints on kinetic orders; and logical constraints for the presence, absence, or magnitude of some of the regulatory loops in the system. In several examples, the authors answered questions about the optimal configuration of networks, such as

1. Which regulatory loops should be active for optimal functioning and which not?
2. Should the type of a particular regulation be an activation or inhibition?
3. Which metabolites should inhibit or activate each reaction?
4. How strong should the regulation be?
5. Which of the manipulated parameters should be altered?
6. How much should the manipulated parameters be altered?
7. Which enzymes should be overexpressed and by how much?
8. What is the optimal combination of regulatory structures?

The approach does not account for dynamical features, such as transition characteristics, but permits the collation of a sequence of alternative regulatory structures, which in some sense are optimal and whose dynamics can be tested in separate simulation analyses. By formulating additional constraints that exclude former solutions, the method leads to a hierarchy of models with respect to the value of the objective function. These can be tested with respect to stability, robustness, and dynamics, and the solution deemed best is then chosen for experimental implementation.

The method of Hatzimanikatis and collaborators is critically dependent on the linearity of the steady-state equations, which is the result of formulating the metabolic network in S-system form. However, using arguments similar to those for the indirect optimization method (Torres et al. 1997), the authors suggest: "...every kinetic description of metabolic systems can be transformed into an approximate S-system representation. Then this MILP formulation can be applied to the transformed systems to suggest promising strategies for achieving a metabolic engineering objective" (Hatzimanikatis et al. 1996a).

REFERENCES

Alves, R., and M. Savageau: Comparing systemic properties of ensembles of biological networks by graphical and statistical methods. *Bioinformatics* 16, 527–33, 2000a.

Alves, R., and M. Savageau: Systemic properties of ensembles of metabolic networks: Application of graphical and statistical methods to simple unbranched pathways. *Bioinformatics* 16, 534–47, 2000b.

Alves, R., and M. Savageau: Extending the method of mathematically controlled comparison to include numerical comparisons. *Bioinformatics* 16, 786–98, 2000c.

Alves, R., and M. Savageau: Effect of overall feedback inhibition in unbranched biosynthetic pathways. *Biophys. J.* 79, 2290–304, 2000d.

Archer, D.B., D.A. MacKenzie, and D.J. Jeenes: Genetic engineering: Yeasts and filamentous fungi. In: C. Ratledge and B. Kristiansen (Eds.), *Basic Biotechnology* (Chapter 5). Cambridge University Press, Cambridge, U.K., 2001.

Arkun, Y., and G. Stephanopoulos: Optimizing control of industrial chemical processes: State of the art review. *Proc. Joint Autom. Contr. Conf.*, WP5–A, San Francisco, CA, 1980.

Bailey, J.: Toward a science of metabolic engineering. *Science* 252, 1668–74, 1991.

Bailey, J.: Mathematical modeling and analysis in biochemical engineering: Past accomplishment and future opportunities. *Biotechnol. Prog.* 14(1), 8–20, 1998.

Bertsekas, D.P.: *Network Optimization: Continuous and Discrete Models*. Athena Scientific, Belmont, MA, 1998.

Brown, G.C., R.P. Hafner, and M.D. Brand: A "top-down" approach to the determination of control coefficients in metabolic control theory. *Eur. J. Biochem.* 188, 321–5, 1990.

Candler, W., and R. Norton: Multilevel programming and development policy. World Bank Staff Working Paper No. 258, *IBRD*, Washington, D.C., 1977.

Candler, W., and R. Townsley: A linear two-level programming problem. *Comp. Ops. Res.* 9, 59–76, 1982.

Charnes, A., and W.W. Cooper: *Management Models and Industrial Applications of Linear Programming* (Vols. I and II). John Wiley & Sons, New York, 1961.

Clark, P.A.: Bilevel programming for steady-state chemical process design – II. Performance study for nondegenerate problems. *Comp. Chem. Eng.* 14, 99–109, 1990.

Clark, P.A., and A.W. Westerberg: Bilevel programming for steady-state chemical process design – I. Fundamentals and algorithms. *Comp. Chem. Eng.* 14, 87–97, 1990.

Cochrane, J.L., and M. Zeleny (Eds.): *Multiple Criteria Decision Making*. University of South Carolina Press, Columbia, SC, 1973.

Conejeros, R., and V. S. Vassiliadis: Analysis and optimization of biochemical process reaction pathways. 1. Pathway sensitivities and identification of limiting steps. *Ind. Eng. Chem. Res.* 37, 4699–708, 1998a.

Conejeros, R., and V.S. Vassiliadis: Analysis and optimization of biochemical process reaction pathways. 2. Optimal selection of reaction steps for modification. *Ind. Eng. Chem. Res.* 37, 4709–14, 1998b.

Conejeros, R., and V.S. Vassiliadis: Dynamic biochemical reaction process analysis and pathway modification predictions. *Biotechnol. Bioeng.* 68(3), 285–97, 2000.

Dantzig, G.B.: *Linear Programming and Extensions*. Princeton University Press, Princeton, NJ, 1963.

Duncan, T.M., and J.A. Reimer: *Chemical Engineering Design and Analysis: An Introduction*. Cambridge University Press, Cambridge, U.K., 1998.

Ebenhöh, O., and R. Heinrich: Evolutionary optimization of metabolic pathways. Theoretical reconstruction of the stoichiometry of ATP and NADH producing systems. *Bull. Math. Biol.* 63, 21–55, 2001.

Fell, D.A.: Metabolic control analysis – a survey of its theoretical and experimental development. *Biochem. J.* 286, 313–30, 1992.

Fell, D.A., and J.R. Small: Fat synthesis in adipose tissue. An examination of stoichiometric constraints. *Biochem. J.* 238, 781–6, 1986.

French, S., R. Hartley, L.C. Thomas, and D.J. White (Eds.): *Multi-Objective Decision Making*. Academic Press, London, 1983.

Garcia, C.E., and M. Morari: Optimal operation of integrated processing systems. *AIChE J.* 27, 960–8, 1981.

Garfinkel, R.S., and G.L. Nemhauser: *Integer Programming*. John Wiley & Sons, New York, 1972.

Gass, S.I.: *Linear Programming: Methods and Applications*. McGraw-Hill, New York, 1985.

Gavalas, G.R.: *Nonlinear Differential Equations of Chemically Reacting Systems*. Springer-Verlag, Berlin, 1968.

Glansdorff, P., and I. Prigogine: *Thermodynamic Theory of Structure, Stability, and Fluctuations*. Wiley-Interscience, London, UK, 1971.

Górak, A., A. Krasławski, and A. Vogelpohl: Simulation und optimierung der mehrstoff-rektifikation. *Chem. Ing. Tech.* 59, 95–106, 1987.

Groetsch, C.W., and J.T. King: *Matrix Methods and Applications*. Prentice-Hall, Englewood Cliffs, NJ, 1988.

Hadley, G.: *Linear Programming*. Addison-Wesley, Reading, MA, 1962.

Harmon, J., S.A. Svoronos, and G. Lyberatos: Adaptive steady-state optimization of biomass productivity in continuous fermentors. *Biotechnol. Bioeng.* 30, 335–44, 1987.

Hartley, R.: Survey of algorithms for vector optimisation problems. In: S. French, R. Hatley, L.C. Thomas, and D.J. White (Eds.), *Multi-Objective Decision Making* (pp. 1–34). Academic Press, London, 1983.

Harwood, C.R., and A. Wipat: Genome management and analysis: Prokaryotes. In: C. Ratledge and B. Kristiansen (Eds.), *Basic Biotechnology* (Chapter 4). Cambridge University Press, Cambridge, U.K., 2001.

Hatzimanikatis, V., and J.E. Bailey: MCA has more to say. *J. Theor. Biol.* 182, 233–42, 1996.

Hatzimanikatis, V., C.A. Floudas, and J.E. Bailey: Optimization of regulatory architectures in metabolic reaction networks. *Biotechnol. Bioeng.* 52(4), 485–500, 1996a.

Hatzimanikatis, V., C.A. Floudas, and J.E. Bailey: Analysis and design of metabolic reaction networks via mixed-integer linear optimization. *AIChE J.* 42(5), 1277–92, 1996b.

Heinrich, R., and S. Schuster: *The Regulation of Cellular Systems*. Chapman and Hall, New York, 1996.

Heinrich, R., S.M. Rapoport, and T.A. Rapoport: Metabolic regulation and mathematical models. *Prog. Biophys. Mol. Biol.* 32, 1–82, 1977.

Hlavacek, W.S., and M.A. Savageau: Subunit structure of regulator proteins influences the design of gene circuitry: Analysis of perfectly coupled and completely uncoupled circuits. *J. Mol. Biol.* 248, 739–55, 1995.

Hlavacek, W.S., and M.A. Savageau: Rules for coupled expression of regulator and effector genes in inducible circuits. *J. Mol. Biol.* 255(1), 121–39, 1996.

Hlavacek, W.S., and M.A. Savageau: Completely uncoupled and perfectly coupled gene expression in repressible systems. *J. Mol. Biol.* 266, 538–58, 1997.

Horn, F., and R. Jackson: General mass action kinetics. *Arch. Rational Mech. Anal.* 47, 81–116, 1972.

Irvine, D. H.: The method of controlled mathematical comparison. In: E. O. Voit (Ed.), *Canonical Nonlinear Modeling. S-System Approach to Understanding Complexity* (Chapter 7). Van Nostrand Reinhold, New York, 1991.

Irvine, D.H., and M.A. Savageau: Network regulation of the immune response: Alternative control points for suppressor modulation of effector lymphocytes. *J. Immunol.* 134, 2100–16, 1985a.

Irvine, D.H., and M.A. Savageau: Network regulation of the immune response: Modulation of suppressor lymphocytes by alternative signals including contra-suppression. *J. Immunol.* 134, 2117–30, 1985b.

Kacser, H., and L. Acerenza: A universal method for increase in metabolite production. *Eur. J. Biochem.* 216, 361–7, 1993.

Koffas M., C. Roberge, K. Lee, and G. Stephanopoulos: Metabolic engineering. *Ann. Rev. Biomed. Eng.* 1, 535–57, 1999.

Kolman, B. Introductory *Linear Algebra with Applications* (6th ed.). Prentice Hall, Upper Saddle River, NJ, 1997.

Kreyszig, E.: *Advanced Engineering Mathematics* (7th ed.). John Wiley & Sons, New York, 1993.

Liao, J.C., and J. Delgado: Advances in metabolic control analysis. *Biotechnol. Prog.* 9, 221–33, 1993.

Lübbert, A., and R. Simutis: Measurement and control. In: C. Ratledge and B. Kristiansen (Eds.), *Basic Biotechnology* (Chapter 10). Cambridge University Press, Cambridge, U.K. 2001.

Luenberger, D.G.: *Linear and Nonlinear Programming*. Addison-Wesley, Reading, MA, 1984.

Majewski, R.A., and M.M. Domach: Simple constrained-optimization view of acetate overflow in *E. coli. Biotechnol. Bioeng.* 35, 732–8, 1990.

Marín-Sanguino, A., and N.V. Torres: Optimization of tryptophan production in bacteria. Design of a strategy for genetic manipulation of the tryptophan operon for tryptophan flux maximization. *Biotechnol. Prog.* 16(2), 133–45, 2000.

Mavrovouniotis, M.L., G. Stephanopoulos, and G. Stephanopoulos: Computer-aided synthesis of biochemical pathways. *Biotechnol. Bioeng.* 36, 1119–32, 1990.

Meléndez-Hevia, E.: The game of the pentose phosphate cycle: A mathematical approach to study the optimization in design of metabolic pathways during evolution. *Biomed. Biochim. Acta* 49, 903–16, 1990.

Meléndez-Hevia, E., and A. Isidoro: The game of the pentose phosphate cycle. *J. Theor. Biol.* 117, 251–63, 1985.

Meléndez-Hevia, E., and N.V. Torres: Economy of design in metabolic pathways: Further remarks on the game of the pentose phosphate cycle. *J. Theor. Biol.* 132, 97–111, 1988.

Meléndez-Hevia, E., T.G. Waddell, and F. Montero: Optimization of metabolism: The evolution of metabolic pathways toward simplicity through the game of the pentose phosphate cycle. *J. Theor. Biol.* 166, 201–19, 1994.

Meléndez-Hevia, E., T.G. Waddell, and M. Cascante: The puzzle of the Krebs cycle: Assembling the pieces of chemically featusible reactions and opportunism in design of metabolic pathways during evolution. *J. Mol. Evol.* 43, 293–303, 1996.

Meléndez-Hevia, E., T.G. Waddell, R. Heinrich, and F. Montero: Theoretical approaches to the evolutionary optimization of glycolysis. *Eur. J. Biochem.* 244, 527–43, 1997.

Mittenthal, J.E., A. Yuan, B. Clarke, and A. Scheeline: Designing metabolism; alternative connectivities for the pentose-phosphate pathway. *Bull. Math. Biol.* 60, 815–56, 1998.

Mittenthal, J.E., B. Clarke, T.G. Waddell, and G. Fawcett: A new method for assembling metabolic networks, with application to the Krebs citric acid cycle. *J. Theor. Biol.* 208(3), 361–82, 2001.

Oldenburger, R.: *Optimal Control.* Holt, Rinehart and Winston, New York, 1966.

Papoutsakis, E.T.: Equations and calculations for fermentations of butyric acid bacteria. *Biotechnol. Bioeng.* 26, 174–87, 1984.

Papoutsakis, E.T., and C.L. Meyer: Equations and calculations of product yields and preferred pathways for butanediol and mixed-acid fermentations. *Biotechnol. Bioeng.* 27, 50–66, 1985a.

Papoutsakis, E.T., and C.L. Meyer: Fermentation equations for propionic-acid bacteria and production of assorted oxychemicals from various sugars. *Biotechnol. Bioeng.* 27, 67–80, 1985b.

Pareto, V.: *Cours d'Econome Politique.* Rouge, Lausanne, 1896.

Petersen, J.N., and G.A. Whyatt: Dynamic on-line optimization of a bioreactor. *Biotechnol. Bioeng.* 35, 712–18, 1990.

Pissara, P.N., J. Nielsen, and M.J. Bazin: Pathway kinetics and metabolic control analysis of a high-yielding strain of *Penicillium chrysogenum* during fed-batch cultivations. *Biotechnol. Bioeng.* 51, 168–76, 1996.

Quant, P.A.: Experimental application of top-down analysis to metabolic systems. *Trends Biochem. Sci.* 18, 26–30, 1993.

Regan, L., I.D.L. Bogle, and P. Dunhill: Simulation and optimization of metabolic pathways. *Comp. Chem. Eng.* 17(5/6), 627–37, 1993.

Rodríguez-Acosta, F., C.M. Regalado, and N.V. Torres: Non-linear optimization of biotechnological processes by stochastic algorithms: Application to the maximization of the production rate of ethanol, glycerol and carbohydrates by *Saccharomyces cerevisiae. J. Biotechnol.* 68, 15–28, 1999.

Rolf, M.J., and H.C. Lim: Experimental adaptive on-line optimization of cellular productivity of a continuous bakers' yeast culture. *Biotechnol. Bioeng.* 27, 1236–45, 1985.

Rustem, B.: Multiple objectives with convex constraints. In: S. French, R. Hartley, L.C. Thomas, and D.J. White (Eds.), *Multi-Objective Decision Making* (pp. 127–43). Academic Press, London, 1983.

Savageau, M.A.: Biochemical systems analysis, II. The steady-state solutions for an n-pool system using a power-law approximation. *J. Theor. Biol.* 25, 370–9, 1969.

Savageau, M.A.: Concepts relating behaviour of biochemical systems to their underlying molecular properties. *Arch. Biochem. Biophys.* 145, 612–21, 1971a.

Savageau, M.A.: Parameter sensitivity as a criterion for evaluating and comparing the performance of biochemical systems. *Nature* 229 (5286), 542–4, 1971b.

Savageau, M.A.: The behavior of intact biochemical control systems. *Curr. Top. Cell Reg.* 6, 63–130, 1972.

Savageau, M.A.: Comparison of classical and autogenous systems of regulation in inducible operons. *Nature (London)* 252, 546–9, 1974a.

Savageau, M.A.: Genetic regulatory mechanisms and the ecological niche of *Escherichia coli*. *Proc. Natl. Acad. Sci. USA* 71, 2354–455, 1974b.

Savageau, M.A.: Optimal design of feedback control by inhibition: Dynamic considerations. *J. Mol. Evol.* 5, 199–222, 1975.

Savageau, M.A.: *Biochemical Systems Analysis. A Study of Function and Design in Molecular Biology.* Addison-Wesley, Reading, MA, 1976.

Savageau, M.A.: A theory of alternative designs for biochemical control systems. *Biomed. Biochim. Acta* 44, 875–80, 1985.

Savageau, M.A.: Are there rules governing patterns of regulation? In: B.C. Goodwin and P.T. Saunders (Eds.), *Theoretical Biology – Epigenetic and Evolutionary Order* (pp. 42–66). Edinburgh University Press, Edinburgh, U.K., 1989.

Savinell, J.M.: *Analysis of Stoichiometry in Metabolic Networks.* Ph.D. dissertation, University of Michigan, 1991.

Savinell, J.M., and B.Ø. Palsson: Network analysis of intermediary metabolism using linear optimization. I. Development of mathematical formalism. *J. Theor. Biol.* 154(4), 421–54, 1992a.

Savinell, J.M., and B.Ø. Palsson: Network analysis of intermediary metabolism using linear optimization. II. Interpretation of *hybridoma* cell metabolism. *J. Theor. Biol.* 154(4), 455–73, 1992b.

Savinell, J.M., and B.Ø. Palsson: Optimal selection of metabolic fluxes for *in vivo* measurement. I. Development of mathematical methods. *J. Theor. Biol.* 155(2), 201–14, 1992c.

Savinell, J.M., and B.Ø. Palsson: Optimal selection of metabolic fluxes for *in vivo* measurement. II. Application to *Escherichia coli* and *hybridoma* cell metabolism. *J. Theor. Biol.* 155(2), 215–42, 1992d.

Semones, G.B., and H.C. Lim: Experimental multivariable adaptive optimization of the steady-state cellular productivity of continuous baker's yeast culture. *Biotechnol. Bioeng.* 33, 16–25, 1989.

Seressiotis, A., and J.E. Bailey: MPS: An artificially intelligent software system for the analysis and synthesis of metabolic pathways. *Biotechnol. Bioeng.* 31, 587–602, 1988.

Simpson, T.W., G.E. Colon, and G. Stephanopoulos: Two paradigms of metabolic engineering applied to amino acid biosynthesis. *Biochem. Soc. Trans.* 23(2), 381–7, 1995.

Simpson, T.W., B.D. Follstad, and G. Stephanopoulos: Analysis of the pathway structure of metabolic networks. *J. Biotechnol.* 71, 207–23, 1999.

Stephanopoulos, G.: Metabolic engineering. *Biotechnol. Bioeng.* 58(2–3), 119–20, 1998.

Stephanopoulos, G., and T.W. Simpson: Flux amplification in complex metabolic networks. *Chem. Eng. Sci.* 52(15), 2607–27, 1997.

Stephanopoulos, G.N., A.A. Aristidou, and J. Nielsen: *Metabolic Engineering: Principles and Methodologies.* Academic Press, San Diego, CA, 1998.

Steuer, R.E.: *Multiple Criteria Optimization: Theory, Computation, and Application.* John Wiley & Sons, New York, 1986.

Strang, G.: *Introduction to Applied Mathematics.* Wellesley-Cambridge Press, Wellesley, MA, 1986.

Torres, N.V., E.O. Voit, and C.H. Alcón: Optimization of nonlinear biotechnological processes with linear programming. Application to citric acid production in *Aspergillus niger. Biotechnol. Bioeng.* 49, 247–58, 1996.

Torres, N.V., E.O. Voit, C. Glez-Alcón, and F. Rodriguez: An indirect optimization method for biochemical systems: Description of method and application to the maximization of the rate of ethanol, glycerol, and carbohydrate production in *Saccharomyces cerevisiae. Biotechnol. Bioeng.* 55(5), 758–72, 1997.

Torres, N.V., E.O. Voit, C. Glez-Alcón, and F. Rodríguez: A novel approach to design of overexpression strategy for metabolic engineering. Application to the carbohydrate metabolism in the citric acid producing mould *Aspergillus niger. Food Technol. Biotechnol.* 36(3), 177–84, 1998.

Vagners, J.: Optimization techniques. In: C.E. Pearson (Ed.), *Handbook of Applied Mathematics. Selected Results and Methods* (2nd ed., pp. 1140–216). Van Nostrand Reinholt Company, New York, 1983.

Voit, E.O. (Ed.): *Canonical Nonlinear Modeling. S-System Approach to Understanding Complexity* (xi + 365 pp.). Van Nostrand Reinhold, New York, 1991.

Voit, E.O.: Optimization in integrated biochemical systems. *Biotechnol. Bioeng.* 40, 572–82, 1992.

Voit, E.O.: *Computational Analysis of Biochemical Systems. A Practical Guide for Biochemists and Molecular Biologists* (xii + 532 pp.). Cambridge University Press, Cambridge, U.K., 2000.

Voit, E.O., and M. Del Signore: Assessment of effects of experimental imprecision on optimized biochemical systems. *Biotechnol. Bioeng.* 74(5), 443–8, 2001.

CHAPTER SIX

Optimization of Citric Acid Production in *Aspergillus niger*

INTRODUCTION

Optimization of yield and productivity has been a major goal since the very beginnings of industrial citric acid production. The first significant step in this regard is to be credited to Currie (1916, 1917), who observed that a large number of black *Aspergillus* strains were very active in the production of citric acid. With his extensive studies of culture media for the mold and his advances toward competitive production of citric acid through fermentation, Currie may also be considered the originator of efforts toward optimizing the biotechnological production process. In fact, Currie's success allowed the United States to become self-sufficient and even to export citric acid to Europe. Since Currie's seminal work, probably a thousand or more scientific studies have been published and several hundred patents issued.

Today, we are at a threshold of further significant improvements. The widespread accessibility of recombinant DNA technology and the possibility of altering phenotypes in specific directions and magnitudes are creating novel and powerful means of microbial manipulation (Archer, MacKenzie, and Jeenes 2001; Harwood and Wipat 2001). Combined with effective mathematical methods, these experimental tools are poised to propel microbial biotechnology to a new level of success. Mathematical guidance is necessary, because it is often no longer sufficient to overexpress the gene of one "rate-limiting" step. Instead, several molecular changes have to be introduced simultaneously and sometimes in very specific proportions, if any real improvements in yield or productivity are to be obtained.

A good example of this emerging situation was recently debated quite emphatically in the literature (Ratlegde 2000; Ruijter 2000). Ruijter, Panneman, and Visser (1997, 1998; Ruijter et al. 2000) had studied the effects of independently overproducing both glycolytic and mitochondrial *A. niger* enzymes. In the first study, the two key glycolytic enzymes phosphofructokinase and pyruvate kinase were overexpressed. Against their expectation, it turned out that moderate overexpression (three- to five-fold the wild-type level), in either one enzyme or simultaneously in both, did not increase citric acid production. More recently, the same group found that overproduction of citrate syntase (up to eleven-fold) did not increase the rate of citric acid production by the fungus either. Even though all three enzymes are known to be crucial for citric acid production, the overall effect on yield was found to be negligible, and the topic of discussion in the literature was whether the negative results could have been foreseen.

Similarly, Stephanopoulos and collaborators (Stephanopoulos 2001) attempted to activate the alleged key step in the production of lysine in *Corynebacterium glutamicum*. To the authors' disappointment, activation of just the crucial conversion of pyruvate into oxalacetate did not improve flux. Increasing the "second most important step," degradation of oxalacetate, alone did not change yield either. Simultaneous activation of both steps led to significant enhancements of flux.

A general discussion and explanation of these and similar phenomena is the topic of this chapter. We will use the comprehensive S-system model of citric acid production in *A. niger* during idiophase, which we developed in Chapter 3, and subject it to some of the optimization methods introduced in Chapter 5. The optimization results will show that alterations in one or a few steps cannot necessarily be expected to result in significant improvements in yield or productivity. In fact, the specific results will indicate that quite a few enzymes and transport steps need to be altered in prescribed magnitudes to achieve a worthwhile rerouting of material toward citric acid production. To what degree of accuracy this type of specific manipulation is experimentally possible is presently an unanswered question. The results are nevertheless important, because they make it clear that "simple" alterations in the metabolic structure of a pathway have not much of a chance of succeeding, especially if the pathway had been the subject of extensive experimental optimization, for instance, with traditional mutagenesis and selection.

During idiophase growth, citric acid accumulation is the main metabolic activity in *A. niger*. The S-system model of the process (Chapter 3) was parameterized with data from the literature, and steady-state and dynamic analyses supported its validity. Furthermore, the model was shown to be fairly robust, as measured by gain and sensitivity coefficients with respect to independent variables, rate constants, and kinetic orders. The steady-state results suggested that the range of accuracy of the S-system representation should be wide enough to model realistic deviations from the nominal steady state. The dynamic analysis indicated reasonable response times, thereby providing further support for the model structure and its numerical implementation.

The successful assessments of model quality and reliability allow us now to address questions of pathway optimization with respect to increased citrate production. With this step, we are leaving the realm of pure biochemistry and enter the modern field of metabolic engineering. Thus, our goal in this chapter is the following: prescribe how some or all steps of citric acid metabolism should be altered such that the rate of citric acid excretion to the medium during idiophase is increased as much as possible. The specific alterations (probably to be implemented with recombinant DNA techniques) are expected to result in a targeted redirection of some or all metabolic fluxes and an amplification of those fluxes that ultimately lead to citric acid.

The overall strategy is to define a linear program with citric acid production as the objective function. The objective function is to be optimized under steady-state conditions and additional constraints that are dictated by the biological idiosyncrasies of metabolic systems in vivo and define boundary conditions that guarantee cell viability. Some of the constraints prevent intermediate and/or enzyme concentrations from overburdening the biochemical balance of the organism, whereas others limit the magnitudes of those fluxes that strongly affect other parts of the cellular metabolism, which in turn are necessary for overall proper functioning. The control variables, whose values are to be optimized, are enzyme activities and magnitudes of transport steps, which can be accessed and manipulated experimentally, at least in principle. Expressed differently, we will assume that the pathway structure is fixed, but that we are able to amplify the rates of some of the reactions leading to the citric acid excretion.

We will execute the optimization of the entire pathway model and also show how one can systematically identify smaller numbers of alterations in enzymes and transport steps that still lead to some improvements in yield. Furthermore, we will discuss the consequences of imprecise control over the control variables during the experimental implementation. We will not pursue other strategies of metabolic engineering, such as the introduction of new biochemical reactions or pathways, that are possible with methods of transfection or transformation (Koffas et al. 1999).

A LINEAR PROGRAM FOR OPTIMIZED CITRIC ACID PRODUCTION IN *ASPERGILLUS NIGER*

The key advantage of formulating the biochemical pathway as an S-system model is the fact that the steady state in this representation is characterized by a system of linear algebraic equations, even though the dynamic model itself is nonlinear (Savageau 1969; see also Chapter 2). As the steady-state equations, typical objective functions, and constraints on metabolites, enzyme activities and fluxes can be formulated as linear equations or inequalities, such that the entire optimization problem reduces to one of straightforward linear programming (Voit 1992). Thus, even though the underlying dynamical model is highly nonlinear, optimization of yields and fluxes can be executed with algorithms that are readily available from the field of operations research (see Chapter 5).

The system of interest is the amended model of citric acid metabolism in *A. niger* during idiophase (Figure 6.1; Eqs. 3.5, 3.6, and 3.7). The objective function of the linear program consists of the elimination flux of the citric acid pool; this function is to be maximized. There are three types of constraints, which any optimized solution must satisfy. First, the system should operate under steady-state conditions, which are given by a set of linear equations in the logarithms of the original variables: $y_i = \ln(X_i)$. Second, the system contains two types of variables: dependent metabolite concentrations and independent control variables, which are usually enzyme activities or magnitudes of transport steps. Both types of variables should remain within certain

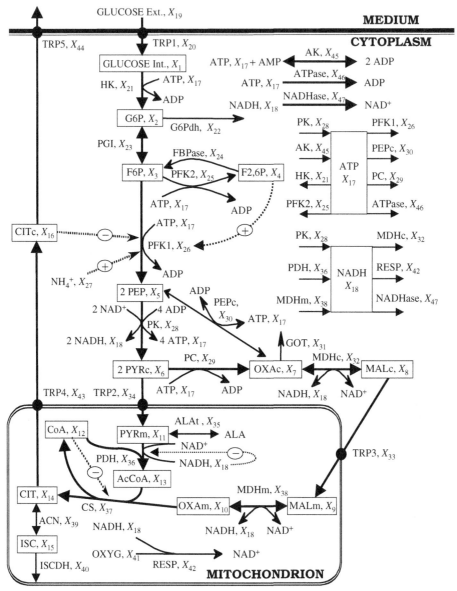

Figure 6.1. Citric acid metabolism in *A. niger* during idiophase. Redrawn from Figure 3.11, where abbreviations are explained.

ranges. Transformation to logarithms does not change the relationships between actual concentration values and upper or lower limits, but makes the resulting y_i variables consistent with those in the steady-state constraints. Third, we are adding a type of constraint that was not discussed before. It pertains to the total cellular protein mass within the cell, to the fact that simultaneous overexpression of many genes and subsequent high amounts of enzymes cause stress to the cell, and to the expectation that the cell presumably has a threshold for the total enzyme content. We formulate this constraint in such a fashion that the total amount of enzyme remains below some threshold.

Objective Function

The rate of citric acid excretion is represented in our model by the degradation term V_{16}^- of the sixteenth differential equation in model (3.7) (see Chapter 3). It has the form

$$V_{16}^- = 0.7621 \cdot X_{16}^{0.1135} X_{44}. \tag{6.1}$$

After taking logarithms and rearranging, we obtain as the objective function

$$0.1135 y_{16} + y_{44}. \tag{6.2}$$

This function is to be maximized. Note that the constant term $\ln(0.7621) = (-0.2717)$ is omitted from Eq. (6.2). It is irrelevant for the search of the location of the optimum, because it is simply added to any potential solution. Nonetheless, we need to record the value for later, because it is part of the actual value of the optimized excretion rate.

Constraints

Steady-state constraints. The steady-state equations are obtained from the model in Chapter 3 as shown in Chapter 2. Each differential equation is set equal to zero, its beta-term is moved to the left-hand side, and both sides are transformed to linearity by taking logarithms. The result for the citric acid model consists of sixteen linear equations and reads:

$$-0.9915 y_1 - 0.5798 y_{17} + 0.0125 y_{19} + y_{20} - y_{21} = -9.9071$$

$$0.9760 y_1 - 2.2280 y_2 + 1.3806 y_3 + 0.5707 y_{17} + 0.9843 y_{21}$$
$$- 0.1968 y_{22} - 0.7875 y_{23} = -3.2385$$

$$2.2587 y_2 - 2.0165 y_3 + 0.1594 y_4 + 0.2462 y_{16} - 0.1212 y_{17} + 0.8208 y_{23}$$
$$+ 0.1628 y_{24} - 0.1628 y_{25} - 0.8208 y_{26} - 0.5335 y_{27} = 5.7267$$

$$0.96 y_3 - 0.9852 y_4 + 0.5833 y_{17} - y_{24} + y_{25} = 6.959$$

$$0.5004 y_3 + 0.0012 y_4 - 0.9124 y_5 + 0.0267 y_7 - 0.2926 y_{16} + 3.3289 y_{17}$$
$$+ 0.0047 y_{18} + 0.9756 y_{26} + 0.6341 y_{27} - 0.9893 y_{28} + 0.0137 y_{30}$$
$$- 3.7481 y_{48} - 0.1014 y_{49} = -5.1162$$

$$0.909 y_5 - 0.835 y_6 - 3.382 y_{17} - 0.0047 y_{18} + y_{28} - 0.5193 y_{29} - 0.4807 y_{34}$$
$$+ 3.7011 y_{48} + 0.1025 y_{49} = 5.3766$$

$$0.0243 y_5 + 0.6733 y_6 - 1.6989 y_7 + 0.7150 y_8 + 0.0904 y_{17} - 1.6332 y_{18}$$
$$+ 0.9619 y_{29} - 0.0257 y_{30} - 0.0507 y_{31} - 0.8854 y_{32} + 0.1618 y_{48}$$
$$+ 0.0196 y_{49} = -4.7104$$

$$1.7690y_7 - 1.5839y_8 + 1.8075y_{18} + 0.9799y_{32} - 0.9799y_{33}$$
$$- 0.0217y_{49} = 1.2499$$

$$0.7926y_8 - 1.0367y_9 + 0.0573y_{10} + 0.1077y_{18} + 0.9799y_{33} - 0.9799y_{38}$$
$$-1.0497y_{49} = -16.4067$$

$$1.0367y_9 - 0.7378y_{10} - 0.9799y_{12} - 0.2449y_{13} - 0.1077y_{18} - 0.9799y_{37}$$
$$+ 0.9799y_{38} + 1.0497y_{49} = 17.7229$$

$$0.9648y_6 - 0.0923y_{11} - 0.0821y_{12} + 0.0868y_{18} + 0.9834y_{34} - 0.0055y_{35}$$
$$- 0.9779y_{36} - 0.0117y_{49} = 5.0047$$

$$y_{10} - 0.0640y_{11} - 0.0839y_{12} + 0.25y_{13} + 0.0887y_{18} - y_{36} + y_{37}$$
$$- 0.0120y_{49} = 1.0582$$

$$-0.3472y_{10} + 0.0320y_{11} + 0.1501y_{12} - 0.125y_{13} - 0.0443y_{18}$$
$$+ 0.5y_{36} - 0.5y_{37} + 0.006y_{49} = 0.9045$$

$$0.6020y_{10} - 0.1875y_{12} + 0.2167y_{13} - 0.65y_{14} + 0.4463y_{15} + 0.8669y_{37}$$
$$- 0.0443y_{39} - 0.8225y_{43} = -2.3786$$

$$2.7186y_{14} - 2.5307y_{15} + 0.25y_{39} - 0.25y_{40} = 10.8635$$

$$0.2038y_{14} - 0.1135y_{16} + y_{43} - y_{44} = 2.0069$$

$$-0.2329y_1 - 0.1322y_3 - 0.0002y_4 + 0.6980y_5 - 0.13855y_6 - 0.01031y_7$$
$$+ 0.05638y_{16} - 4.7147y_{17} - 0.0036y_{18} - 0.2349y_{21} - 0.0372y_{25}$$
$$- 0.1879y_{26} - 0.1221y_{27} + 0.7622y_{28} - 0.1979y_{29} - 0.0052y_{30}$$
$$+ 0.1557y_{45} - 0.2546y_{46} + 4.8902y_{48} + 0.0781y_{49} = 4.6901$$

$$0.4601y_5 - 0.4367y_7 + 0.1953y_8 + 0.2559y_9 - 0.0141y_{10} + 0.0154y_{11}$$
$$+ 0.0203y_{12} - 1.629y_{17} - 0.97622y_{18} + 0.5061y_{28} - 0.2419y_{32}$$
$$+ 0.2419y_{36} + 0.2419y_{38} - 0.0025y_{41} - 0.2495y_{42} - 0.4985y_{47}$$
$$+ 1.8734y_{48} + 0.0771y_{49} = 3.3495.$$

$$(6.3)$$

Constraints on control variables. Constraints on control variables limit by how much any of the enzyme activities or transport steps are allowed to vary. Recombinant DNA techniques suggest that an enzyme may be experimentally implemented with an activity in a range between about 0.1 and 50 times its basal value (Guarante et al. 1980). The introduction of (controllable) plasmids (Mattanovich et al. 1998), along with promoters that are being collected in promoter libraries (Jensen and Hammer 1998), allows some fine-tuning within this range. It is noted that the upper limit of fifty is an extreme value that should not be attained by very many enzyme simultaneously, because a large collection of high values could potentially impede cell viability. This concern is addressed in the section on cellular protein economy.

It is prudent to consider three groups of exceptions to the range of [0.1, 50].

1. Glucose 6-phosphate dehydrogenase (X_{22}), aspartate aminotransferase (X_{31}), alanine transaminase (X_{35}), and isocitrate dehydrogenase (X_{40}) are maintained constant at their basal values. These enzymes divert flux from the pathway that ultimately leads to citrate (see Figure 6.1), and it would be tempting to lower their activity. However, the concern is that drastic changes in the side branch fluxes – for instance toward the pentose shunt – could negatively affect the well-being of the cell and subsequently lead to lower productivity or even compromised cell viability.

2. External glucose (X_{19}), oxygen (X_{41}), and ammonium (X_{27}; a positive effector of phosphofructokinase) are also kept constant at their basal value to retain the observed conditions in the normal mode of operation as well as possible.

3. The total amount of adenine nucleotides (X_{48}) and total amount of pyridine nucleotides (X_{49}) are allowed to change only within a range of about 10% of their base values. These constraints are consistent with the small range of variation reported for these variables in vivo (Führer, Kubicek, and Röhr 1980).

Accordingly, the mathematical formulation of constraints on control variables, expressed on a logarithmic scale, is as follows:

$$
\begin{aligned}
y_{19} &= 5.67 & y_{35} &= -0.69 \\
-5.76 \le y_{20} &\le 0.46 & -3.83 \le y_{36} &\le 2.38 \\
-5.12 \le y_{21} &\le 1.10 & -7.26 \le y_{37} &\le -1.05 \\
y_{22} &= -4.07 & -0.06 \le y_{38} &\le 6.15 \\
0.43 \le y_{23} &\le 6.64 & -5.88 \le y_{39} &\le 0.34 \\
-2.78 \le y_{24} &\le 3.43 & y_{40} &= -4.02 \\
-3.86 \le y_{25} &\le 2.35 & y_{41} &= 5.01 \\
-4.89 \le y_{26} &\le 1.32 & -3.91 \le y_{42} &\le 2.30 \\
y_{27} &= 0.85 & -7.00 \le y_{43} &\le -0.79 \\
-4.20 \le y_{28} &\le 2.01 & -7.60 \le y_{44} &\le -1.39 \\
-6.50 \le y_{29} &\le -0.29 & -1.27 \le y_{45} &\le 4.94 \\
-3.00 \le y_{30} &\le 3.22 & -0.67 \le y_{46} &\le 5.54 \\
y_{31} &= -3.04 & -2.30 \le y_{47} &\le 3.91 \\
-0.58 \le y_{32} &\le 5.63 & -0.44 \le y_{48} &\le -0.23 \\
-2.92 \le y_{33} &\le 3.30 & -0.39 \le y_{49} &\le -0.19. \\
-5.40 \le y_{34} &\le 0.81
\end{aligned}
$$

$$(6.4)$$

Constraints on metabolites. To guarantee a physiologically meaningful solution, it is necessary to limit the range of operation for all metabolites. The overexpression of enzymes that are associated with the overproduction of some product often evokes changes in the intermediate metabolite pools that might have deleterious effects on other pathways in the cell. This in turn tends to lead to poor yield in product synthesis. It is unknown and difficult to estimate what magnitudes of deviations from the normal concentrations would be inconsequential. In fact, these magnitudes presumably vary greatly among metabolites and conditions. To operate on the safe side, we limit the ranges of these metabolite pools to a modest 20% of their steady-state levels. Metabolite variation within this restrictive range seems small enough to be readily tolerated by all pathways of the cell. Obviously, if better information about tolerable ranges becomes available, it is straightforwardly incorporated in the optimization routine, without the need for structural changes.

The mathematical formulation of the metabolite constraints in logarithmic coordinates is

$$
\begin{aligned}
-6.12 &\le y_1 \le -5.72 \\
-2.53 &\le y_2 \le -2.12 \\
-3.91 &\le y_3 \le -3.51 \\
-6.73 &\le y_4 \le -6.32 \\
-4.83 &\le y_5 \le -4.42 \\
-2.34 &\le y_6 \le -1.94 \\
-7.68 &\le y_7 \le -7.27 \\
0.73 &\le y_8 \le 1.14 \\
2.34 &\le y_9 \le 2.75 \\
-6.34 &\le y_{10} \le -5.94 \\
-0.80 &\le y_{11} \le -0.40 \\
-2.45 &\le y_{12} \le -2.04 \\
-4.65 &\le y_{13} \le -4.24 \\
3.22 &\le y_{14} \le 3.62 \\
-0.03 &\le y_{15} \le 0.37 \\
1.50 &\le y_{16} \le 1.91 \\
-0.92 &\le y_{17} \le -0.51 \\
-3.58 &\le y_{18} \le -3.17.
\end{aligned}
$$

$$(6.5)$$

Constraints against metabolic burden. It is our declared goal to maximize productivity of citric acid production, and it is reasonable to expect that this goal can only be attained if the organism is not being altered to a point of compromised viability. The first two sets of constraints, discussed previously, limit the maximally permitted overexpression of individual enzymes involved in the biosynthesis of citric acid and the variations in intermediate metabolites. The present section discusses an additional constraint that is designed to keep the organism within a functional range. This constraint poses an upper limit on the total changes in all altered enzymes combined. One can imagine that a cell might tolerate a fifty-fold overexpression in one or two enzymes. However, if the results of an optimization analysis called for fifty-fold overexpression of many or all involved enzymes, one could reasonably assume that the cellular metabolism would be overburdened in terms of general functioning, maintenance, and metabolite solubility.

Several authors have discussed these thoughts in the literature and coined the terms *metabolic burden, metabolic load, protein burden,* and *cellular protein economy* to summarize the efforts of the cell geared toward managing and limiting its biosynthetic investments in metabolic material and energy. Issues of cellular protein economy have important consequences for metabolic engineering, in that strong overexpression of the enzymes of a target pathway does not necessarily lead to the expected enhancements in flux. As a very relevant example, it was found that the growth rate in recombinant bacteria decreases monotonically, in one case in an almost linear fashion, with an increase in the number of introduced plasmid copies (Bentley et al. 1990; Snoep et al. 1995; Smits et al. 2000). The authors of these studies suggest that the additional plasmids create a competitive disadvantage and lead to subsequent instability of the host strain within the microbial population. The disadvantage may be caused by the additional stress placed on the host by plasmid replication, rDNA transcription, and plasmid-coded mRNA translation. One consequence of this physiological stress is a growth rate differential in comparison to plasmid-free cells, which may be so severe that a bioreactor with plasmid-containing cells may become unproductive in just a few generations. The development and implementation of controllable promoters can counteract the decrease in yield to some degree (Bentley et al. 1990; Mattanovitch et al. 1998).

Protein burden may also be ascribed to the dilution of essential enzymes necessary for protein production and to an increase in energy demand for the additional production of protein. As a consequence, the effects of protein burden become limiting when the overproduced enzymes constitute a substantial part of the total protein content. As an example, consider the gram-negative bacterium *Zymomonas mobilis*, in which 50% of the cytoplasmic enzymes are associated with glycolysis (An et al. 1991). Snoep et al. (1995) engineered simultaneous overexpression of key glycolytic operons, expecting increased yield; however, the overall glycolytic flux was reduced, presumably due to the increased protein burden.

A related issue associated with very high degrees of overexpression that are achievable through multiple introduced promoter fragments is a possible competition for important global transcription factors. The unmet demand for these factors is thought to cause general changes in transcription rates with a reduced expression of several housekeeping genes, which ultimately would also lead to reduced growth rates. A recent discussion with numerous references on issues associated with metabolic burden is available in Görgens et al. (2001).

These reasons of metabolic burden and cellular protein economy suggest that we should impose some constraint on the total concentration of all involved enzymes, in addition to the maximally permissible amplification factor for each enzyme. There are several options for a global constraint. For instance, one could force the total enzyme amount to remain at the original level, while trying to optimize production through a targeted redirection of fluxes. Expressed differently, this strategy would correspond to an optimal rearrangement of the profile of enzyme activities. A variation on this strategy is the permission for the entire enzyme pool to increase by a certain factor, such as 2, 5, or 10. A formulation of these types of constraints is

$$\sum_i X_i \leq X_T, \tag{6.6}$$

where X_T is defined as

$$X_T = p \cdot \sum_i (X_i)_{\text{Basal}}. \tag{6.7}$$

The index i covers all enzymes ($i = $ 20–21, 23–26, 28–30, 32–34, 36–39, 42–47), and p is the "cellular protein economy factor" with a value of 0.5, 1, 2, 5, or 10. To retain formal consistency within the linear program in logarithmic coordinates, it is beneficial to approximate the constraint (6.6) with the corresponding power-law function

$$X_T \geq \gamma_t \prod_i X_i^{f_{iT}}, \tag{6.8}$$

where the exponents f_{iT} are analogous to the kinetic orders g_{ij} and h_{ij}. They are computed just like any other kinetic orders, namely by partial differentiation of $\ln(X_T)$ with respect to each $\ln(X_i)$:

$$f_{iT} = \frac{\partial X_T}{\partial X_i} \cdot \frac{X_i}{X_T}, \tag{6.9}$$

which is computed at the operating point. The multiplier γ_t resembles the rate constants α_i and β_i. Taking the logarithm in Eq. (6.8) yields for the case of all twenty-two independent variables and $p = 1$:

$$\begin{aligned}
0.0007 y_{20} &+ 0.0014 y_{21} + 0.3646 y_{23} + 0.0147 y_{24} + 0.005 y_{25} + 0.0017 y_{26} \\
&+ 0.0035 y_{28} + 0.0003 y_{29} + 0.01192 y_{30} + 0.1334 y_{32} + 0.0128 y_{33} \\
&+ 0.001 y_{34} + 0.0007 y_{36} + 0.0001 y_{37} + 0.2240 y_{38} + 0.0006 y_{39} \\
&+ 0.0047 y_{42} + 0.0002 y_{43} + 0.0001 y_{44} + 0.0729 y_{45} + 0.1215 y_{46} \\
&+ 0.0238 y_{47} \leq 1.9394,
\end{aligned} \tag{6.10}$$

where again $y_i = \ln(X_i)$.

The constraint in Eq. (6.10) is a reformulation of Eq. (6.8), which in turn is an approximation of Eq. (6.6). It is not difficult to see that the constraint in power-law form (Eq. 6.8) is actually more stringent than the original constraint in Cartesian coordinates (Eq. 6.6). As an illustration, consider the sum of just two variables X_1 and X_2, which must not exceed X_T. Thus, the constraint in Cartesian coordinates is $X_1 + X_2 < X_T$. For simplicity (but without loss of generality of the argument), suppose the operating point is chosen arbitrarily but such that $X_{10p} = X_{20p}$. The power-law approximation, according to Eq. (6.9), therefore is

$$2 \cdot \sqrt{X_1 X_2} \approx X_1 + X_2. \tag{6.11}$$

At the operating point, both sides of the equation are equal by definition. For other points, divide both sides by 2, which shows that Eq. (6.11) compares the geometric mean on the left-hand side and the arithmetic mean on the right-hand side. Elementary operations demonstrate that the geometric mean is at most as large as the arithmetic mean. Thus,

$$2 \cdot \sqrt{X_1 X_2} \leq X_1 + X_2 \tag{6.12}$$

and if the system variables are forced to remain below the function on the left of the inequality (6.12), they certainly remain below the right-hand side.

Because optimization itself does not address questions of stability, the steady-state solution of the optimized S-system should be checked with respect to stability and possibly with respect to robustness. Both types of analyses can in principle be executed analytically, but are much more easily assessed in PLAS© (Ferreira 2000). Should the optimization routine result in an unstable steady-state solution, caution is necessary. The solution itself is normally discarded, because the actual fermentation system should be stable. The instability may furthermore indicate that even the baseline system is at the edge of normal functioning, and/or that some of the constraints on metabolites, enzyme activities, or fluxes were specified too loosely and need to be revisited (Torres et al. 1997).

OPTIMAL SOLUTIONS

The linear program, with all previously detailed constraints, was executed with the software package LINDO/PC 6.01©.[1] Entering all kinetic information and debugging a program of this size typically requires a little bit of time and effort, but once this is accomplished, the optimization algorithm itself produces optimal solutions very quickly. The main results are summarily shown in Table 6.1.

The table contains the following information. The first column lists all independent variables, which are assumed to be available for experimental manipulation. It also includes the scaled rate of citrate excretion (SRCE), which is the subject of greatest interest here. The second column exhibits the basal value of each independent variable and, for easy comparison with other results, the baseline citrate excretion, which is set as "1." The baseline values for enzyme activities were taken from the literature, as discussed in the original descriptions of the optimization task (Alvarez-Vasquez 2000; Alvarez-Vasquez, González-Alcón, and Torres 2000). These values also correspond to the steady state described and analyzed in Chapter 3. The body of the table shows the changes in each independent variable and the resulting scaled rate of citrate excretion for any given protein economy factor. Except for the cellular protein economy factor, the remaining constraints were the same in all optimizations, with the specifics described in the previous sections.

The solutions in Table 6.1 show several interesting features. Consider the column representing the optimal solution under the condition that the cellular protein economy factor equals 1. A factor of 1 corresponds to the requirement that the total amount of enzyme must not be increased; this scenario reflects a mere redirection of fluxes throughout the network. Even without a concomitant overall increase in metabolic activity, the optimized solution yields five times as much citric acid production. This might be surprising at first, but is readily explained with two arguments. First, natural selection favors the genotype with the best chance for survival of the organism, which clearly is a different objective than maximizing citric acid production. A strain with the characteristics of this column would possibly not survive in the outside world, but under the appropriate experimental conditions would be

---

[1] LINDO/PC 6.01©: LINDO Systems, Inc. 1415 North Dayton St., Chicago IL, 60622 USA.

Table 6.1. Normalized Optimum Enzyme Profiles [$(X_i)_{optimum}/(X_i)_{basal}$] and Scaled Rate of Citrate Excretion [SRCE $= (V_{16}^-)_{optimum}/(V_{16}^-)_{basal}$] for Different Values of Cellular Protein Economy[a]

Enzyme	Basal Value [μmM/min]	$(X_i)_{optimum}/(X_i)_{basal}$ Cellular Protein Economy Factor				
		0.5	1	2	5	10
X_{20}	0.032	1.5	3	5.6	13	16.5
X_{21}	0.06	1.1	2.2	4.2	9.7	13.3
X_{23}	15.3	0.6	1.6	3.5	9.9	26.1
X_{24}	0.62	0.1	0.1	0.1	0.1	1
X_{25}	0.21	0.1	0.1	0.1	0.1	1
X_{26}	0.07	1.7	4.1	9	25.3	31.8
X_{28}	0.15	4	12.5	27.2	50	50
X_{29}	0.01	1.1	2.7	5.7	15.2	26.7
X_{30}	0.5	0.1	0.1	0.1	0.1	0.1
X_{32}	5.6	0.6	1.5	3.5	10.1	13.6
X_{33}	0.54	1.7	4.3	9.7	28.1	38.1
X_{34}	0.04	1.6	3.9	8.7	25.1	50
X_{36}	0.21	2	4.9	11.2	32.3	42.6
X_{37}	0.007	1.5	3.8	8.7	25.1	37.7
X_{38}	9.4	1.1	2.7	6.2	18	31.5
X_{39}	0.02	0.1	0.1	0.1	0.1	1
X_{42}	0.2	50	50	50	50	10.1
X_{43}	0.009	2	5.1	12.1	37.1	50
X_{44}	0.005	2	5.3	12.4	38	50
X_{45}	2.8	0.1	0.1	0.1	0.1	0.1
X_{46}	5.1	0.1	0.1	0.1	0.4	1
X_{47}	1.0	0.2	0.9	3.1	14.5	50
SRCE	1	2.01	5.17	12.15	37.14	51.05

[a] Enzyme activities are given in μmM/min.

expected to produce noticeably more citric acid. Secondly, suppose the artificial task of maximizing citric acid coincided with the genuine survival tasks of the organism. In a sense, this supposition is not entirely untrue, given that the experimental conditions over the past century have been selecting for citric acid production and that the parent strain in this analysis is no longer a "natural" wildtype either. Even if citric acid production is the dominant task, evolution toward the optimum has proceeded essentially as a sequence of single mutations. This is in stark contrast to our modeling approach, which allows systematic determinations of optimal solutions even if they require numerous simultaneous changes. During natural or experimental stepwise selection, each intermediate step might be disadvantageous, but the combination of several alterations is ultimately superior. This alleged need for several simultaneous alterations provides at least one possible explanation of why the flux redistribution yielding the highest level of citric acid production may not yet have been identified experimentally. Having said that, one must caution that theoretical predictions of the type shown will not always come true, because they potentially ignore factors outside the model that could have an unanticipated, significant impact on the viability and optimal functioning of the organism.

Even if the total enzyme activity is reduced by 50% (economy factor 0.5), it is possible to reorganize fluxes in a fashion that would lead to higher citric acid production and excretion. More interesting, however, is the rest of the table, in which the total enzyme activity is permitted to rise to two-, five-, or ten-fold the baseline level. The theoretical predictions in these cases are that one could improve yield quite considerably, if one could alter all enzyme steps in the proportions shown in Table 6.1 and if the assumptions of the model hold. Even if not all of the model assumptions are satisfied, these theoretical predictions provide some guidance as to where to look for new avenues of yield optimization. Experimental testing of theoretical enzyme profiles will either yield improved fluxes as predicted or provide further insights into the structure of the pathway, which can be used for the next iteration of modeling and optimization.

The optimum profiles contain other interesting and potentially useful information for optimization purposes. It can be seen that all transport processes, namely, the glucose carrier (X_{20}), the mitochondrial malate transport system (X_{33}), the mitochondrial pyruvate transport system (X_{34}), the mitochondrial citrate carrier (X_{43}), and the cytosolic citrate carrier (X_{44}) should be amplified by significant factors if an increase in the rate of citrate production is to be observed. This finding points in the same direction as previous experimental and theoretical studies (Torres et al. 1996a; Netik et al. 1997). Moreover, the respiratory system (X_{42}) is the process requiring the highest overexpression, in most cases suggesting an expression level at the permitted upper limit. This observation reflects the importance of oxygen and the mechanisms involved in the overflow metabolism of *A. niger*, when the organism operates under conditions of manganese deficiency (Wallrath, Schmidt, and Weiss 1991; Schmidt et al. 1992). Other metabolic steps whose catalytic activity should be increased are the glycolytic steps, and in particular the reactions catalyzed by hexokinase (X_{21}), phosphofructokinase I (X_{26}), and pyruvate kinase (X_{28}). These results are also in agreement with the available experimental evidence, which has pointed out the important role of the glycolytic pathway in citric acid synthesis (Schreferl-Kunar et al. 1989; Steinbock et al. 1991). By contrast, the enzymes involved in the substrate cycle, leading to the synthesis and degradation of the PFKI effector fructose 2,6-bishosphate, PFK II (X_{25}), and fructose bishosphatase (X_{24}), actually contribute more strongly to the increase in citric acid production rate, if their activities are reduced to about 10% of their basal values. This indicates that a high rate of fructose 6-phosphate recycling has a negative effect on the rate of citrate synthesis. The situation is similar in the other substrate cycle of the model, the phosphoenolpyruvate carboxykinase (X_{30}) cycle.

Apart from the respiratory system, only two mitochondrial enzyme activities need to be increased, namely pyruvate dehydrogenase (X_{36}) and citrate synthase (X_{37}). Finally, there is a set of processes, outside the abovementioned substrate cycles, whose activities should be decreased. These processes are involved in the transformation of the adenylate phosphate pools and are catalyzed by adenylate kinase (X_{45}) and

ATPase (X_{46}). Also to be decreased are the activities of NADHase (X_{47}) and aconitase (X_{39}). The latter catalyzes an equilibrium reaction transforming the mitochondrial citrate (X_{14}) into isocitrate (X_{15}). Its activity should be decreased because it negatively affects the rate of citrate excretion.

Overall, the enzyme and transport processes whose activities should be increased belong to the direct pathway from external glucose (X_{19}) to citrate (X_{16}). The remaining enzymes change minimally or are even reduced in activity. Examination of the optimal profile without increase in overall metabolic activity (economy factor 1) furthermore shows that at least thirteen steps need to be upmodulated for attaining significant increases in rate of citrate efflux. The situation is quite similar for other economy factors, and the profiles of enzyme activities follow the same pattern. When no limits are imposed on the total enzyme concentration (cellular protein economy factor unlimited), the results do not deviate significantly from the solution obtained for economy factor 10. The reason for this is that other constraints become dominant and render the constraint on cellular protein economy mathematically irrelevant.

SYSTEMATIC SEARCH FOR THE BEST SEQUENCE OF STEPS TOWARD THE OPTIMAL SOLUTION

Although it is very important to determine the optimum profile, it is almost equally important for practical purposes to determine which particular alterations in expression would have the strongest potential of improvements in yield. In other words, one might ask: If it is not possible or not feasible to alter the entire system, which less complex changes would be most efficacious? After all, each alteration in gene expression and/or enzyme activity requires considerable experimental effort, either in terms of the introduction of strong promoters or the use of additional copies of the relevant genes, and may not even be possible in exactly the prescribed proportions. To respond to these practical issues, it is useful to execute two types of analyses. The first is a set of restricted optimizations, in which only smaller subsets of all possible independent variables are accessible to optimization. This analysis follows. The results suggest a hierarchy of steps that would gradually improve yield and ultimately lead to the optimum solution. The second type of analysis addresses the uncertainties that accompany all experimental implementations by asking: What happens if the prescribed enzyme profiles can only be implemented with a certain degree of inaccuracy? This type of analysis is discussed in the subsequent section of this chapter.

Table 6.2 shows the results of such a systematic search for optimal solutions in which subsets of one, two, three, up to all twenty-two catalytic and transport steps were available for alteration. All constraints remained exactly as they were for the full optimization. For the results in the table, the cellular protein economy factor was specified as 2; a discussion of other factors follows. For each number of enzymes changed (first column), the table shows only the one combination of enzyme activities that yields the largest increase in citrate efflux. For instance, consider a subset size of six. The enzyme profile generating the highest yield identifies the variables X_{20}, X_{34}, X_{36}, X_{37}, X_{43}, and X_{44} be altered and specifies in the corresponding columns the

Table 6.2. Normalized Optimum Enzyme Profiles for Different Subsets of Enzymes (1–22) That Were Available for Modulation[a]

Normalized Modulated Enzyme Activities: $X_i/(X_i)_{basal}$

Enzymes	X_{20}	X_{21}	X_{23}	X_{24}	X_{25}	X_{26}	X_{28}	X_{29}	X_{30}	X_{32}	X_{33}	X_{34}	X_{36}	X_{37}	X_{38}	X_{39}	X_{42}	X_{43}	X_{44}	X_{45}	X_{46}	X_{47}	SRCE
1													1.03										1.02
2													1.04						1.06				1.03
3													1.04				1.7		1.06				1.04
4												1.42	1.2					1.22	1.2				1.17
5	1.06											1.32	1.25					1.29	1.27				1.24
6	1.10											1.31	1.35	1.05				1.36	1.31				1.3
7	1.13											1.35	1.39	1.08			1.11	1.4	1.38				1.34
8	1.15						1.58					1.30	1.43	1.11				1.44	1.42			1.21	1.38
9	1.25					1.56						1.30	1.56	1.24			1.86	1.62	1.52		1.74		1.56
10	1.28					1.60	1.85				1.44	1.32	1.61	1.42			2.01	1.68	1.58				1.61
11	1.30					1.64	1.84		0.1		1.48	1.41	1.7	1.47			1.99	1.74	1.63				1.67
12	1.42	1.73				1.57		1.52			1.61	2.18	1.81	1.44			4.55	1.9	1.79	0.24			1.82
13	2.07	2.32				2.88	4.03	2.37			2.64	3.56	3.05	2.36	3.22			3.2	3.01			3.69	3.07
14	2.18	1.96	1.59			2.45		1.93			2.80	2.56	3.14	2.77	2.22		15.08	3.4	3.34		13.23		3.26
15	3.53	2.65	1.95			4.97	9.67	3.41		1.89	5.27	4.75	6.06	4.71	3.37			6.63	6.51			9.21	6.35
16	4.19	3.30	2.42			6.18	6.84	4.16		2.36	6.58	5.92	7.56	5.88	4.21			8.37	8.23	0.1		12.52	7.9
17	5.36	4.02	3.29		0.13	8.38	26.14	5.71		3.24	9.03	8.11	10.38	8.07	5.78			11.68	11.49	0.1	0.13	20.4	11.2
18	5.50	4.13	3.4		0.13	8.65	26.14	5.54	0.1	3.35	9.33	8.38	10.73	8.34	5.97			12.1	11.89	0.1	0.13	20.72	11.6
19	5.64	4.23	3.51			8.93	26.97	5.71	0.1	3.46	9.65	8.66	11.08	8.62	6.17		50	12.52	12.31	0.1	0.14	3.07	12
20	5.66	4.25	3.52	0.75		8.97	27.1	5.73	0.1	3.48	9.69	8.7	11.14	8.66	6.2		50	12.59	12.38	0.1	0.14	3.09	12.07
21	5.68	4.27	3.54	0.1	0.13	9.01	27.22	5.76	0.1	3.50	9.74	8.74	11.19	8.7	6.23		50	12.65	12.44	0.1	0.19	3.11	12.13
22	5.69	4.27	3.55	0.1	0.13	9.03	27.27	5.77	0.1	3.50	9.76	8.76	11.21	8.72	6.24	0.1	50	12.14	12.46	0.1	0.19	3.12	12.15

[a] The cellular protein economy factor considered in this analysis was 2.

degree of overexpression. In this particular case, the increase in citrate excretion is predicted to be 30%. No other combination of alterations of six steps yields a higher rate of citric acid production, while still satisfying all constraints.

The first observation from Table 6.2 worth noting is the fact that simultaneous alterations in just a few enzymes have essentially no effect on citric acid production. For instance, even the best possible alteration of three steps would under ideal conditions still only improve yield by 4%. An intuitive explanation may be the following. Strains with improved productivity have mostly been achieved with the traditional approach of random mutagenesis and selection. A viable mutant strain is in itself optimized for the present conditions, and each additional random alteration in one more gene is usually detrimental, or neutral at best. To lead to drastic increases in a product like citric acid, simultaneous changes in several catalytic or transport steps are necessary, and these have to occur in specific magnitudes. An improved combination of two or even three genes might result from trial and error. However, complex, specific changes involving six, seven, or more steps are highly unlikely because of the inherent randomness of the experimental approach. As a consequence, each new strain occupies a local optimum, but none of these optima may be the global optimum for citric acid production.

The Conclusions chapter discusses two studies that have not been investigated experimentally as thoroughly as citric acid production in molds. One study addresses the optimization of carnitine production in *E. coli*, and the other discusses microbial tryptophan production. In both cases, alterations in just one or two steps are shown to improve yield significantly, and relatively modest experimental effort promises several-fold increases. In the carnitine case, some of the predictions were already tested and confirmed experimentally (Alvarez-Vasquez et al., forthcoming). Chapter 7 also suggests that well-studied pathways are more difficult to improve. There, we first study ethanol production, which has been investigated for very long and does not allow simple means of improvement. The same model is then used to optimize the production of secondary products, which is possible with reasonable effort.

The requirement of simultaneous overexpression of many enzymes for improved yield that we observe here is in line with earlier studies. For instance, Niederberger et al. (1992) and Stephanopoulos, Aristidou, and Nielsen (1998) demonstrated that significant increases in fluxes were attained only when multiple enzymes were up-modulated (see also Torres et al. 1997). Indirect support for the need of multiple, simultaneous alterations can also be seen in the natural design of bacterial operons, in which several enzymes are under the control of a single promoter and consequently expressed in equal amounts in response to relevant signals. It has been argued that this design ensures metabolite homeostasis and is advantageous in terms of cellular protein economy (Thomas and Fell 1998).

The sequence of combinations of enzyme alterations and their effects reveals that no really significant increase in citrate flux is obtained until at least thirteen enzymes are modulated. Under ideal conditions, this modulation is predicted to yield a citrate production rate three times larger than that at the basal steady state. All involved enzymes are overexpressed, with activities that are increased about two to four times.

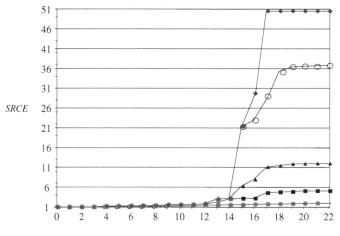

Number of enzymes available for modulation

Figure 6.2. Trends in optimized scaled rates of citrate excretion, $SRCE = (V_{16}^-)_{opt}/(V_{16}^-)_{basal}$, against the number of steps available for modulation. The symbols identify different cellular protein economy factors [0.5 (●); 1 (■); 2 (▲); 5 (O); 10 (♦)].

Enzymes requiring modulation include all transport processes and all glycolytic enzymes except for the phosphoglucose isomerase (X_{23}). These enzymes account for eight of the total thirteen steps to be altered. This finding is consistent with features of the optimum profile shown in Table 6.1. It has been shown that these steps are among those with a high overexpression factor. Thus, these results suggest that simultaneous overexpression of the transport processes and key glycolytic enzymes might be the best strategy toward the optimal solution (however, see Discussion). The same pattern was observed when the cellular protein economy factor was specified as five or ten.

Beyond simultaneous changes in thirteen enzymes, citric acid production increases steadily from three-fold to eleven-fold with only four additional alterations. These alterations include two enzymes that are to be overexpressed, namely phosphoglucose isomerase (X_{23}) and cytosolic malate dehydrogenase (X_{32}), and two to be down-modulated, namely adenylate kinase (X_{45}) and ATPase (X_{46}). Additional alterations, beyond these seventeen enzymes, are essentially ineffective.

Figure 6.2 shows the evolution of flux amplification with each optimal combination of enzymes to be altered. In all cases investigated (cellular protein economy factor of two, five, and ten), the pattern is essentially the same. Changes of up to five enzymes have almost no effect on the citric acid flux. Simultaneous alterations of six to nine enzymes lead to moderate improvements (up to 50%). Additional changes of one, two, or three enzymes, for a total between ten and twelve, do not much improve flux. Truly significant increases in citric acid flux occur once thirteen enzymes are upmodulated. Between thirteen to seventeen steps, a steady increase in flux amplification is achieved with every additional enzyme. The slope within this range is more pronounced as the cellular protein economy factor increases. Beyond seventeen enzymes, a plateau is reached where no significant increase in flux is observed with any further enzyme modulation.

Figure 6.3 visualizes some key expression scenarios. Modest increases in flux are predicted by simple activation of key steps of glycolysis, as well as citrate production and excretion. More significant results are obtained when the cytosolic branch toward oxalacetate and malate and the transport of malate into mitochondria are overexpressed. The maximal production is reached if all intermediate steps are also adjusted.

ANALYSIS OF THE EFFECTS OF IMPRECISION DURING IMPLEMENTATION

Optimization of a subset of enzymes yields lower improvements in comparison to the overall optimal solution. At the same time, alterations of a limited number of variables have the obvious advantage of significantly reduced effort in terms of experimental implementation. This poses the immediate challenge of deciding whether the effort of adjusting more rather than fewer control variables is worthwhile. A component of this cost-benefit decision is the inaccuracy accompanying any experimental implementation of altered profiles. This inaccuracy can be expected to grow with the number of necessary adjustments, and one may ask how each optimal solution (for a limited number of permissible alterations) is affected if the proposed profile of independent variables cannot be implemented experimentally with mathematical exactness. For instance, if the optimal solution calls for an 8.15-fold increase in some enzyme activity, one might ask whether an actual eight-fold increase would significantly alter the optimal yield and to what degree it would violate one or several constraints.

If "error" distributions are attached to the input values of the control variables, the output variables (metabolite concentrations and fluxes) are distributed as well. The particularities of the output distributions are difficult to predict. The logarithmic gains indicate how strongly a metabolite or flux is affected if one of the control variables is changed, so that high logarithmic gains may be expected to translate into relatively wide spreads of the output distributions. If two or more control variables are changed simultaneously, one could in principle assess the spread of the output distributions with *synergism coefficients*, which are tensor-based analogues of logarithmic gains (Salvador 2000a,b). One could also approach the effects of inaccuracies with statistical methods, as Petkov and Maranas (1997) proposed for uncertainties in parameter values. For practical purposes, involving five, ten, or more simultaneous changes, these methods become overwhelmingly cumbersome.

An alternate – admittedly brute force – fashion of addressing this issue is the use of Monte-Carlo simulations, in which each control variable is drawn from a statistical distribution that is centered about the optimal "target" value and has some specified range and shape (Voit and Del Signore 2001). We will illustrate this method here with the *A. niger* system.

No general information is available about the shape of the true or best-suited input distributions that describe inaccuracies in input, and we will use uniform and triangular distributions as likely extremes. The triangular distribution corresponds

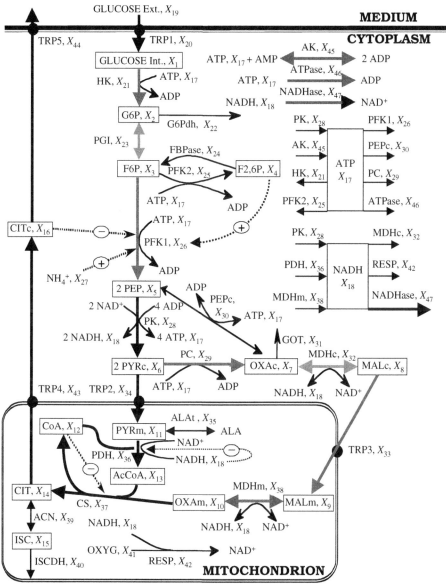

Figure 6.3. Visualization of some of the optimization results shown in Table 6.2. To achieve close to a 40% increase in citric acid flux, eight steps (heavy black arrows) need to be modulated. They constitute a rather direct path from external glucose uptake to pyruvate and citric acid production with subsequent excretion. Note that some of the steps of the glycolytic pathway need not be overexpressed, which indicates that they were originally not operating at maximal capacity ("rate limiting"). Further activation of five additional steps (dark gray arrows) increases the citric acid flux to roughly three times the baseline value. The mechanism to accomplish this increase is an activation of the cytosolic oxalacetate-malate branch. A total of seventeen enzyme alterations (all heavy arrows) results in a predicted twelve-fold increase in yield. The additionally activated steps (light gray arrows) "fill in" the gaps of the former profile. Notice that the reversible steps in the pathway are among the last to require alteration, which makes intuitively sense. See Figure 3.11 for nomenclature.

to an implementation of the optimal control profile that is most likely to achieve the target value but also accounts for close-by values, though with decreasing probability. At the other end of the spectrum, the uniform distribution models an implementation of each control variable that is equally likely within a certain range around the target. Obviously, there are uncounted ways of specifying distributions and ranges. For streamlined discussion, we select representative scenarios consisting of five, nine, thirteen, and seventeen enzyme alterations, which correspond to noticeable improvements over fewer alterations and respective increases in yield of about 23%, 54%, three-fold, and eleven-fold (Table 6.2). Also, we generally assume ±20% ranges for all uncertainty distributions. For the actual implementation of the simulations, the Microsoft Excel® add-on Crystal Ball® (Sargent and Wainwright 1996) is most convenient. Crystal Ball® allows the specification of numerous input distributions and offers a large spread of graphical and statistical representations of the simulation output.

Shapes of Output Distributions

The simulations demonstrate that different metabolites respond to the same changes in input in distinctly different ways. For instance, if five control variables are sampled from triangular distributions with ±20% ranges, some metabolite distributions (X_6, X_7, X_8, X_9, X_{12}) are more or less Gaussian, others look more or less lognormal (X_3, X_4, X_{14}, X_{15}, X_{16}, X_{17}), and yet others resemble triangular (X_1, X_5, X_{18}) or "log-triangular" (X_2, X_{13}) distributions. X_{10} is close to normal but exhibits slight skewness to the left, whereas X_{11} has an extremely long right tail. If seventeen control variables are sampled from triangular distributions, similar results are obtained, but the distributions lose some of the triangular character and are more rounded out, approaching normality or lognormality. A few of the distributions obtained from triangular input are shown in Figure 6.4.

If the input distributions are uniform, the resulting output distributions also are shaped in different ways. For five input variables, some dependent variables have shapes that still reflect uniformity, whereas others are almost normal or lognormal. For seventeen input variables, the uniform character is almost entirely lost. A few of the distributions obtained from uniform input are displayed in Figure 6.5.

Representative statistical features of some of the output distributions are presented in Tables 6.3 and 6.4. If the distributions of the dependent variables are more or less symmetrical, their means and medians are very similar to that of the optimal solution. In cases of strong skewness (in particular, X_{11}; see also Figures 6.4 and 6.5), the means and medians differ, and it seems that the median reflects the optimal solution better than the mean. Given the particularities of the simulations and the input distributions, this is probably not surprising.

Citrate Production

The citrate flux distributions for all scenarios resemble normal distributions (Figure 6.6), even though the metabolites contributing to this flux are distributed in a

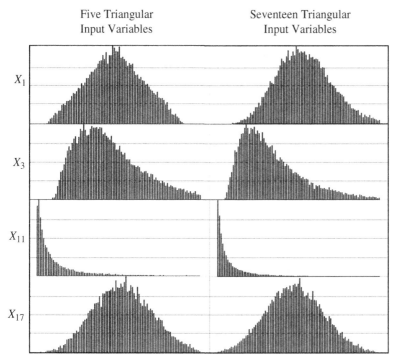

Figure 6.4. Distributions of four representative dependent variables that result from *triangular* input distributions of five or seventeen independent variables.

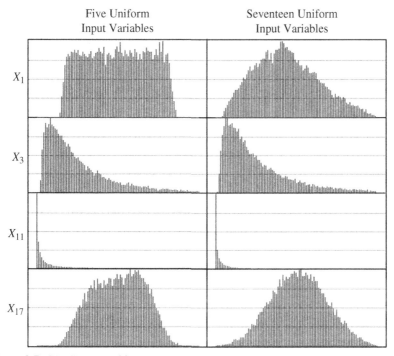

Figure 6.5. Distributions of four representative dependent variables that result from *uniform* input distributions of five or seventeen independent variables.

Table 6.3. Means and Medians of Representative Variables Under a Simulation Scenario Where Five Independent Variables are Distributed

Variable	Optimal Solution	Triangular Input Distributions		Uniform Input Distributions	
		Mean	Median	Mean	Median
X_1	0.0029	0.0029	0.0029	0.0029	0.0029
X_3	0.0286	0.0314	0.0286	0.0391	0.0287
X_{11}	0.5127	1.2572	0.5002	2.8668	0.4895
X_{17}	0.5077	0.5078	0.5078	0.5076	0.5079

number of different ways (Figures 6.3 and 6.4). The means and medians of the citrate flux distributions are essentially identical with those of the corresponding optimal solutions. In response to the ±20% ranges in input distributions, the citrate flux distributions have ranges in slight excess of ±20%. However, in all cases tested, the tenth and ninetieth percentiles of the output distribution fall within ±20% of the optimal citrate value. Thus, inaccuracies in input are not amplified much in the output of interest.

The citrate flux distributions overlap in scenarios where five or nine enzyme activities are subject to optimization. This implies that low-performing cases of nine adjustments would not necessarily yield more citric acid than high-performing cases of only five adjustments. This may not be surprising, because the improvements in citrate yield for the two scenarios do not differ much (1.23-fold and 1.54-fold, respectively, according to Alvarez-Vasquez et al. 2000). For adjustments of thirteen and seventeen enzymes, the distributions are separated. For more widely spread input distributions, one might expect the citrate flux distributions to be more dispersed and their overlap more pronounced.

Constraint Violations

The optimization task is constrained not only by the steady-state equations, but also by limits to the values of the dependent variables. The actual limits in vivo are unknown, and it is unclear whether the violation of one or two constraints would actually compromise cell viability. In fact, it is to be expected that the tolerance of the

Table 6.4. Means and Medians of Representative Variables Under a Scenario Where Seventeen Independent Variables are Distributed

Variable	Optimal Solution	Triangular Input Distributions		Uniform Input Distributions	
		Mean	Median	Mean	Median
X_1	0.0032	0.0032	0.0032	0.0032	0.0032
X_3	0.0201	0.0243	0.0203	0.0288	0.0201
X_{11}	0.4995	1.4535	0.4839	3.6678	0.4766
X_{17}	0.6012	0.6011	0.6010	0.6012	0.6011

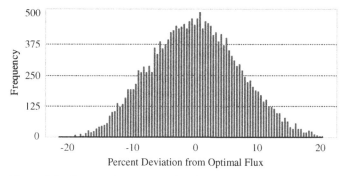

Figure 6.6. Citric acid production flux in a simulation with 20,000 iterations of seventeen triangularly distributed independent variables (thirty-one outliers not shown). Values on the x axis are shown as $100 \cdot (Flux_S - Flux_O)/Flux_O$, where $Flux_S$ is the simulated flux and $Flux_O$ is the flux in the optimal solution. The distribution for uniformly distributed input looks rather similar.

organism depends very strongly on the particular excess metabolite. On one hand, concentrations of metabolites are tightly buffered. For instance, the concentrations of serum metabolites in humans usually remain in a close vicinity of their nominal values (Geigy 1960; Wallach 1986). On the other hand, concentrations may change very drastically. For example, yeast contains, under normal growth conditions, only traces of the disaccharide trehalose. With increases in temperature, however, trehalose may rise to huge intracellular amounts of 1 gram per gram protein (Hottiger, Schmutz, and Wiemken 1987).

Without a systematic approach, it is difficult to estimate how many combinations of input values for n control variables would violate one or more constraints. If one would simplistically assume that all altered enzyme activities of the optimal solution caused the system to be located at one of the boundaries of the admissible simplex (see Chapter 5), then one could surmise that half of all random draws would violate some constraint. If ten enzymes were involved, one would deduce that only one out of $2^{10} = 1,024$ combinations would be admissible. For seventeen enzymes, the odds would be reduced to one in $2^{17} = 131,072$, which would be very discouraging for practical implementations of the optimal solution.

Indeed, the simulations show that very many random combinations of independent variables violate one or several constraints if the conservative limits on metabolites ($\pm 20\%$) are retained (Table 6.5). Even if these limits are relaxed to half and twice the nominal metabolite concentrations, respectively, the majority of iterations yield results that violate one or more constraints (Table 6.6). Two general patterns become evident. First, constraint violations are more prominent for uniformly distributed input. This is not surprising, given that relatively more of the randomly drawn values of the independent variables are farther away from the target values of the optimal solution. Second, as one might expect, simulations with five enzyme variations perform slightly better in terms of satisfying constraints than simulations with more possible enzyme alterations. However, there is not much difference in performance between simulations in which nine, thirteen, or seventeen independent variables were subject to uncertainty. The reason may be that among these latter scenarios roughly the

Table 6.5. Numbers of Simulation Results (out of 20,000) Satisfying All Constraints or Showing at Most One, Two, or Three Constraint Violations[a]

Scenario	All ±20% Constraints Satisfied	At Most One Violation (±20%)	At Most Two Violations (±20%)	At Most Three Violations (±20%)	All ½/2X Constraints satisfied	At Most One Violation (½/2X)	At Most Two Violations (½/2X)
T5	20	122	389	947	2,209	6,172	7,107
T9	1	10	44	206	1,187	4,088	6,095
T13	0	0	6	77	1,322	4,254	6,095
T17	0	0	2	28	1,270	4,089	5,794
U5	1	40	108	299	487	2,328	3,806
U9	0	2	10	48	329	1,752	3,432
U13	0	0	0	10	319	1,702	3,098
U17	0	0	1	5	353	1,668	3,139

[a] Terminology for scenarios: T = triangular, U = uniform; following number denotes number of independent variables allowed to be optimized; "±20%" refers to the constraint that all variables remain with 20% of their nominal value; "1/2/2X" refers to the relaxed constraint that all variables remain within half and twice their nominal value.

same number of constraints is actually active and many dependent and independent variables are not positioned at a boundary.

If the constraints are relaxed, a larger percentage of simulation results is admissible. For instance, if all metabolites are allowed to decrease to half or increase to twice their nominal values, 2,209 out of 20,000 combinations of five triangularly distributed inputs are within limits (Table 6.5). The corresponding citrate flux distributions more or less reflect the distributions of all outputs in the same category. They have about the same means and medians as the parent distributions, but smaller standard deviations and ranges, which reflects that predominantly extreme combinations violate constraints on metabolites (Table 6.6). Although more combinations are admissible, their percentage among all simulations is still small.

The main reason for the large percentages of constraint violations is presumably a small, but important set of relatively high logarithmic gains. Most prominent are the gains of mitochondrial pyruvate (X_{11}) with respect to changes in glucose transport (X_{20}) and pyruvate dehydrogenase (X_{36}), which are $+12.5$ and -11.1, respectively. These gains suggest that a 1% change in glucose transport would approximately evoke a 12.5% change in mitochondrial pyruvate. Consequently, if a value for X_{20} is drawn randomly from within a ±20% range about its nominal value, the response in X_{11} is magnified roughly ten times, thus, more often than not, violating the corresponding constraints. In fact, inspection of the output distributions for X_{11} and their long right tails reflects this amplification effect (Figures 6.3 and 6.4). In the optimized solution, the positive and negative gains keep each other in balance, but this does not happen if random inaccuracies are introduced. It is probably no coincidence that the independent variables with high gains are included in most subsets of enzymes that lead to improved citrate yield.

Table 6.6. Comparison of Statistical Features of Citrate Flux Distributions for Different Scenarios and Either All Simulation Results or Only Those Results That Are Admissible (adm.) with Respect to Relaxed Constraints on Dependent Variables (Range Equals Half and Twice the Nominal Concentration)[a]

Scen.	Target	Mean (all)	Mean (adm.)	Median (all)	Median (adm.)	σ (all)	σ (adm.)	Range (all)	Range (adm.)
T5	0.0058	0.0058	0.0058	0.0058	0.0058	0.0004	0.0003	0.0046, 0.0070	0.0053, 0.0062
U5	0.0058	0.0058	0.0058	0.0057	0.0058	0.0005	0.0005	0.0043, 0.0074	0.0047, 0.0070
T9	0.0072	0.0072	0.0072	0.0072	0.0072	0.0005	0.0003	0.0057, 0.0088	0.0062, 0.0084
U9	0.0072	0.0072	0.0073	0.0071	0.0073	0.0007	0.0006	0.0054, 0.0091	0.0059, 0.0087
T13	0.0142	0.0142	0.0144	0.0141	0.0144	0.0010	0.0007	0.0104, 0.0177	0.0123, 0.0170
U13	0.0142	0.0141	0.0150	0.0141	0.0151	0.0015	0.0118	0.0102, 0.0190	0.0119, 0.0172
T17	0.0517	0.0516	0.0524	0.0516	0.0524	0.0037	0.0027	0.0393, 0.0660	0.0445, 0.0615
U17	0.0517	0.0515	0.0527	0.0514	0.0528	0.0054	0.0458	0.0373, 0.0693	0.0438, 0.0639

[a] Terminology as in Table 6.5.

As Alvarez-Vasquez et al. (2000) correctly pointed out, the vast majority of logarithmic gains in the system are low, indicating overall robustness. However, it seems that the two high gains mentioned, six additional gains with a magnitude above five, and nine gains with magnitudes between three and five are sufficient to cause the observed large percentages of constraint violations. Whether these constraint violations would compromise cell viability or be easily buffered by the organism cannot be predicted without actual experimental investigation.

It is worth noting that the relatively high gains are not randomly distributed throughout the system but are mostly clustered around the glucose and citrate transporters X_{20}, X_{43}, and X_{44} and the dependent variable X_{11} (mitochondrial pyruvate). Thus, model amendments surrounding these clusters could possibly alleviate many of the constraint violations observed here, if the high gains are actually artifacts.

CONCLUSIONS

In a well-studied system like citric acid production in *A. niger*, simultaneous alterations in just a few enzymes have essentially no positive effect on yield. Even under ideal conditions, the best possible alteration of three steps is predicted to improve yield by only by 4%. As discussed in the introduction, an intuitive explanation of this result may be that the intense experimental effort in optimizing this system may already have identified a local optimum that cannot be improved by simple alterations. For instance, the fine-tuning of fermentation conditions, including medium composition, cofactors, pH, and temperature, has co-evolved with new, better performing strains, so that the baseline system in our analysis is difficult to improve in a gradual fashion as it results from the traditional approach of random mutagenesis and selection. Instead, novel, improved solutions have to be determined with methods that "leap" away from the current state. Of course, such leaps can occur in almost unlimited ways and, without strong guidance, most would result in inferior production or impaired viability. Only a rigorously structured approach can overcome this challenge.

The optimization methods proposed here and in other chapters are certainly not the final word on improvements in microbial or biotechnological production, but they do open new and promising avenues for exploration that are not possible with strictly experimental means. The combination of a proven nonlinear modeling structure with a straightforward optimization method facilitates large numbers of explorations in a short period of time. For instance, it permits analyses of subsets of enzyme alterations and greatly simplifies multi-objective optimizations (see Chapters 5 and 7). These streamlined methods, which are an immediate consequence of the S-system structure, are even useful, in an indirect manner, if the pathway of interest is initially modeled in a different mathematical structure (see Chapter 7).

Beyond the types of optimization discussed here, the power and flexibility of the linear optimization of an intrinsically nonlinear dynamic metabolic model have permitted the optimization of the model structure itself and the assessment of uncertainties in structure or input. For instance, Hatzimanikatis, Floudas, and Bailey (1996a,b) used the S-system formulation to optimize the regulatory structure of pathways. The S-system formulation facilitates this type of analysis, because each regulatory effect is uniquely characterized by a kinetic order and a rate constant, so that different modes of regulation simply differ in the values of these parameters. Petkov and Maranas (1997) studied the effects of variations in parameter values with strictly statistical means. Finally, the previous sections of this chapter showed how imprecision in input may be evaluated with simulations. Again, the explicit linear form of the steady-state equations in log space was of great benefit, because it avoided computationally expensive, nonlinear searches for the steady state within each iteration of the simulation.

Focusing on the pathway of this chapter, we saw that imprecision in input frequently causes the optimization results to violate some of the artificially imposed constraints. This result is somewhat disappointing, but also leads to specific proposals for future research. The first calls for closer scrutiny of the dynamics of the glucose and citrate transporters X_{20}, X_{43}, and X_{44} and the dependent variable X_{11} (mitochondrial pyruvate). The high gains and sensitivities associated with these variables could be a true reflection of reality. However, they could also be artifacts, in which case the model could potentially be improved through further amendments.

The second proposal suggests fixing the steps with very high gains and optimizing the system in terms of all other control variables. The problematic steps may be kept at baseline or at any other actually implementable level. As mentioned earlier, this strategy has been used for branches that divert flux to avoid negative effects on cell viability.

The third proposal is a hierarchical optimization toward the best achievable solution. Although this is not always the case, the required adjustment of n enzymes typically includes as a subset all enzymes of the previous step with $n - 1$ adjustments. Furthermore, in many cases, the magnitude of adjustment of a particular enzyme changes monotonically with the number of enzymes being adjusted. This suggests a stepwise optimization strategy. In the first phase, one selects a target level of increased production that is deemed significant enough to warrant the experimental effort. In the case of citric acid production, this level may be set at four or five enzymes, leading to 15% or 20% increases, respectively, as opposed to a mere 4% predicted improvement from the adjustment of only three enzymes. Using the optimal enhancement factors for the five-enzyme control profile as targets, one attempts to implement the system experimentally. If this is possible with some accuracy, one may stop or enter the second phase, identifying the next level of significant improvement and proceeding as in the first phase. However, if the optimal control profile cannot be established experimentally, it might be wise to use the relatively best scenario and to execute a new optimization, fixing the problematic enzyme activities at implementable levels and allowing other enzyme activities to float in the linear programming process. Under these new constraints, the step function of improved yields is likely to change, thereby defining new target values for subsequent optimizations. These options for dealing with imprecision remain to be explored for different pathways on a case-by-case basis.

Even if the theoretically predicted optimal results are strongly affected by idealized assumptions or issues of uncertainty and imprecision, the analysis of the *A. niger* pathway shown here offers interesting insights. The most important observation is that any really significant improvement of citric acid production would require the upmodulation of a large number of steps, unless the mold is transfected in a fashion that would create entirely new pathways. If this prediction is true, it implies that random approaches will become less and less likely to yield improvement. Indeed, this conclusion is consistent with increasingly frequent observations made on well-studied organisms.

REFERENCES

Alvarez-Vasquez, F.: Modelización, análisis y optimización del metabolismo del hongo *Aspergillus niger* en condiciones de producción de ácido cítrico. Ph.D. dissertation, University of La Laguna, Tenerife, Spain, 2000.

Alvarez-Vasquez, F., C. González-Alcón, and N.V. Torres: Metabolism of citric acid production by *Aspergillus niger*: Model definition, steady state analysis and constrained optimization of the citric acid production rate. *Biotechnol. Bioeng.* 70(1), 82–108, 2000.

Alvarez-Vasquez, F., M. Cánovas, J.L. Iborra, and N.V. Torres: Modelling and optimization of continuous L-(-)-carnitine production by high-density *Escherichia coli* cultures. Forthcoming.

An, H., R.P. Scopes, M. Rodríguez, K.-F. Kesvav, and L.O. Ingram: Gel electrophoretic analysis of *Zymomonas mobilis* glycolytic and fermentative enzymes: Identification of an alcohol dehydrogenase II as a stress protein. *J. Bacteriol.* 173, 5975–82, 1991.

Archer, D.B., D.A. MacKenzie, and D.J. Jeenes: Genetic engineering: Yeasts and filamentous fungi. In: C. Ratledge and B. Kristiansen (Eds.), *Basic Biotechnology* (Chapter 5). Cambridge University Press, Cambridge, U.K., 2001.

Bentley, W.E., N. Mirjalili, D.C. Andersen, R.H. Davis, and D.S. Kompala: Plasmid-encoded protein: The principal factor in the "metabolic burden" associated with recombinant bacteria. *Biotechol. Bioeng.* 35, 668–81, 1990.

Currie, J.N.: On the citric acid production of *Aspergillus Niger*. *Science* 44, 215–16, 1916.

Currie, J.N.: The citric acid fermentation. *J. Biol. Chem.* 31, 15–21, 1917.

Ferreira, A.E.N.: *PLAS*©: http://correio.cc.fc.ul.pt/~aenf/plas.html, 2000.

Führer, L., C.P. Kubicek, and M. Röhr: Pyridine nucleotide levels and ratios in *Aspergillus niger*. *Can. J. Microbiol.* 26, 405–08, 1980.

Geigy, A.G.: Documenta Geigy: *Wissenschaftliche Tabellen* (6th ed.). Pharmazeutische Abteilung der Firma J.R. Geigy A.G., Basel, 1960.

Görgens, J.F., W.H. van Zyl, J.H. Knoetze, and B. Hahn-Hägerdal: The metabolic burden of the *PGK1* and *ADH2* promoter systems for heterologous xylanase production by *Saccharomyces cerevisiae* in defined medium. *Biotechnol. Bioeng.* 73(3), 238–45, 2001.

Guarante, G., L. Gail, T.M. Roberts, and M. Ptashane: Improved methods for maximizing expression of a cloned gene: A bacterium that synthesizes rabbit β-globin. *Cell* 20, 543–53, 1980.

Harwood, C.R., and A. Wipat: Genome management and analysis: Prokaryotes. In: C. Ratledge and B. Kristiansen (Eds.), *Basic Biotechnology* (Chapter 4). Cambridge University Press, Cambridge, U.K., 2001.

Hatzimanikatis, V., C.A. Floudas, and J.E. Bailey: Optimization of regulatory architectures in metabolic reaction networks. *Biotechnol. Bioeng.* 52(4), 485–500, 1996a.

Hatzimanikatis, V., C.A. Floudas, and J.E. Bailey: Analysis and design of metabolic reaction networks via mixed-integer linear optimization. *AIChE J.* 42(5), 1277–92, 1996b.

Hottiger, T., P. Schmutz, and A. Wiemken: Heat-induced accumulation and futile cycling of trehalose in *Saccharomyces cerevisiae. J. Bacteriol.* 169, 5518–22, 1987.

Jensen, P.R., and K. Hammer: Artificial promoters for metabolic optimization. *Biotechnol. Bioeng.* 58, 191–5, 1998.

Koffas, M., C. Roberge, K. Lee, and G. Stephanopoulos: Metabolic engineering. *Annu. Rev. Biomed. Eng.* 1, 535–57, 1999.

Mattanovitch, D., W. Kramer, C. Lüttich, R. Weik, K. Bayer, and H. Katinger: Rational design of an improved induction scheme for recombinant *Escherichia coli. Biotechnol. Bioeng.* 58, 296–8, 1998.

Netik, A., N.V. Torres, J.M. Riol, and C.P. Kubicek: Uptake and export of citric acid by *Aspergillus niger* is reciprocally regulated by manganese ions. *Biochim. Biophys.* Acta 1326, 287–94, 1997.

Niederberger, P., R. Prasad, G. Miozzari, and H. Kacser: A strategy for increasing an in vivo flux by genetic manipulations. The tryptophan system of yeast. *Biochem. J.* 287, 473–9, 1992.

Petkov, S.B., and C.D. Maranas: Quantitative assessment of uncertainty in the optimization of metabolic pathways. *Biotechnol. Bioeng.* 56, 145–61, 1997.

Ratledge, C.: Look before you clone. Letter to the editor. *FEMS Microbiol. Lett.* 189, 317–18, 2000.

Ruijter, C.J.G.: Life is not that simple. Letter to the editor. *FEMS Microbiol. Lett.* 189, 318–19, 2000.

Ruijter, C.J.G., H. Panneman, and J. Visser: Overexpression of phosphofructokinase and pyruvate kinasse in citric acid producing *Aspergillus niger. Biochim. Biophys.* Acta 1334, 317–23, 1997.

Ruijter, C.J.G., H. Panneman, and J. Visser: Metabolic engineering of the glycolytic pathway in *Aspergillus niger. Food Technol. Biotechnol.* 36(3), 185–8, 1998.

Ruijter, C.J.G., H. Panneman, X. Ding-Bang, and J. Visser: Properties of *Aspergillus niger* citrate synthase and effects of citA overexpression on citric acid production. *FEMS Microbiol. Lett.* 184, 35–40, 2000.

Salvador, A.: Synergism analysis of biochemical systems. I. Conceptual framework. *Math. Biosc.* 163(2), 105–29, 2000a.

Salvador, A.: Synergism analysis of biochemical systems. II. Tensor formulation and treatment of stoichiometric constraints. *Math. Biosc.* 163(2), 131–58, 2000b.

Sargent, R., and E. Wainwright (Eds): *Crystal Ball: Forecasting & Risk Analysis for Spreadsheet Users,* Version 4.0, CG Press, Broomfield, CO, 1996.

Savageau, M.A.: Biochemical systems analysis II. The steady state solutions for an *n*-pool system using a power-law approximation. *J. Theor. Biol.* 25, 370–9, 1969.

Savageau, M.A. and E.O. Voit: Recasting nonlinear differential equations as S-systems: A canonical nonlinear form. *Math. Biosci.* 87, 83–115, 1987.

Schmidt, M., J. Wallrath, A. Dörner, and H. Weiss: Disturbed assembly of the respiratory chain NADH: Ubiquinone reductase (complex I) in citric-acid-accumulating *Aspergillus niger* strain B60. *Appl. Microbiol. Biotechnol.* 36, 667–72, 1992.

Schreferl-Kunar, G., M. Grotz, M. Röhr, and C.P. Kubicek: Increased citric acid production by mutants of *Aspergillus niger* with increasased glycolytic capacity. *FEMS Microbiol. Lett.* 59, 297–300, 1989.

Smits, H.P., J. Hauf, S. Müller, T.J. Hobley, F.K. Zimmermann, B. Hahn-Hägerdal, J. Nielsen, and L. Olsson: Simultaneous overexpression of enzymes of the lower part of glycolysis can enhance the fermentative capacity of *Saccharomyces cerevisiae. Yeast* 16, 1325–34, 2000.

Snoep, J.L., L.P. Yomano, H.V. Westerhoff, and L.O. Ingram: Protein burden in *Zymomonas mobilis*: Negative flux and growth control due to overproduction of glycolytic enzymes. *Microbiology* 141, 2329–37, 1995.

Steinbock, F.A., I. Held, S. Chojun, H. Harsen, M. Rohr, E.M. Kubicek-Pranz, and C.P. Kubicek: Regulatory aspects of carbohydrate metabolism in relation to citric acid accumulation by *Aspergillus niger*. *Acta Biotechnol.* 11(6), 571–81, 1991.

Stephanopoulos, G.: A platform of flux and gene expression measurements for metabolic engineering and drug discovery (Presentation). *Biol. Inf. Proc. Syst.* Workshop, Clemson University, Clemson, SC, January 19–20, 2001.

Stephanopoulos, G.N., A.A. Aristidou, and J. Nielsen: *Metabolic Engineering*. Academic Press, San Diego, 1998.

Thomas, S., and S. Fell: The role of multiple enzyme activation in metabolic flux control. *Adv. Enzyme Reg.* 38, 65–85, 1998.

Torres, N.V., J.M. Riol-Cimas, M. Wolschek, and C.P. Kubicek: Glucose transport by *Aspergillus niger*: The low-affinity carrier is only formed during growth on high glucose concentrations. *Appl. Microbiol. Biotechnol.* 44, 790–4, 1996a.

Torres, N.V., E.O. Voit, and C. González-Alcón: Optimization of nonlinear biotechnological processes with linear programming: Application to citric acid production by *Aspergillus niger*. *Biotechnol. Bioeng.* 49, 247–58, 1996b.

Torres, N.V., E.O. Voit, C. González-Alcón, and F. Rodríguez: An indirect optimization method for biochemical systems: Description of method and application to the maximization of the rate of ethanol, glycerol, and carbohydrate production in *Saccharomyces cerevisiae*. *Biotechnol. Bioeng.* 49, 247–58, 1997.

Voit, E.O.: Optimization in integrated biochemical systems. *Biotechnol. Bioeng.* 40, 572–82, 1992.

Voit, E.O., and M. Del Signore: Assessment of effects of experimental imprecision on optimized biochemical systems. *Biotechnol. Bioeng.* 74(5), 443–8, 2001.

Wallach. J.: *Interpretation of Diagnostic Tests* (4th ed.). Little, Brown, and Company, Boston, 1986.

Wallrath, J., J. Schmidt, and H. Weiss: Concomitant loss of respiratory chain NADH:ubiquinone reductase (complex I) and citric acid accumulation in *Aspergillus niger*. *Appl. Microbiol. Biotechnol.* 36, 76–81, 1991.

## CHAPTER SEVEN

Maximization of Ethanol Production in *Saccharomyces cerevisiae*

INTRODUCTION

The previous chapter demonstrated how a fairly complex, nonlinear pathway model may be optimized with methods of linear programming, once it is formulated as an S-system. As an illustration we used the pathway of citric acid production in the mold *A. niger*, which we had modeled step by step in Chapter 3. Strictly following the tenets of the power-law formalism, we had formulated S-system equations from the known topology of the pathway, estimated parameter values from various available sources, and also suggested which methods of analysis and diagnostics should be applied to validate the model or to pinpoint problem areas that needed numerical refinements or structural amendments.

The present chapter discusses in some detail another biotechnologically interesting example. In fact, it is probably the paradigm of fermentation, namely the conversion of glucose into ethanol, glycerol, trehalose, and glycogen in the brewer's or baker's yeast *Saccharomyces cerevisiae*. Obviously, fermentation in yeast has been known well for at least 5,000 years, as the practice of brewing beer and baking bread in antiquity attests (Enfors 2001). Fermentation may also constitute the first example of a truly biomathematical joint venture: In the late eighteenth century, two famous researchers, the biochemist Antoine Lavoisier and the mathematician Pierre Simon Laplace, collaborated on respiration processes and suggested the fermentation equation

must of grapes = carbonic acid + alcohol.

As Lavoisier himself explained, this equation was not an algebraic curiosity but had purpose and applicability: "We may consider the substance submitted to fermentation, and the products resulting from that operation, as forming an algebraic equation, and, by successively supposing each of the elements in this equation unknown, we can calculate their values in succession and thus verify our experiments by calculation and our calculations by experiments reciprocally" (cited in Leicester 1974, p. 139).

In 1998, the United States used 550 million bushels of corn to produce fuel ethanol and exceeded the 1995 record production of 1.4 billion gallons (Bothast, Nichols, and Dien 1999). Due to its enormous economic and industrial importance, uncounted articles have been written about the various aspects of alcohol fermentation. A comprehensive review is not the purpose of this chapter, and the interested reader may consult the recent accounts by Bothast et al. (1999), Gong et al. (1999) and Pretorius (2000) for an entry to the extensive literature. Instead, the chapter uses the fermentation pathway for a demonstration of three aspects of metabolic pathway analysis and optimization.

First, it shows how an existing biochemical systems model from the literature may be reformulated as an S-system model, whose structure facilitates analysis and, ultimately, optimization. This demonstration does not represent a singular special case. To the contrary, the formulation of enzyme-catalyzed reactions within the mathematical framework of Michaelis–Menten equations and their generalizations is still so prevalent that the construction of a power-law model through reformulation of existing models is currently more often the rule than the exception. It would be desirable to construct power-law models de novo and base parameter estimates directly on experimental data. The future will probably allow us to pursue this strategy; however, at this point in time, we do not have the luxury of a complete set of directly applicable data and must make best use of all that is available, and that is typically expressed in the literature in terms of kinetic parameters like K_M and V_{max} that are intimately tied to the concepts of the Michaelis-Menten mechanism.

To be sure, models of Michaelis–Menten or other types could, in principle, address many of the same relevant questions that we intend to ask and answer with power-law models. For instance, one could assess local stability with eigenvalue analysis of the linearized system at steady state or ask which enzymes should be altered to improve a desired product or flux. However, the rational functions underlying the Michaelis–Menten approach are simply mathematically too inconvenient to permit comprehensive, streamlined analyses of even moderately sized pathways (see discussions in Savageau 1976; Heinrich, Rapoport, and Rapoport 1977; Shiraishi and Savageau 1992a). In particular, the structure of these models has confined attempts to optimize kinetic systems to rather small pathways (Heinrich, Hoffman, and Holzhütter 1990; Heinrich and Hoffman 1991; Heinrich, Schuster, and Holzhütter 1991; Schuster and Heinrich 1991; Schuster, Schuster, and Heinrich 1991; Pettersson 1992; Heinrich and Schuster 1996).

The second goal of this chapter is the introduction of the *Indirect Optimization Method*. This method allows the optimization of a metabolic model, for instance in Michaelis–Menten form, in a stepwise fashion that consists of reformulation as an S-system, optimization, and comparison with results of the original model.

Thirdly, the chapter discusses issues of multi-objective optimization that were mentioned in general terms in Chapter 5. Here, some of these methods are applied to an actual system, namely fermentation in yeast.

BACKGROUND

Over the past decades, industrial ethanol has become synonymous with chemical energy. As a "biofuel," commercial ethanol is a viable alternative to traditional petrochemical products that are dependent on fossils fuels, which obviously constitute a limited resource. Ethanol is also less polluting and therefore attractive from an environmental point of view.

Yeast and other microbial organisms can utilize a number of different raw materials as substrates for the production of ethanol. In the free market economy, three basic raw materials have emerged as particularly efficient: molasses, maize, and cassava. In addition, ligno-cellulose materials, obtained either directly as forestry products or indirectly from common wastes such as straw or paper, are so abundant and cheap that the existing supplies could support a significant part of the world demand for transportation fuels (Jeffries and Shi 1999). Although acid hydrolysis of cellulose to fermentable sugars is technically possible and was in fact used in times of war, it is expensive and always compromised by the presence of by-products like hemicellulose and lignin. Recent research has therefore been directed toward the development of microbial strains and pretreated media that permit the efficient fermentation of xylose and other hemicellulosic constituents, which could become the basis of economically viable biomass production of suitable fermentation substrate.

Yeasts are the most commonly used microorganisms for fermentation purposes. They are able to produce ethanol with high selectivity and with only traces of undesired by-products. Of particular biotechnological importance are high-producing strains from the species *Saccharomyces cerevisiae*, *S. carlsbergensis*, and *Candida utilis*. An immense amount of detailed biochemical and physiological knowledge about fermentation in yeasts has been accumulated over the years. Uncounted studies have investigated different experimental set-ups and conditions under which yeasts produce ethanol, and much is known about rates of glucose uptake and of glycerol and ethanol formation in vivo. The metabolic pathway itself is fairly easily accessible and therefore rather well understood in terms of metabolites and kinetic processes. The rich body of qualitative and quantitative information has led to extensive experimental manipulation and to mathematical models with the ultimate goal of yield improvement and optimization.

In this chapter, we focus on one relatively simple fermentation model that includes only the most important kinetic control points of nongrowing yeast cells and was originally proposed by Galazzo and Bailey (1989, 1990, 1991). The pathway describes how yeast cells use glucose to produce ethanol, as well as glycerol, glycogen, and trehalose. More detailed models were proposed, for instance, by Schlosser, Riedy, and Bailey (1994) and by the Sørensen group (Hynne, Dano, and Sørensen 2001).

Merging relevant information from the literature with results from their own experimental studies, Galazzo and Bailey (1990) proposed the simplified reaction scheme shown in Figure 7.1 and formulated corresponding kinetic equations in the traditional form of Michaelis–Menten reactions and their derivatives.

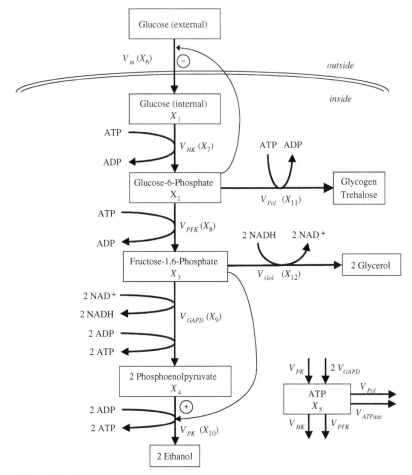

Figure 7.1. Simplified model of anaerobic fermentation of glucose to ethanol, glycerol, and polysaccharides in the yeast *Saccharomyces cerevisiae*. Heavy arrows represent reactions and light arrows show modulations.

In the first step of the model, external glucose is transported into the cell. This step must not be ignored, because it has a direct bearing on fermentation experiments, in which the supply of external glucose can easily be manipulated. Also, this transport step is inhibited by glucose 6-phosphate and thus constitutes an important point of regulation.

Hexokinase phosphorylates internal glucose to glucose 6-phosphate. In vivo, glucose 6-phosphate is used as a starting point for different pathways. Galazzo and Bailey consider only two. One is the main glycolytic branch that leads toward ethanol, whereas the other branch is responsible for the production of the polysaccharides glycogen and trehalose. The flow into the oxidative and nonoxidative pentose pathways is considered insignificant under the experimental conditions of interest and, therefore, omitted. Glucose 6-phosphate and fructose 6-phosphate are assumed to be at equilibrium. Because they are very readily converted into each other, they are pooled in the model.

In the next step, phosphofructokinase phosphorylates fructose 6-phosphate. If the enzyme fructose 1,6-diphosphatase is active, a futile cycle between fructose 6-phosphate and fructose 1,6-diphosphate may be of importance. However, under the experimental conditions of interest, this enzyme is typically inactive, and the futile cycle is therefore not included in the model. Fructose 1,6-diphosphate can be used to produce glycerol or phosphoenolpyruvate. Finally, pyruvate kinase catalyzes the production of ethanol. This step is activated by fructose 1,6-diphosphate. In addition to these reactions of the main pathway, Galazzo and Bailey consider conversions between ADP and ATP.

GALAZZO AND BAILEY'S MODEL

Mass Balance Equations

As is typical for metabolic modeling, the starting point of Galazzo and Bailey's model is a set of mass balance equations. In our notation, these take the form

$$\dot{X}_1 = V_{in} - V_{HK}$$
$$\dot{X}_2 = V_{HK} - V_{PFK} - V_{Pol}$$
$$\dot{X}_3 = V_{PFK} - V_{GAPD} - 0.5\, V_{Gol} \qquad\qquad (7.1)$$
$$\dot{X}_4 = 2 \cdot V_{GAPD} - V_{PK}$$
$$\dot{X}_5 = 2 \cdot V_{GAPD} + V_{PK} - V_{HK} - V_{Pol} - V_{PFK} - V_{ATPase}.$$

The variables X_i represent intermediate metabolite concentrations. Specifically: X_1 is the intracellular glucose concentration (Glc_{in}), X_2 represents glucose 6-phosphate (G6P), X_3 codes for fructose 1,6-diphosphate (FDP), X_4 is phosphoenolpyruvate (PEP), and X_5 represents ATP. The indexed quantities V represent the following fluxes: V_{in} denotes the sugar transport into the cells, V_{HK} summarizes all hexokinases, V_{PFK} is the phosphofructokinase reaction, V_{GADP} represents glyceraldehyde 3-phosphate dehydrogenase, V_{PK} represents pyruvate kinase, V_{Pol} describes glycogen synthetase, V_{Gol}, the glycerol 3-phosphate dehydrogenase is proportional to V_{PK}, and V_{ATPase} summarizes collectively the use of ATP. The rate of ethanol production is given directly by the flux through the pyruvate kinase reaction, V_{PK}. We will refer to the ethanol production flux also as V_{Eth} or V_4^-.

KINETIC MODEL

Clearly, internal glucose (X_1), glucose 6-phosphate (X_2), fructose 1,6-diphosphate (X_3), phosphoenolpyruvate (X_4), and ATP (X_5) are dependent variables, whereas the external glucose concentration is under the control of the experimenter and thus independent. Other independent variables include enzyme activities and quantities that affect the pathway but are not themselves changed by the dynamics of the system. These considerations lead to the steady-state metabolite concentrations and enzyme activities shown in Tables 7.1 and 7.2.

Table 7.1. Steady-State Concentrations of the Fermentation Pathway (Galazzo and Bailey 1990)

Metabolite	Acronym	Symbol	Concentration (mM)
Cytosolic glucose	Glc(in)	X_1	0.035
Glucose 6-phosphate	G6P	X_2	1.01
Fructose 1,6-diphosphate	FDP	X_3	9.14
Phosphoenolpyruvate	PEP	X_4	0.0095
Adenosine triphosphate	ATP	X_5	1.13
Nicotinamide adenine dinucleotide ratio	NADH/NAD$^+$	X_{14}	0.042 (unitless)

The intracellular concentrations of glucose 6-phosphate, fructose 6-phosphate, fructose 1,6-diphosphate, and 3-phosphoglycerate were estimated individually from broad sugar phosphate resonance measurements obtained through in vivo ^{31}P NMR (Shanks and Bailey 1988). Uptake of glucose and its conversion into various end products were measured through NMR with ^{13}C-glucose.

Secondary substrates like ADP and NAD$^+$ are not explicitly represented in the model but contribute actively to the dynamics of the pathway and need to be included in the model. Exchanges between NAD$^+$ and NADH and among the adenylates are quantified indirectly. Because the ratio of NAD$^+$ and NADH seems to be fairly constant, it is considered an independent variable (see Table 7.1). Experimental evidence furthermore suggests that the sum of NAD$^+$ and NADH equals about 2 mM. Thus, there are two equations, from which the individual concentrations are obtained as NAD$^+$ = 1.919 mM and NADH = 0.081 mM.

AMP, ADP, and ATP change over time, but the total pool of adenylates does not fluctuate much. Galazzo and Bailey assumed it to have a constant value of AMP + ADP + ATP = 3 mM. The adenylates are catalyzed by adenylate kinase reaction with

$$K_{eq} = \frac{\text{ADP}^2}{\text{ATP} \cdot \text{AMP}} = 1 \tag{7.2}$$

Table 7.2. Maximal Enzyme Activities, Fluxes (mM/min) Through Reactions, and Transport Steps of the Fermentation Pathway (Galazzo and Bailey 1990)

Enzyme/Step	Acronym	Symbol	Maximal Activity	Net Flux
Glucose transport	V_{in}^M	X_6	19.7	15.96
Hexokinase	V_{HK}^M	X_7	68.5	15.96
Phosphofructo-1-kinase	V_{PFK}^M	X_8	31.7	15.94
Glyceraldehyde Dehydrogenase	V_{GAPD}^M	X_9	49.9	15.06
Pyruvate kinase	V_{PK}^M	X_{10}	3440	30.11
Polysaccharide biosynthesis	V_{Pol}^M	X_{11}	14.31	0.014
Polyol biosynthesis	V_{Gol}^M	X_{12}	203	1.777
ATPase	V_{ATPase}^M	X_{13}	25.1	28.31

(Su and Russel 1968). In this case, there are three unknowns, namely the concentrations of AMP, ADP, and ATP, and two relationships among them. If one concentration is known, the other two can be computed from these constraint relationships. Expressed explicitly, AMP and ADP depend on ATP in the following fashion:

$$ADP = \frac{1}{2}\left(\sqrt{12 \cdot ATP - 3 \cdot ATP^2} - ATP\right)$$

$$AMP = 3 - ATP - ADP. \tag{7.3}$$

Galazzo and Bailey used the steady-state concentration ATP $= 1.1278$ mM. Curto, Sorribas, and Cascante (1995) considered ATP a dependent variable and defined AMP and ADP according to the constraint relationships among the adenylates. Given the concentration of ATP, the solution to Eqs. (7.2) and (7.3) is ADP $= 0.9948$ mM and AMP $= 0.8774$ mM.

Similar constraints exist between other metabolites. It is important to account for them, because they allow us to keep the number of explicitly modeled, dependent variables to a minimum. Specifically, fructose 6-phosphate, glyceraldehyde 3-phosphate, 3-phosphoglycerate, and UDPG do not need to be modeled explicitly, because they are involved in equilibrium relationships with some of the model variables. Galazzo and Bailey (1989) measured the following constraints:

$$[F6P]/[G6P] = 0.3$$
$$[G3P]/[FDP] = 0.01$$
$$[PEP]/[3PG] = 0.1$$
$$[G6P]/[UDPG] = 1.444. \tag{7.4}$$

Using the experimental methods outlined and accounting for the constraints among metabolites, Galazzo and Bailey (1990) were able to translate the mass balance Eq. (7.1) into a complete set of rate equations within the Michaelis–Menten formalism. The individual flux rates in their model are

$$V_{Glc(in)} = V_{IN}^M - 3.7 \cdot X_2$$

$$V_{HK} = \frac{V_{HK}^M}{\dfrac{6.2 \cdot 10^{-4}}{X_1 \cdot X_5} + \dfrac{0.11}{X_1} + \dfrac{0.1}{X_5} + 1}$$

$$V_{PFK} = \frac{50 \cdot V_{PFK}^M \cdot X_2 \cdot X_5 \cdot R_1}{R_1^2 + 3.342 \cdot 10^3 \cdot L_1^2 \cdot T_1^2}$$

$$R_1 = 1 + 0.3 \cdot X_2 + 16.67 \cdot X_5 + 50 \cdot X_2 \cdot X_5$$

$$L_1 = \frac{1 + 0.76 \cdot AMP}{1 + 40 \cdot AMP}$$

$$T_1 = V_1 + 1.5 \cdot 10^{-4} \cdot X_2 + 16.67 \cdot X_5 + 0.0025 \cdot X_2 \cdot X_5$$

$$V_{Pol} = \frac{1.1 \cdot V_{Pol}^M}{\left(1 + \left(\dfrac{2}{X_2}\right)^{8.25}\right)\left(\dfrac{1.1}{0.7 \cdot X_2} + 2.43\right)}$$

$$V_{GADP} = \frac{V_{GADP}^M}{1 + \dfrac{0.25}{X_3} + 0.0938 \cdot \left(1 + \dfrac{67.428}{X_3}\right) \cdot \left(1 + \dfrac{AMP}{1.1} + \dfrac{ADP}{1.5} + \dfrac{X_5}{2.5}\right)}$$

$$V_{PK} = \frac{V_{PK}^M \cdot X_4 \cdot ADP \cdot \left(2.519 \cdot R_2 + 0.656 \cdot T_2 \cdot L_2^2\right)}{1.0832 \cdot \left(R_2^2 + 164.084 \cdot L_2^2 \cdot T_2^2\right)}$$

$$R_2 = 1 + 125.94 \cdot X_4 + 0.2\, ADP + 2.1519 \cdot X_4 \cdot ADP$$

$$T_2 = 1 + 0.024 \cdot X_4 + 0.2 \cdot ADP + 0.004 \cdot X_4 \cdot ADP$$

$$L_2 = \frac{1 + 0.05 \cdot X_3}{1 + 5 \cdot X_3}$$

$$V_{Gol} = \frac{V_{Gol}^M}{V_{PK}^M} V_{PK}$$

$$V_{ATPase} = V_{ATPase}^M. \tag{7.5}$$

The parameter value for V_{HK} was taken from Wilkinson and Rose (1979); V_{PFK} from Hess and Plesser (1978); V_{Pol} from Rothman and Cabib (1967); V_{GAPD} from Yang and Deal (1969); and V_{PK} from Hess and Plesser (1978) and Brown, Taylor, and Chan (1985).

S-SYSTEM MODEL

Given the mass balance equations in Eqs. (7.1), the constraints in Eqs. (7.2–7.5), and the specifications in Tables 7.1 and 7.2, the corresponding S-system model is obtained in a straightforward manner. Curto, Sorribas, and Cascante (1995) described the reformulation process for four experimental situations. Focusing strictly on one of these situations, namely suspended cells at pH 4.5, Voit (2000) used the same model as a case study example and showed intermediate manipulations not explicitly stated by Curto and co-workers. Cascante, Curto, and Sorribas (1995) and Sorribas, Curto, and Cascante (1995) analyzed the steady-state and dynamic properties of the model under different sets of experimental conditions.

The transition from the Michaelis–Menten to the S-system model consists of three steps. First, one establishes the symbolic S-system model by setting up one product of power-law functions for all fluxes entering a given pool and one product for all fluxes leaving a given pool. Each product contains those and only those variables that directly affect the process under consideration. Each variable obtains a kinetic order, whose indices refer to the affected process and the affecting variable. For instance, the power-law term corresponding to V_{HK} contains the independent variables X_1 (internal glucose) and X_5 (ATP), and the independent variable X_7 (V_{HK}^M). The symbolic S-system resulting from these simple rules is

$$\dot{X}_1 = \alpha_1 X_2^{g_{12}} X_6^{g_{16}} - \beta_1 X_1^{h_{11}} X_5^{h_{15}} X_7^{h_{17}}$$

$$\dot{X}_2 = \alpha_2 X_1^{g_{21}} X_5^{g_{25}} X_7^{g_{27}} - \beta_2 X_2^{h_{22}} X_5^{h_{25}} X_8^{h_{28}} X_{11}^{h_{2,11}}$$

$$\dot{X}_3 = \alpha_3 X_2^{g_{32}} X_5^{g_{35}} X_8^{g_{38}} - \beta_3 X_3^{b_{33}} X_4^{b_{34}} X_5^{b_{35}} X_9^{b_{39}} X_{12}^{b_{3,12}} X_{14}^{b_{3,14}}$$

$$\dot{X}_4 = \alpha_4 X_3^{g_{43}} X_5^{g_{45}} X_9^{g_{49}} X_{14}^{g_{4,14}} - \beta_4 X_3^{b_{43}} X_4^{b_{44}} X_5^{b_{45}} X_{10}^{b_{4,10}}$$

$$\dot{X}_5 = \alpha_5 X_3^{g_{53}} X_4^{g_{54}} X_5^{g_{55}} X_9^{g_{59}} X_{10}^{g_{5,10}} X_{14}^{g_{5,14}}$$

$$- \beta_5 X_1^{b_{51}} X_2^{b_{52}} X_5^{b_{55}} X_7^{b_{57}} X_8^{b_{58}} X_{11}^{b_{5,11}} X_{13}^{b_{5,13}}. \tag{7.6}$$

In the second step, one selects an appropriate operating point, at which the S-system and the Michaelis–Menten model are exactly equivalent. Typically, this point is chosen to be the nominal steady state, whose characteristics in the present case are given in Tables 7.1 and 7.2.

The final step consists of computing values for the kinetic orders and rate constants that transform the symbolic S-system model into a numerical S-system model. The kinetic orders are computed as partial derivatives of the logarithm of the flux of interest (in Michaelis–Menten form) with respect to the logarithm of the variable of interest, as it was shown in Chapter 2. Each rate constant is subsequently computed by equating the original flux term with the corresponding power-law term of the S-system and solving for the rate constant. These steps do not require biological insight, but are strictly a matter of mathematical manipulation. The result for Galazzo and Bailey's fermentation model is

$$\dot{X}_1 = 0.8122 X_2^{-0.2344} X_6 - 2.8632 X_1^{0.7464} X_5^{0.0243} X_7$$

$$\dot{X}_2 = 2.8632 X_1^{0.7464} X_5^{0.0243} X_7 - 0.5239 X_2^{0.735} X_5^{-0.394} X_8^{0.999} X_{11}^{0.001}$$

$$\dot{X}_3 = 0.5232 X_2^{0.7318} X_5^{-0.3941} X_8 - 0.0148 X_3^{0.584} X_4^{0.03} X_5^{0.119}$$

$$\times X_9^{0.944} X_{12}^{0.056} X_{14}^{-0.575}$$

$$\dot{X}_4 = 0.022 X_3^{0.6159} X_5^{0.1308} X_9 X_{14}^{-0.6088} - 0.0945 X_3^{0.05} X_4^{0.533} X_5^{-0.0822} X_{10}$$

$$\dot{X}_5 = 0.0913 X_3^{0.333} X_4^{0.266} X_5^{0.024} X_9^{0.5} X_{10}^{0.5} X_{14}^{-0.304}$$

$$- 3.2097 X_1^{0.198} X_2^{0.196} X_5^{0.372} X_7^{0.265} X_8^{0.265} X_{11}^{0.0002} X_{13}^{0.47}. \tag{7.7}$$

QUALITY ASSESSMENT OF THE MODEL

Before the model is used for optimization, it is necessary to test its validity. By the very nature of modeling, no mathematical description of a complex biological phenomenon is absolutely complete, "true," or "valid." Nonetheless, the model should be able to make realistic predictions or give reasonable explanations, and a necessary – though not sufficient – prerequisite for that is that the model responds to perturbations in an adequate manner. It is not possible to execute an all-encompassing battery of tests of such responses, but it is feasible to assess key properties of the model with some representative analyses. These include local stability analysis, robustness analysis, and an assessment of the dynamics of the model. We will discuss these briefly, but note that comprehensive analyses are found in Cascante, Curto, and Sorribas (1995) and Sorribas, Curto, and Cascante (1995).

Local Stability of the Steady State

A reasonable model, situated at the observed steady state, is expected to tolerate moderate perturbations. In other words, if one of the variables is slightly altered for a short period of time, the model will react in some fashion that may be difficult to predict. However, the model should return to the steady state after a reasonable period of time. Whether it actually does so is largely determined by its eigenvalues. If their real parts are all negative, the system is able to tolerate small perturbations (Chapter 2).

In principle, the steady state of the S-system and its stability could be computed by hand with standard methods of linear algebra. But, of course, it is much easier to compute them with the canonical modeling software PLAS (Ferreira 2000). For the given pathway model, PLAS indeed confirms that the model reproduces the observed steady state and also provides the eigenvalues of the system, namely, -1.846; $-13.6 + 7.84\,i$; $-13.6 - 7.84\,i$; -1687; and -341.6. All eigenvalues have negative real parts, which indicates local stability of the pathway model. The numerical values of the real parts furthermore indicate the relative time scales of the processes. The smallest values are related with slow processes, whereas the larger values are associated with the fast reactions. Two eigenvalues form a pair of complex conjugates, which alludes to the potential of oscillations. Overall, stability analysis presents no reason to doubt the quality of the model. Further details are found in Sorribas, Curto, and Cascante (1995).

Robustness of the Model

The sensitivities and logarithmic gains indicate how robust the system is. As the mathematical representation of a real metabolic pathway, the model should not change drastically if either a system parameter or an independent variable is slightly altered. As discussed in Chapter 2, these types of responses are captured with sensitivities and logarithmic gains. Again, sensitivities and gains could in theory be computed by hand, but a numerical characterization with PLAS is clearly the method of choice.

Sensitivities. The sensitivities of rate constants for both fluxes and intermediate metabolites are all below 2, and most of them (90%) are below 1. The situation is rather similar with regard to the kinetic order sensitivities. Among a total of 115 kinetic order sensitivities with respect to metabolites, only four are above 1, and the highest sensitivity, namely $S(X_4, h_{4,4})$, has a value of about 4. The kinetic order sensitivities with respect to fluxes are similarly well constrained. The overall interpretation of the sensitivity analysis is that the system is robust, yet responsive. For further details see Cascante, Curto, and Sorribas (1995).

Logarithmic gains. The gains are of twofold – and in some sense opposite – interest. On one hand, large gains may be an indication that the model was not constructed appropriately, because a small alteration in an independent variable would cause unreasonably strong responses in some metabolites or gains. On the other hand, it is the goal of biotechnological manipulation to evoke significant changes in a

Table 7.3. Logarithmic Gains L(V_i, X_j) of the Fermentation Pathway with Respect to Fluxes (Cascante, Curto, and Sorribas 1995)

	V_1	V_2	V_3	V_4	V_5
$L(V_i, X_6)$	0.694	0.694	0.689	0.689	0.688
$L(V_i, X_7)$	-5.04×10^{-5}	-5.04×10^{-5}	-5.10×10^{-5}	-5.12×10^{-5}	-5.14×10^{-5}
$L(V_i, X_8)$	0.2201	0.220	0.224	0.224	0.223
$L(V_i, X_9)$	9.59×10^{-5}	9.59×10^{-5}	9.73×10^{-5}	-0.000266	-0.000423
$L(V_i, X_{10})$	-0.0105	-0.0107	0.0456	0.0464	0.0464
$L(V_i, X_{11})$	0.000392	0.000392	-0.000602	-0.000602	-0.000601
$L(V_i, X_{12})$	0.0103	0.0103	0.0105	-0.0455	-0.0454
$L(V_i, X_{13})$	0.0866	0.0866	0.0879	0.0880	0.0883
$L(V_i, X_{14})$	-0.000186	-0.000186	-0.000189	0.000325	0.000820

metabolite or a flux, and if all gains were very small, the pathway would not allow any real improvements with reasonable effort.

The logarithmic gains with respect to fluxes are presented in Table 7.3. It is interesting to note that their magnitudes vary widely across the pathway. Glucose transport (X_6) has the strongest effect on the system, followed by phosphofructokinase (X_8); none of the other activities has an appreciable effect on any of the fluxes. In addition to the effect exerted by enzyme activities, the effect of the [NADH]/[NAD$^+$] ratio (X_{14}) may be considered, but is negligible in this case. Cascante, Curto, and Sorribas (1995) provide further detail.

The logarithmic gains with respect to metabolites are presented in Table 7.4. Examination of the results shows that no independent variable has a particularly strong effect on any parts of the system. At the same time, the degree of impact among affected metabolites varies by orders of magnitude. The effect of glucose transport (X_6) on the concentration of phosphoenolpyruvate (X_4) is noticeable (1.31), whereas the effect of hexokinase activity (X_7) on phosphoenolpyruvate (X_4) is essentially negligible (3.78×10^{-6}). The highest value in magnitude is L(X_4, X_{10}) = -1.77, which implies that a 1% increase in pyruvate kinase activity leads to an approximate

Table 7.4. Logarithmic Gains L(X_i, X_j) of the Fermentation Pathway with Respect to Metabolites (Cascante, Curto, and Sorribas 1995)

	X_1	X_2	X_3	X_4	X_5
X_6	0.907	1.31	0.975	1.31	0.679
X_7	-1.34	0.000215	-0.000195	3.78e-06	0.000529
X_8	0.287	-0.937	0.314	0.425	0.229
X_9	0.000161	-0.000409	-1.62	0.152	-0.00101
X_{10}	-0.0177	0.0449	0.0505	-1.78	0.111
X_{11}	0.000577	-0.00167	-0.000642	-0.00131	-0.00158
X_{12}	0.0173	-0.0440	-0.0509	-0.0973	-0.108
X_{13}	0.146	-0.369	0.336	-0.00649	-0.909
X_{14}	-0.000313	0.000793	0.989	-0.0918	0.00195

decrease in phosphoenolpyruvate that is still less than 2%. All other responses are even smaller. The system is thus responsive to perturbations, but it does not react in a disastrous fashion, where a small perturbation would compromise the balance of metabolites. The [NADH]/[NAD$^+$] ratio (X_{14}) generally has little influence on the system. A notable exception is the effect on X_3 (FDP), which is understandable because X_3 is the substrate of the reaction that is directly modulated by X_{14}. The same influence is observed for X_9, but in the opposite direction. The reason here is that X_9 catalyzes the flux from X_3 to X_4, a process that is inhibited by X_{14}.

Dynamic Behavior

Even if a model is locally stable and robust in terms of sensitivities and gains, it may be unreasonable in its dynamics. For instance, one could imagine that the model would take hours to return to the steady state, even if it were perturbed only very slightly. Shiraishi and Savageau (1992c) discussed such a situation for a model of the tricarboxylic acid cycle, where an extremely long relaxation time indicated flaws in the model. Further diagnostics of factors contributing to the time constants of the model led to a much improved model.

Sorribas, Curto, and Cascante (1995) executed many simulations with the fermentation model discussed here. They analyzed system responses after increases and decreases in the dependent variables X_1, X_4, X_5 and studied the effects of alterations in the independent quantities X_{10} and X_{14}. Overall, the authors came to the conclusion that the model responds dynamically to perturbations in a very reasonable fashion: none of the transients were found to be extraordinary, no sustained oscillations were observed, and the time for the system to return to the steady state was generally short.

The analysis of stability, gains, sensitivities, and dynamics demonstrates that the S-system model is robust. Small changes in initial values, independent variables, or parameters do not alter the steady state much and the dynamic responses to perturbations seem reasonable. The robustness of the model is no guarantee that the model is in some sense "correct," but it does strengthen our confidence when we proceed to different optimization tasks.

OPTIMIZATION: THE INDIRECT OPTIMIZATION METHOD (IOM)

Our model system for ethanol fermentation in *Saccharomyces cerevisiae* fulfills four basic requirements that make a biotechnological process interesting for optimization by means of mathematical modeling. First, it deals with the production of an important commodity. Second, its basic microbiology and biochemistry are well known. Third, a sound mathematical description is available for integrating this basic information. And fourth, a considerable body of scientific and technical expertise renders the actual implementation of DNA vectors in modified strains of *S. cerevisiae* feasible (Heinish 1986; Schaaff, Heinish, and Zimmermann 1989; Davies, and Brindle 1992; Archer, Mackenzie, and Jeenes 2001).

The optimization of the fermentation process proceeds in a fashion similar to that of citric acid production in Chapter 5. However, the fact that the original model was presented in terms of generalized Michaelis–Menten rate laws affords us the opportunity to compare direct S-system optimization with an indirect linear optimization strategy that uses the S-system formulation as an intermediate and with a direct nonlinear optimization of the Michaelis–Menten model.

When a biochemical pathway is represented with rational functions, such as Michaelis-Menten rate laws, linear optimization is not directly possible. A major stumbling block is the fact that these models are not amenable to algebraic methods for computing steady states. Steady-state conditions are crucial in many types of optimization, and the lack of analytical computability and the corresponding need for repeated numerical solutions makes optimization computationally expensive. Thus, if one insists on optimizing the rational function model, one has to make the decision either to forego the advantages of linear optimization or to make the pathway model somehow adhere to the S-system form. In the former case, problems of nonlinear programming are involved, which may or may not be solvable with reasonable effort (see Chapter 4). The latter case calls for a stepwise and sometimes iterative process, in which the rational functions are approximated by power-law terms, optimization is performed with the resulting S-system, the optimization results are transferred back to the original Michaelis–Menten model representation, and the results are evaluated within this representation. This stepwise procedure has been coined the *Indirect Optimization Method* (Torres et al. 1997). In the following, we will describe this method and compare the optimization results from the IOM with those obtained from the direct S-system optimization and with results obtained from a nonlinear optimization routine based on a multistart stochastic algorithm.

Direct S-System Optimization

The linear program for optimizing a flux under typical constraints consists of two major components: the objective function and a set of constraints. As shown in Chapter 5, the objective function for a typical S-system model reads

> maximize ln(flux).

Maximizing the logarithm of the flux instead of the flux itself converts the product of power-law functions to linearity in the logarithms of the involved dependent and independent variables.

The constraints fall into three categories. First, the *steady-state constraints* ensure that the optimized solution corresponds to a steady state, no matter how the altered enzyme activities are selected. Secondly, constraints on dependent and independent variables force metabolites and enzyme activities to stay within preset limits. Again, these constraints are expressed in terms of logarithms to maintain linearity (in logarithmic coordinates) throughout the entire optimization program. These constraints thus have the general form

$$c_{\text{lower bound}} \leq \ln(\text{dependent or independent variable}) \leq c_{\text{upper bound}}.$$

As a special case, the setting $c_{\text{lower bound}} = c_{\text{upper bound}}$ forces the corresponding variable to be fixed at a given value. The third category contains *stoichiometric constraints*. In the optimization process, the system deviates from the original steady-state operating point, and as a consequence of the particular approximation that underlies the S-system model, some of the stoichiometric relationships that hold for the base solution are no longer strictly satisfied. To avoid significant stoichiometric deviations, one may add an additional restriction that reflects the required dependency among fluxes. This type of constraint may seem unusual at first but reflects the price that must be paid for the luxury of using linear programming. Much has been written about the deviation of S-system models from precise flux stoichiometry, but experience has demonstrated that these deviations are often negligible. Later in this chapter, we will have the opportunity to assess this question first hand. When pondering these constraints, one must not forget that all mathematical models are approximations that balance mathematical and computational convenience with inaccuracies in representation. In the particular case of flux aggregation at branch points, which distinguishes GMA and S-systems, the flux stoichiometry is better represented by GMA models, but the metabolite concentrations are more accurately modeled with S-systems, when the two alternatives are compared to Michaelis–Menten models (Voit and Savageau 1987).

A typical stoichiometric constraint limits the ratio of two fluxes. Again in logarithmic coordinates, such a constraint reads

$$\ln(\text{flux}_1/\text{flux}_2) \leq \text{constant}.$$

The next sections describe the implementation of the linear program for the specific case of the fermentation pathway in yeast.

Objective Functions

The objective function clearly depends on the particular goals of the optimization. As an illustration, we consider three optimization tasks that address the maximization of different output fluxes. The first task is maximization of the flux through the pyruvate kinase reaction, V_4^- (Eq. 7.1). This task corresponds to the typical fermentation goal of maximizing ethanol production. The second and third tasks will be maximization of the fluxes toward glycerol and carbohydrates, respectively. These fluxes are quantitatively marginal, but serve here as additional illustration of different aspects of the indirect optimization approach.

Maximization of the rate of ethanol production. The S-system term representing ethanol flux V_4^- corresponds to the pyruvate kinase step and is given as V_4^- (V_{PK} in Eq. 7.1). It reads

$$V_4^- = V_{Eth} = 0.0945 \cdot X_3^{0.05} X_4^{0.533} X_5^{-0.0822} X_{10}. \tag{7.8}$$

Taking logarithms and rearranging terms, we obtain as the objective function

$$0.05 y_3 + 0.533 y_4 - 0.0822 y_5 + y_{10}, \tag{7.9}$$

where $y_i = \ln(X_i)$. As discussed in Chapter 5, the constant term $\ln(0.0945)$ is irrelevant for the optimization algorithm, but must be remembered later for the evaluation of the optimized rate.

Maximization of the rates of glycerol (V_{Gol}) and carbohydrate (V_{Pol}) production.
These production fluxes do not explicitly appear in the current S-system model, even though they are implicitly part of it. The reason they are not apparent is that glycerol and the carbohydrate pool are not modeled as dependent variables and that their production fluxes are aggregated within the overall degradation of fructose 6-phosphate and glucose 6-phosphate, respectively. Nonetheless, these fluxes are directly obtained from Galazzo and Bailey's model just like the other fluxes. In particular, the power-law terms are computed from

$$V_{Gol} = \frac{X_{12} \cdot X_4 \cdot ADP \cdot (2.519 \cdot R_2 + 0.656 \cdot T_2 \cdot L_2^2)}{1.0832 \cdot (R_2^2 + 164.084 \cdot L_2^2 \cdot T_2^2)}$$

$$\approx \beta_{Gol} \cdot X_3^{h_{Gol,3}} \cdot X_4^{h_{Gol,4}} \cdot X_5^{h_{Gol,5}} \cdot X_{12}^{h_{Gol,12}} \tag{7.10}$$

(see Eq. 7.5 for definitions of R_2, T_2, and L_2) and

$$V_{Pol} = \frac{1.1 \cdot X_{11}}{\left(1 + \left(\dfrac{2}{X_2}\right)^{8.25}\right)\left(\dfrac{1.1}{0.7 \cdot X_2} + 2.43\right)} \approx \beta_{Pol} \cdot X_2^{h_{Pol,2}} \cdot X_{11}^{h_{Pol,11}}, \tag{7.11}$$

where the kinetic orders are again computed through partial differentiation, and the rate constants are obtained from equating the rational functions and the power-law terms at the steady state (see Chapter 2). Curto, Sorribas, and Cascante (1995) executed these operations and obtained

$$V_{Gol} = 0.093 X_3^{0.05} X_4^{0.533} X_5^{-0.0822} X_{12} \tag{7.12}$$

$$V_{Pol} = 8.9 \cdot 10^{-4} X_2^{8.6107} X_{11}. \tag{7.13}$$

Again taking logarithms, rearranging terms, and omitting the constant parts yields the linear objective functions

$$0.05 y_3 + 0.533 y_4 - 0.0822 y_5 + y_{12} \tag{7.14}$$

and

$$8.6107 y_2 + y_{11}, \tag{7.15}$$

respectively, for the two optimization tasks.

Constraints

The steady-state constraints are directly derived from the model Eqs. (7.1), as was shown in Chapter 2. Expressed in terms of the logarithms of the dependent and independent variables, these constraints take the following form:

$$-0.7464 y_1 - 0.2344 y_2 - 0.024 y_5 + y_6 - y_7 = 1.261$$

$$0.7464 y_1 - 0.739 y_2 + 0.418 y_5 + y_7 - y_8 - 0.001 y_{11} = -1.699$$

$$0.731 y_2 - 0.58 y_3 + 0.03 y_4 - 0.513 y_5 + y_8 - 0.94 y_9 - 0.056 y_{12}$$
$$+ 0.575 y_{14} = -3.561$$

$$0.5659 y_3 - 0.533 y_4 + 0.213 y_5 + y_9 - y_{10} - 0.608 y_{14} = 1.456$$

$$-0.198 y_1 - 0.196 y_2 + 0.333 y_3 + 0.266 y_4 - 0.348 y_5 - 0.265 y_7 - 0.265 y_8$$
$$+ 0.5 y_9 + 0.5 y_{10} - 0.002 y_{11} - 0.47 y_{13} - 0.304 y_{14} = 33.039. \qquad (7.16)$$

Within the category of variable constraints, we consider two types, those referring to metabolite concentrations (dependent variables) and those related to enzyme activities and other independent variables. We begin with the independent variables.

Constraints on independent variables allow us to determine which enzyme activities may vary and by how much. In line with biotechnological feasibility, a rule of thumb is to allow enzymes to vary between 1 and 50 times their base values. However, to ensure that metabolism outside the fermentation processes remains relatively unperturbed, those enzymes diverting flux from the target pathway are kept constant at the base steady-state values. With these settings, we obtain the following constraints on the control variables for each of the flux maximization tasks.

Ethanol maximization. For this task, enzymes whose activities are kept constant at their base values are glycogen synthetase (X_{11}) and glycerol 3-phosphate dehydrogenase (X_{12}). It might be helpful to identify the role of these processes in Figure 7.1. The constraints on independent variables are thus

$$2.980 \leq y_6 \leq 6.892$$
$$4.226 \leq y_7 \leq 8.138$$
$$3.456 \leq y_8 \leq 7.368$$
$$3.910 \leq y_9 \leq 7.822$$
$$8.143 \leq y_{10} \leq 12.055$$
$$y_{11} = 2.66$$
$$y_{12} = 5.313$$
$$3.222 \leq y_{13} \leq 7.134.$$

Glycerol maximization. Enzymes whose activities are kept constant at their base values in this case are glyceraldehyde 3-phosphate dehydrogenase (X_9), pyruvate kinase (X_{10}), and glycogen synthetase (X_{11}). As in the previous case, we originally allowed the other enzymes to vary in activity between 1 and 50 times their basal value. However, subsequent analysis of the optimized solution revealed significant constraint violations when the optimized setting was implemented in the Michaelis-Menten model. These discrepancies disappear when the upper limits of enzyme variations are set to 10 times rather than 50 times the base value (see section on Quality Assessment) suggesting ten as the maximum increase in enzyme activity. The corresponding constraints are therefore

$$2.980 \leq y_6 \leq 5.283$$
$$4.226 \leq y_7 \leq 6.529$$
$$3.456 \leq y_8 \leq 5.758$$
$$y_9 = 3.91$$
$$y_{10} = 8.143$$
$$y_{11} = 2.66$$
$$5.313 \leq y_{12} \leq 7.615$$
$$3.222 \leq y_{13} \leq 5.525.$$

Carbohydrate maximization. As in the case of ethanol production, enzyme activities are allowed to vary between 1 and 50 times their base value, which yields

$$2.980 \leq y_6 \leq 6.892$$
$$4.226 \leq y_7 \leq 8.138$$
$$3.456 \leq y_8 \leq 7.368$$
$$y_9 = 3.91$$
$$y_{10} = 8.143$$
$$2.66 \leq y_{11} \leq 6.57$$
$$y_{12} = 5.313$$
$$3.222 \leq y_{13} \leq 7.134.$$

Enzymes whose activities are kept constant at base value are glyceraldehyde 3-phosphate dehydrogenase (X_9), pyruvate kinase (X_{10}), and glycerol 3-phosphate dehydrogenase (X_{12}).

Constraints on dependent variables. Analogous to the constraints on enzymes, we limit the range of variation in the metabolite concentrations (i.e., the dependent variables). For all three optimization tasks, the lower and upper limits of the dependent variables, X_1 to X_5, are set to 0.8 and 1.2 times their base values, which corresponds to 20% variation about the steady-state levels. This range is presumed small enough to avoid significant changes in the overall functioning of the organism.

The ratio NADH/NAD$^+$ (X_{14}) may be treated as an independent control variable but should probably be kept constant at the value proposed by Galazzo and Bailey (1990). Fluorescence measurements performed by these authors indicated small variations in the ratio, which may be explainable by the participation of NAD$^+$ and NADH in other pathways, which tends to buffer fluctuations. Accordingly, the mathematical formulation in logarithmic coordinates leads to the following constraints on the metabolite concentrations:

$$-3.604 \leq y_1 \leq -3.199$$
$$-0.213 \leq y_2 \leq 0.192$$
$$1.989 \leq y_3 \leq 2.395$$
$$-4.879 \leq y_4 \leq -4.474$$
$$-0.109 \leq y_5 \leq 0.304.$$

Stoichiometric constraints. The only stoichiometric conservation constraint evoked here takes the form

$$(V_1^+)/(V_4^-) > 0.5, \tag{7.17}$$

which ensures that the ethanol production rate V_4^- cannot be greater than twice the input flux. The factor 2 accounts for the splitting of each fructose diphosphate molecule (X_3) into two molecules of phosphoenolpyruvate (X_4).

After substituting the power-law terms for V_1^+ and V_4^- from Eq. (7.1) and taking logarithms, the flux constraint becomes

$$-0.2344y_2 - 0.05y_3 - 0.533y_4 + 0.0822y_5 + y_6 - y_{10} > -2.842. \tag{7.18}$$

It should be noted that other stoichiometric conservation constraints could be imposed instead of, or in addition to, Eq. (7.18). For instance, we could require that the flux through the branchpoint fructose diphosphate (X_3) cannot be greater than the glucose input flux ($V_3^-)/(V_1^+) < 1$) or that the flux toward glycerol should be less than twice the glucose uptake ($V_{Gol})/(2 \cdot V_1^+) < 1$). Again, these constraints become linear, upon logarithmic transformation. Detailed analyses indicate that constraint (7.18) prevents significant deviations in flux stoichiometry and that additional constraints do not change the results obtained with (7.18) as the sole constraint.

RESULTS

In contrast to the results of the optimization of citric acid production in *A. niger*, which focused exclusively on optimization with the S-system model, the results here are comparative. They show the optimal profiles of control variables that are computed through linear programming with the S-system model and, for comparison, implement these profiles in the corresponding Michaelis–Menten model that was proposed by Galazzo and Bailey (1990, 1991). Because the S-system and the Michaelis–Menten model are two different mathematical representations, the results are generally different. Which of the two models provides a representation closer to the truth is an unanswered question.

Maximization of Ethanol Production

As mentioned in previous sections, one purpose of this chapter is to use different objective functions for illustration of the indirect optimization method. The first task is maximization of ethanol production.

The lower and upper limits of the dependent variables, the value of the NADH/NAD$^+$ ratio (X_{14}), the ranges of admissible variations in metabolites and enzyme concentrations, and the flux stoichiometric conservation constraints are defined as stated in the constraint sections. Results are shown in Table 7.5.

The first block of results shows the computed optimal profile of enzyme activities. One notices that the optimized solution requires only moderate levels of overexpression of the enzymes involved (factors ranging from 3.15 to 4.25). These levels are attainable in yeast without major difficulties (see Heinish 1986; Schaaff et al. 1989; Davies and Brindle 1992; Smits et al. 2000).

The second and third blocks of results in Table 7.5 show the resulting steady-state levels of metabolite concentrations and of the fluxes of primary interest. For the S-system, which was used to optimize the enzyme profile, these levels are a direct consequence of the optimization procedure. By contrast, the levels in the Michaelis-Menten model are indirectly obtained. Specifically, they are computed by replacement of the original enzyme levels with those computed for the optimal S-system and by subsequent computation of the steady state of the Michaelis-Menten model. Consequently, their values differ from those of the S-system model. In some cases, the differences are relatively small, but in other cases they are noticeable. For instance, internal glucose (X_1) is predicted to decrease by a factor of 0.8 in the S-system model and by a factor of 0.96 in the Michaelis–Menten model.

Overall, the solution of the Michaelis–Menten model corresponds to a steady state that leads to an increased rate of ethanol production that is 3.48 times higher than the base steady-state rate and is comparable with the optimal solution obtained directly with the S-system. As is to be expected, some of the variables are close to the imposed constraints, and because the enzyme activities are only to be altered slightly, the dominating constraints are found among the metabolites. In fact, because of the mathematical differences between the S-system and the Michaelis–Menten model, three intermediate metabolites actually exceed their limits: X_2 (glucose 6-phosphate) shows the strongest deviation observed (73%); X_3 (fructose-bisphosphate) and X_4 (phosphoenolpyruvate) follow with 42% and 57%, respectively. It is interesting to note that most fluxes are only slightly elevated in the Michaelis–Menten model, but that V_{Pol} is very much higher than predicted by the S-solution. Whether these differences are a cause for concern is unknown. If so, it would be advisable to revisit the constraints imposed on the various quantities in the optimization task.

A related, but different issue is the absolute quality of the results for the Michaelis–Menten model. After all, these results are obtained indirectly by optimizing the corresponding S-system form, which reduces the nonlinear optimization problem to a much simpler linear task. A comparison between the results here and the results from a nonlinear optimization is presented in a later section of this chapter, but it might be noted here that the results are generally rather similar.

Changing the lower or upper limits of the enzyme concentrations (with ranges varying from 0.01 to 100) does not alter the observed enhancements, and the control variables remain unaltered. This is not surprising, because none of the control variables in the previous scenario were even close to the allowed limits. By contrast, when one raises the limits for the metabolite concentrations, for instance, allowing up to 90% variation about the base levels, the ethanol production increases five-fold over the original value (details not shown).

The yield of ethanol production (expressed as $100 \cdot 0.5 \cdot V_4^- / V_1^+$) is 100% for the S-system solution; all rates of synthesis increase, and the rate of ethanol production is enhanced by a factor of 3.2. The overall ethanol yield for the Michaelis–Menten model with S-system optimized enzyme profile is 95.82%, which is comparable to

Table 7.5. Optimization of the Ethanol Rate Production in *Saccharomyces cerevisiae* in Suspended Yeast Cells at pH 4.5[a]

	Base Solution	S-system Solution	Michaelis-Menten Solution
	$(X_i)_{base}$	$X_i/(X_i)_{base}$	$X_i/(X_i)_{base}$
Optimized enzyme level			
X_6	19.7	3.15	3.15*
X_7	68.5	3.58	3.58*
X_8	31.5	2.42	2.42*
X_9	49.9	2.94	2.94*
X_{10}	3440	2.82	2.82*
X_{13}	25.1	4.25	4.25*
	$(X_i)_{base}$	$X_i/(X_i)_{base}$	$X_i/(X_i)_{base}$
Metabolite level in optimized system			
X_1	$3.4 \cdot 10^{-2}$	0.8	0.96**
X_2	1.01	1.2	1.73**
X_3	9.14	1.2	1.42**
X_4	$9.5 \cdot 10^{-3}$	1.2	1.57**
X_5	1.13	0.8	0.85**
	$(V_i)_{base}$	$V_i/(V_i)_{base}$	$V_i/(V_i)_{base}$
Flux level in optimized system			
V_{Gol}	1.77	1.11	1.25**
V_{Pol}	$1.4 \cdot 10^{-2}$	4.8	83.84**
V_{Eth}	30.11	3.2	3.48**

[a] Enzyme activities and fluxes are given in mM/min and concentrations in mM.
* Computed through linear optimization of the corresponding S-system model.
** Computed from the Michaelis–Menten model upon replacing values of the control variables with those optimized in the S-system model.

the yield of the base steady state, but with a rate of ethanol production that is almost three-and-a-half times higher.

A consequence of the aggregation of fluxes in the S-system representation is typically a deviation of the total output flux from the total input flux. This type of deviation has been discussed widely in the literature and used as an argument against using the S-system representation (for a recent critique, see de Atauri et al. 1999). In the present case, the deviation of the total output flux from the total input flux is a mere 2.1%, which demonstrates how small the aggregation error really is. The small magnitude of error has to be seen in light of the advantages of the S-system representation, which is the crucial basis of the linear programming approach.

Inspection of eigenvalues confirms that the optimized S-solution is stable (details not shown).

The solution shown in Table 7.5 requires quantitatively prescribed changes in all enzyme activities. Such an implementation necessitates considerable experimental

Table 7.6. Optimized Solutions Obtained for Subsets of Enzymes Involved in Ethanol Production in *Saccharomyces cerevisiae*

Number of Enzymes	Normalized Modulated Enzyme Activities $X_i/(X_i)_{base}$						$V_{Eth}/(V_{Eth})_{base}$ MM
	X_6	X_7	X_8	X_9	X_{10}	X_{13}	
1	1.15						1.12
2	1.15					1.14	1.13
3	1.06		1.4		1.28		1.13
4	1.09		1.45	1.31	1.35		1.16
5	1.30	1.1		1.5	1.42	1.61	1.29
6	3.15	3.58	2.42	2.94	2.82	4.25	3.48

work and requires either the introduction of strong promoters or the use of additional copies of the relevant genes, for instance, by integration or by using multicopy vectors. A pertinent question, therefore, is whether similar results could be achieved with the modulation of fewer enzymes. Accordingly, it is useful to explore the minimum subset of enzymes necessary to produce a solution that might only be slightly inferior to the optimum. The results for the maximization of ethanol are shown in Table 7.6. For each number of modulated enzymes, the combination of steps shown in Table 7.6 yields a stable, feasible solution that corresponds to the highest possible target flux under the given conditions.

It becomes evident from Table 7.6 that the entire set of six enzymes has to be modulated for significant increases in ethanol flux. This result reflects the same pattern that emerged during the optimization of citric acid production in *Aspergillus niger* (Chapter 6). The results in Table 7.6 also indicate the order in which enzymes should be modulated to obtain progressively better solutions. Foremost is the rate of substrate uptake (X_6), which is followed closely by ATPase activity (X_{13}). These two steps increase flux by about 10%. The next enzymes to be modulated should be phosphofructokinase (X_8) and pyruvate kinase (X_{10}). Although least important when modulated alone, additional modulation of GADP (X_9) and the hexokinase (X_7) leads to significant flux increases.

Maximization of Glycerol and Carbohydrate Production

As an additional illustration of the method, it is instructive to consider improvements in the production rate of glycerol and carbohydrates. Experimental efforts to maximize these secondary outputs are dwarfed by the maximization of ethanol yield. Nevertheless, the entire pathway will be used later as a model system to study other aspects of yield optimization, such as quality assessment, multi-objective optimization, and the design of a given metabolic system with respect to a chosen target, and it is therefore useful to explore alternative optimization objectives.

As before, we limit the dependent variables (X_1 to X_5) to a range between 0.8 and 1.2 of the base values and set the NADH/NAD$^+$ ratio (X_{14}) constant. Also as previously, the enzyme concentrations are generally allowed to vary between 1 and

50 times the base values. A difference is the flux toward glycerol, or carbohydrates, respectively. For the optimization of glycerol production, the controlling enzyme X_{11} of the branch point flux V_{Pol} and the enzymes downstream from fructose diphosphate, namely X_9, X_{10}, and X_{13}, are kept constant. Similarly, for carbohydrate maximization, the enzymes X_8, X_9, X_{10}, and X_{13} are kept constant. This design facilitates the flux toward glycerol (or carbohydrates) without decreasing other fluxes below the base steady-state level. The flux stoichiometric conservation constraint is the same as shown in Eq. (7.17).

Glycerol maximization. As in the previous case of ethanol maximization, the S-system solution is stable, which is indicated by negative real parts of all eigenvalues of the system. Interestingly, the corresponding Michaelis–Menten solutions are not always stable, if some enzyme activities exceed 15 times their base level. An explanation may be the following. With increasing enzyme activities, the intracellular glucose accumulates steadily, while the ATP concentration decreases to almost zero. If the activity of X_{12} exceeds 15 times its base level, most of fructose diphosphate is converted into glycerol. This leaves no chance for ATP to be replenished and, consequently, no glucose phosphorylation is possible at the hexokinase step, which in turn causes instability of the system. Clearly, this is a shortcoming of the model, which simply does not account for the fact that the cell would have numerous other mechanisms of replenishing ATP. After all, except for water, ADP and ATP are the most "popular" metabolites in a wide range of organisms; they are involved in more reactions than any other metabolites (Jeong et al. 2000).

The situation could be fixed in various ways. One could consist of an expanded model that would allow for sufficiently many other mechanisms of ATP metabolism. Secondly, instead of modeling these numerous mechanisms in detail, they could be collectively represented by a so-called _buffer box_ (Voit and Ferreira 1998). A third option is a set of stricter constraints. For instance, if we allow the enzyme activities to vary up to a maximum of ten, the S-system and the Michaelis–Menten model are both stable, even though some of the metabolites in the Michaelis–Menten exceed the artificial constraints. As it stands, X_5 deviates from the preset maximum by 62% and X_2 attains a steady-state value below the permitted 80% limit. With stricter limits on enzyme activities, these deviations also diminish (results not shown).

Overall, these issues arise because of the limited scope of the model. With only five dependent variables, the system has no wiggle room, and extreme changes cannot be absorbed, as they would be in vivo. Thus, even though the glycolysis model proposed by Galazzo and Bailey (1990, 1991) and subsequently by Curto, Sorribas, and Cascante (1995) covers many observations well, it apparently breaks down if it is not used as intended, namely, in this case, for the increased production of glycerol. Nonetheless, it is instructive to study the responses of the models for alterations of smaller magnitude, because they indicate trends that a more detailed model might capture more accurately.

The only enzyme to be overexpressed in the S-system solution is glycerol 3-phosphate dehydrogenase (X_{12}), whose concentration is to be raised to 10 times the base level (see Table 7.7). This alteration is well within the limits of actual feasibility.

Table 7.7. Optimization of Glycerol Production in S. cerevisiae[a]

	Base Level	S-system Solution	MM Solution
	$(X_i)_{base}$	$X_i/(X_i)_{base}$	$X_i/(X_i)_{base}$
Enzyme			
X_6	19.7	1.20	1.20
X_7	68.5	1.49	1.49
X_8	31.7	1.47	1.47
X_{12}	203.0	10	10
X_{13}	25.1	1.0	1.0
	$(X_i)_{base}$	$X_i/(X_i)_{base}$	$X_i/(X_i)_{base}$
Metabolite level in optimized system			
X_1	$3.4 \cdot 10^{-2}$	0.8	0.89
X_2	1.01	0.8	0.7
X_3	9.14	1.2	0.93
X_4	$9.5 \cdot 10^{-3}$	1.2	1.01
X_5	1.13	0.96	0.38
	$(V_i)_{base}$	$V_i/(V_i)_{base}$	$V_i/(V_i)_{base}$
Flux level in optimized system			
V_{Gol}	1.77	10.95	8.81
V_{Pol}	$1.4 \cdot 10^{-2}$	0.15	0.049
V_{Eth}	30.11	1.11	0.88

[a] Enzyme activities and fluxes are given in mM/min and concentrations in mM.

With this change, the optimized glycerol production is predicted to increase about ten-fold, with a concomitant increase in yield from 5.54% to 48.18%. The rate of carbohydrate synthesis decreases to 15% of the base value and glucose uptake rate increases by 20%. Meanwhile, ethanol production (V_{Eth}) changes only very slightly. These trends are a reflection of the manipulated rerouting of fluxes. Similar results hold for the optimized solution of the Michaelis-Menten model: the glycerol production is more than 8 times greater than at the base steady state, and the overall yield of the conversion of glucose into glycerol is 37.29%.

It is worth noting that these increases in fluxes simply require modulation of the step X_{12}, which is responsible for glycerol synthesis from fructose diphosphate. All other enzymes remain almost unaltered. This is very different from the ethanol maximization and the citric acid optimization in the previous chapter, where three, six, or even more enzymes had to be modulated for any noticeable improvement in yield. One probable reason for this discrepancy may be the fact that glycerol has not been a typical target for experimental optimization. Thus, in contrast to the former systems, this system apparently had not been optimized previously through experimental trial and error. If this explanation holds true, one may deduce that lesser established pathways have a better chance of significant improvement, and with less

Table 7.8. Optimization of Carbohydrate Production in S. cerevisiae[a]

	Base	S-system Solution	MM Solution
	$(X_i)_{base}$	$X_i/(X_i)_{base}$	$X_i/(X_i)_{base}$
Enzyme			
X_6	19.7	1.15	1.15
X_7	68.5	1.3	1.3
X_{11}	14.31	50	50
X_{13}	25.1	1.0	1.0
	$(X_i)_{base}$	$X_i/(X_i)_{base}$	$X_i/(X_i)_{base}$
Metabolite level in optimized system			
X_1	$3.4 \cdot 10^{-2}$	0.8	0.86
X_2	1.01	1.19	1.09
X_3	9.14	1.14	1.13
X_4	$9.5 \cdot 10^{-3}$	1.2	1.13
X_5	1.13	1.09	0.97
	$(V_i)_{base}$	$V_i/(V_i)_{base}$	$V_i/(V_i)_{base}$
Flux level in optimized system			
V_{Gol}	1.77	1.07	1.07
V_{Pol}	$1.4 \cdot 10^{-2}$	234.33	106.77
V_{Eth}	30.11	1.10	1.07

[a] Enzyme activities and fluxes are given in mM/min and concentrations in mM.

effort. Recent results on the carnitine pathway (Alvarez-Vasquez et al. 2002) support this tentative conclusion.

Carbohydrate maximization. The situation for this task is quite similar to that observed for glycerol optimization. Table 7.8 shows the optimized profiles. In the S-system solution, the only enzyme that has to be strongly overexpressed is glycogen synthetase (X_{11}), the enzyme that diverts the flux. Its concentration is to be raised to the maximum of 50 times the original steady-state value. Although such an increase is within the limits of technical and physiological feasibility, it is important to note that even much lesser increases in this enzyme can be expected to show improvements in the rate of carbohydrate synthesis.

In response to the 50-fold rise, the production of carbohydrates in the S-system model increases by a factor of 234.33, while ethanol and glycerol production as well as glucose uptake remain almost unaltered. Overall, this solution predicts production of more carbohydrates without a concomitant loss in ethanol synthesis. Thus, if the end products of interest are ethanol and carbohydrates, the S-solution characterizes a system that is more efficient than the baseline model.

Similar results hold for the Michaelis–Menten model, although numerically to a lesser degree. The optimized enzyme profile leads to increased carbohydrate production of 106.77 times the original, which constitutes 8.1% of the total output flux. The

Table 7.9. Optimized Solutions Obtained for Subsets of Enzymes Involved in Glycerol and Carbohydrate Production in S. *cerevisiae*

Number of Manipulated Enzymes	Normalized Modulated Enzyme Activities						$V_{ij}/(V_{ij})_{base}$ MM
	X_6	X_7	X_8	X_{11}	X_{12}	X_{13}	
Glycerol maximization							
1				—	10		5.91
2	1.18			—	10		7.93
3	1.22	1.02		—	10		8.34
4	1.28	1.07	1.07	—	10		8.81
5	1.30	1.49	1.47	—	10	1.0	8.81
Carbohydrate maximization							
1			—	50	—		33.17
2	1.15		—	50	—		106.77
3	1.15	1.30	—	50	—		106.77
4	1.15	1.30	—	50	—	1.0	106.77

other fluxes are rather similar in the S-system and the Michaelis–Menten solution. Furthermore, both solutions are stable, and the concentrations of all intermediates (X_1 to X_5) satisfy the predefined 20% constraints on variation (see Table 7.8). Although the optimized Michaelis–Menten system shows a higher rate of carbohydrate synthesis and better yields than the base system, it is not as efficient in transforming glucose into carbohydrates as the S-system model.

As in the optimization of ethanol biosynthesis, it is useful to explore the gain in production per enzyme alteration. The optimal results for both glycerol and carbohydrate production are shown in Table 7.9. In all cases the search yielded stable, feasible solutions.

The situation here is different from that observed for ethanol optimization (Table 7.6). In the case of glycerol maximization, modulating just one enzyme (X_{12}) already amplifies glycerol production over 5 times. Subsequent modulation of X_6, and X_7 further increases glycerol synthesis to about 8 times the base value. In fact, modulation of only three enzymes produces essentially the same flux as modulation of five enzymes. Optimization of carbohydrate production indicates a similar pattern. Here the modulation of X_{11} alone produces a 33-fold increase in flux, and the simultaneous modulation of X_6 and X_{11} elevates the flux to its maximum (106.77 times the base flux). These results are of practical interest, because they suggest that only a minimal amount of experimental work may be necessary to obtain strains with improved glycerol or carbohydrate production rates. By contrast, it seems that all presently available experimental means have been explored to optimize the system with respect to ethanol production and that further improvements will require rather complicated alterations of several enzymes simultaneously.

Examination of the optimized ethanol profile (Table 7.5) shows that the highest overexpression factor is slightly above 4. This indicates that constraints such as the cellular economy factor (see Chapter 6) are here probably unnecessary. In fact, imposing a cellular economy constraint with a three- to five-fold protein amplification factor yields optimized metabolite and enzyme profiles quite close to the results tabulated (results not shown).

Now that we have reached several solutions that are mathematically optimal with respect to different tasks, it is prudent to revisit issues of biotechnological implementation and of physiological viability of the predicted solutions. Manipulations of enzyme expression levels are not only theoretically possible, they have already been experimentally demonstrated to be effective, leading to viable strains whose amplification factors range from 5 to 50 times the basal activity (Shaaff et al. 1989; Niederberger et al. 1992; Smits et al. 2000).

As an aside that is intriguing in our context, Niederberger et al. (1992) found that the modified strain increased significantly the production rate of the desired metabolite only after overexpression of five consecutive enzymes, in all cases by factors of 40–50 times the basal enzymatic levels, as it had been predicted by the theoretical optimum solution. As a second note, it is mentioned that the optimal solutions developed so far in this chapter require the simultaneous and coordinated overexpression of most of the enzymes of a pathway. The solutions lead to substantial flux increases, but without significant changes in metabolite levels. This result is consistent with other observations actually seen in cells under in vivo conditions (see Stephanoupoulos and Simpson 1997; Thomas and Fell 1998).

Smits et al. (2000) recently overexpressed the enzymes of the lower part of the glycolytic pathway. Interestingly, the recombinant *S. cerevisiae* strains showed the same specific ethanol production rates as the host strain. Also, the physiological behavior of the recombinant and the host strains was similar. Increased glycolytic flux was only observed under conditions of ATP demand. The authors concluded that all glycolytic enzymes need to be overexpressed simultaneously for a substantial enhancement of the glycolytic flux and that such a strategy should be limited by the total amount of protein synthesis. In a sense, this work constitutes an experimental confirmation of the predictions we based on our model and optimization method. Additional assessments of the quality of predictions are the subject of the following section.

QUALITY ASSESSMENT OF THE INDIRECT OPTIMIZATION METHOD

The IOM approach constitutes a compromise. It calls for taking a model, which we assume to be valid for one reason or another, and transforming it into an S-system model. In the example given, the starting model was the Michaelis–Menten type model of glycolysis proposed by Galazzo and Bailey (1990, 1991). The compromise arises because the S-system model is not an exact analogue of the original model but rather an approximation. The particular approximation has the advantage of linearity at steady state. This advantage is considerable because it replaces costly numerical approximations of the steady state, which would be necessary for Michaelis-Menten or most other representations, with analytical methods of linear algebra.

As part of our analyses, we compared the IOM results against those of the direct S-system optimization. This was illustrative but did not answer the question of how

IOM compares with a direct nonlinear optimization of the Michaelis–Menten model. This comparison is explored now. As examples we use again all three optimization tasks, namely synthesis of ethanol, glycerol, and carbohydrates in *S. cerevisiae*.

Many procedures have been developed for the purpose of nonlinear optimization; we show here the results of a multistart stochastic search algorithm. This type of nonlinear algorithm was inspired by ideas of evolution and fitness (Michalewicz and Attia 1994). It has been successfully applied to a wide range of optimization problems, such as wire routing and scheduling, and to a number of optimal control problems. However, applications to the development of improved, viable strains for biotechnological processes have been very limited (Michalewicz 1992).

Formulation of the Direct Optimization Approach (DOA)

The implementation of the direct optimization approach can be divided into three main steps: Steady-state computation and definition of the optimization task, algorithmic computation, and assessment of results.

Computation of the steady-state solution and definition of constraints and objective functions. The steady-state solution may be computed in two very different ways. In the first approach, one equates the differential equations to zero and lets a search algorithm find the solution(s) in terms of values for all dependent variables. The second approach consists of the numerical integration of the differential equations. Making the (usually valid) assumption that the system is stable, one starts the integrator with reasonable initial values and grinds out the numerical solution, which should converge to the steady state.

Once the steady state is obtained, it should be analyzed with respect to stability and robustness. Again, due to the inherent nonlinearity of typical kinetic models, such an assessment is not as straightforward as in the S-system case (for a detailed discussion, see Shiraishi and Savageau 1992a). Except for the steady-state constraints, which cannot be expressed explicitly and must be computed anew for every iteration of the optimization algorithm, other constraints are essentially the same as before. In many cases, stoichiometric constraints are not needed, because a typical kinetic model, for instance, in Michaelis–Menten form, preserves flux stoichiometry automatically. The objective function is typically the same as in the S-system optimization. Throughout the optimization program for the original kinetic model, the variables are retained in Cartesian coordinates.

The optimization engine: A multistart search algorithm. Among different choices available, we are here using a multistart, improved hill-climbing search algorithm. As all hill-climbing routines, this algorithm uses an iterative technique of improvement, which means that during each iteration a new point of the search space is selected from within a close neighborhood of the current state. If the new state provides an improved value of the objective function, and if the metabolite concentrations and enzyme activities are within the imposed constraints, the new state becomes the current solution. Otherwise, a different neighbor is selected randomly and tested against the current solution. The search terminates if no further improvement is achieved.

At steady state, the model reduces to a set of nonlinear algebraic equations, where the independent variables (R_j) are the enzyme activities and the dependent variables (Z_i) are the metabolite concentrations. (Because enzymes and metabolites play different roles, it is useful to distinguish them through the notation R_j versus Z_i). Such system of equations must be solved for each iteration by a suitable method. We use the well-established Levenberg–Marquardt algorithm (Levenberg 1944; Marquardt 1963; Moré, Burton, and Kenneth 1980). The optimization routine has the following form:

```
begin
    define objective function (F)
    select lower and upper bounds for Z_i(i = 1 : n)
repeat
    repeat
        select current array R_j(j = 1 : m) at random
        solve system (Z)
    until lower-bound-i < Z_i < upper-bound-i
    compute F = F(Z, R)
    repeat
        select new array R'_j in the neighborhood of R_j
        solve system (Z)
        compute F' = F(Z, R')
        if F < F' and lower-bound-i < Z_i < upper-bound-i
            then F = F', R = R'
    until itmax
until smax
end
```

The termination flag *smax* represents the number of starting points and *itmax* is the number of iterations in each search. Typical values, which we also use here, are *smax* = 5 and *itmax* = 20,000. We found this combination to be sufficient to guarantee convergence to the optimum solution. The values of the enzyme activities (R_j) are determined randomly within a ±5% range of the current value. Provided that the search space is sufficiently smooth, theory guarantees that the method reaches the optimum solution. Because the topology of the search space is not obvious, it is recommended to start the stochastic search from different initial points (i.e., with a slightly different initial profile of enzyme activities) and check whether the same optimal solution is obtained (Michalewicz 1992).

Quality assessment of the solutions. The optimization itself does not address questions of stability, and this mandates that the steady state of the optimized solution must be checked separately. Because the steady state is only available in numerical form, the easiest manner of checking stability is probably a numerical integration of the differential equations, which is initiated with several slight perturbations close to the optimal steady state. If the system returns to the steady state, it is deemed to be stable. Unstable steady-state solutions are discarded.

Implementation of DOA for Ethanol Production

Objective function. The first task is maximization of the ethanol production rate. The mathematical expression to be optimized is taken directly from Galazzo and Bailey (1990, 1991). It consists of the flux through the pyruvate kinase step, $V_4^+ = V_{PK}$ (see Eq. 7.1) and has the form

$$V_{PK} = \frac{V_{PK}^M \cdot X_4 \cdot ADP \cdot (2.519 \cdot R_2 + 0.656 \cdot T_2 \cdot L_2^2)}{1.0832 \cdot (R_2^2 + 164.084 \cdot L_2^2 \cdot T_2^2)}, \tag{7.19}$$

where R_2, T_2, and L_2 were defined (see Eq. 7.5).

Constraints. Again, we must consider three types of constraints, those ensuring steady-state performance, those regulating enzyme activities, and those limiting metabolites; stoichiometric constraints are not necessary, as was discussed.

The steady-state constraints ensure that the optimized system is in a steady state, no matter what the altered enzyme concentrations are. They are expressed most easily as follows:

$$V_{Glcin} - V_{HK} = 0$$
$$V_{HK} - V_{PFK} - V_{Pol} = 0$$
$$V_{PFK} - V_{GADP} - \tfrac{1}{2} V_{Gol} = 0$$
$$2 V_{GADP} - V_{PK} = 0$$
$$2 V_{GADP} + V_{PK} - V_{HK} - V_{Pol} - V_{PFK} - V_{ATPase} = 0, \tag{7.20}$$

where V_{Glcin}, V_{HK}, V_{PFK}, V_{Pol}, V_{GADP}, V_{Gol}, V_{PK}, and V_{ATPase} are the fluxes shown in Figure 7.1.

These constraints look perfectly linear and, indeed, they correspond to the solution of the corresponding stoichiometric system, which has been used by others for optimization purposes (Savinell and Palsson 1992). However, each flux is a nonlinear function of some of the system variables, and because these are the subject of our optimization, the stoichiometric linearity really does not help us.

As before, the constraints on enzyme concentrations prevent all enzymes from exceeding variations above 50 times their basal activities. Also as in the S-system optimization, the activities of those enzymes diverting flux from their target products (R_6 and R_7) are kept constant at their basal steady-state values.

The constraints on metabolite concentrations limit variations to ranges of 20% of the original steady-state levels, as in former optimizations.

Optimal IOM and DOA Solutions for Ethanol Production

Table 7.10 shows the optimal metabolic profile obtained from DOA and IOM, when the objective is ethanol rate maximization. The results indicate that the indirect method yields outcomes that are quantitatively rather similar to the direct optimization.

First, it is predicted that six enzymes are to be amplified for significant improvements in yield. Second, the amplification factors, $X_i/(X_i)_{base}$, in both cases position all six enzymes in a range between 30 and 50 times the baseline activities. Third, the

Table 7.10. Optimal IOM and DOA Profiles for the Maximization of Ethanol Rate Production[a]

	Base Solution	IOM Solution	DOA Solution
	$(X_i)_{base}$	$X_i/(X_i)_{base}$	$X_i/(X_i)_{base}$
Enzyme			
X_6	19.7	39.9	41.50
X_7	68.5	33.21	49.19
X_8	31.5	36.02	47.98
X_9	49.9	41.27	47.91
X_{10}	3440	43.17	49.93
X_{13}	25.1	50	49.48
	$(X_i)_{base}$	$X_i/(X_i)_{base}$	$X_i/(X_i)_{base}$
Metabolite level in optimized system			
X_1	$3.4 \cdot 10^{-2}$	1.2	1.04
X_2	1.01	1.2	1.2
X_3	9.14	1.2	1.18
X_4	$9.5 \cdot 10^{-3}$	1.2	1.2
X_5	1.13	1.2	1.16
	$(V_i)_{base}$	$V_i/(V_i)_{base}$	$V_i/(V_i)_{base}$
Flux level in optimized system			
V_{Gol}	1.77	1.07	1.08
V_{Pol}	$1.4 \cdot 10^{-2}$	4.8	4.65
V_{Eth}	30.11	47.31	53.94

[a] Enzyme activities and fluxes are given in mM/min and concentrations in mM.

increase in the rate of ethanol production, $V_{Eth}/(V_{Eth})_{base}$ is about 50 times, and the yield of the optimized solution is around 100% (94.33% for the linear optimization and 99.88% for the direct one). Overall, it is interesting to note that the system achieves significant increases in ethanol rate production only when all six enzymes are manipulated, and this prediction is independent of the model representation (S-system or Michaelis–Menten).

Similarity between the two solutions extends to the search for the relatively best solution, given that modulation is limited to a subset of enzymes (Table 7.11). The particular combination of enzymes to be altered in each case, the enhancement factor $X_i/(X_i)_{base}$, and the predicted rates of increase, $V_{Eth}/(V_{Eth})_{base}$, are the same in all scenarios that were analyzed with the two models. As mentioned, the system achieves significant increases in ethanol production rate only when all six enzymes are altered; any other set of alterations is of little practical interest. Nonetheless, from an academic point of view, these scenarios provide yet another comparison. Table 7.11 indicates the strong agreement between the results obtained with the direct optimization and those obtained by the IOM approach. Furthermore, the unconstrained solution and the direct optimization lead to similar answers. This suggests that the stoichiometric constraints are not really necessary, even though deviations

Table 7.11. Subsets of Enzymes Leading to Relatively Optimal Ethanol Production in S. cerevisiae[a]

| | Normalized Modulated Enzyme Activities $X_i/(X_i)_{base}$ and Ethanol Fluxes $V_{Eth}/(V_{Eth})_{base}$ | | | | | | | | | | | | | |
| | IOM | | | | | | | DOA | | | | | | |
N	X_6	X_7	X_8	X_9	X_{10}	X_{13}	V_{Eth}	X_6	X_7	X_8	X_9	X_{10}	X_{13}	V_{Eth}
1	1.15						1.1	1.11						1.09
2	1.15					1.14	1.11	1.11					1.16	1.11
3	1.06		1.40		1.28		1.14	1.09		1.24		1.29		1.13
4	1.09		1.45	1.31	1.35		1.17	1.10		1.17	1.18	1.36		1.16
5	1.3	1.07		1.5	1.42	1.62	1.27	1.20	1.17		1.12	1.35	1.53	1.21
6	39.89	33.21	36.02	41.27	43.17	50	47.31	41.50	49.19	47.98	47.91	49.93	49.48	53.9

[a] N is the maximum number of enzymes available for alteration. The optimal profiles for each subset of enzymes are quite similar in both the IOM and DOA approaches.

from stoichiometry were observed in some of the IOM solutions. Qualitatively, both optimization approaches render the same profile of subsets of enzymes to be modulated. Quantitative discrepancies between the two approaches are due to the fact that the Michaelis–Menten model and the S-system model constitute different mathematical representations. These differences are small when relatively tight stoichiometric constraints are imposed.

If only one enzyme is permitted to change, it should be the sugar transport system, X_6, which is also to be manipulated in all other cases. It is noted that X_6 is associated with the highest logarithmic gains across all metabolites (Table 7.4), which indicates that gains are often good indicators for potentially effective enzyme manipulations.

The optimum profiles can be interpreted in terms of the pathway structure. For instance, consider X_{13} (ATPase), whose strong upmodulation is necessary to achieve maximum production of ethanol (Table 7.10). The upper part of the glycolytic pathway consumes ATP and produces ADP. This ADP is subsequently used in the lower part of the pathway to recoup ATP; thus, the rate of production of ADP is necessary to sustain the steady state. This production is provided in the model by the glycogen/trehalose branch and by the ATPase reaction. However, the glycogen/trehalose branch carries a very small flux and its activity remains essentially fixed, so that the ATPase (X_{13}) reaction is mostly responsible for generating ADP. In fact, the optimum solution (and therefore the best enzyme combination) involving five enzyme modulations includes X_{13}.

If X_{13} is not allowed to vary, the best enzyme combination includes X_8 instead, and the optimum rate of ethanol production (1.19 and 1.2 for the Michaelis–Menten and S-system models, respectively) is slightly inferior (results not shown).

Two main conclusions can be drawn from these comparisons. First it seems that to unravel which enzymatic activities are to be modified and by how much, both the Michaelis–Menten and the S-system models are essentially equivalent. A reason for this similarity lies in the mathematical structure of both representations. In both models – though not in general – the fluxes are proportional to enzyme activities. Therefore, changes in enzyme activities have similar effects on the dependent variables. By contrast, the models differ in their dependence of fluxes on metabolites, at least once the system leaves the operating point, so that changes in metabolites have different effects. However, because the metabolites are forced to remain within prescribed bounds ($\pm 20\%$ of the steady-state value), the consequences of these differences are also small.

A second conclusion from the comparison is that both approaches are even quantitatively quite similar. It is so far unknown how general this result is, but careful comparisons between power-law and Michaelis–Menten models have indicated that the differences are smaller than one might expect (Voit and Savageau 1984; Shiraishi and Savageau 1992a–d; Curto et al. 1997, 1998a,b).

MULTI-OBJECTIVE OPTIMIZATION

The intermediate step of formulating a model as an S-system reduces the optimization of a nonlinear biochemical pathway to a straightforward task of linear programming.

Figure 7.2. Linear pathway converting X_0 into X_1.

$$X_0 \xrightarrow{\;E_1\;} S_1 \xrightarrow{\;E_2\;} S_2 \xrightarrow{\;E_3\;} X_1$$

We are now turning to the much more complicated topic of multi-objective optimization and ask to what degree S-systems could be helpful here as well. It is clear from Chapters 4 and 5 that multi-objective optimization is bound to be difficult if equations and constraints are nonlinear and have no structure that permits simplification. This section outlines how the linear optimization approach, based either on an original S-system itself or on the S-system representation of another kinetic model, can be used for the simultaneous optimization of different objectives. The example again is ethanol production in yeast.

A first way of approaching this question is through the transformation of a multi-objective optimization problem into a mono-objective problem. An example of this approach is the consideration of a flux and one or more metabolite concentrations as the quantities that are to be optimized simultaneously. For instance, one may be interested in the maximization of a particular flux and the minimization of some metabolite. Combining tasks of this nature, the transition time was proposed as a suitable criterion for the optimization of metabolic processes (Torres 1994).

Let us consider a metabolic pathway as depicted in Figure 7.2. The initial substrate X_0 is converted through a linear chain of two intermediates into X_1. E_1, E_2, and E_3 represent the enzymes catalyzing those transformations. Assuming that the system is in steady state, the transition time for this metabolic pathway is defined as the ratio of the sum of metabolite pools over the flux through the system:

$$\tau = \frac{\sigma}{J}, \tag{7.21}$$

where σ represents the sum of the intermediate metabolite concentrations involved in the metabolic pathway,

$$\sigma = \sum_{i=1}^{2} S_i, \tag{7.22}$$

and J denotes the steady-state flux through the metabolic pathway, which coincides with the flux of production of X_1,

$$J = \frac{dX_1}{dt}. \tag{7.23}$$

This definition of the transition time is general and independent of any particular model of metabolism (Easterby 1981). The transition time can be interpreted either as a temporal characterization of the system as it evolves toward its steady state or as the average time a molecule spends inside the system, while this is operating at steady state, during the conversion from substrate to final product. A metabolic system with short transition time reaches the steady state rapidly. It also transforms substrate into final product faster than other systems with longer transition time. It is rather obvious that shortening of the transition time is a desirable goal in biotechnology.

Minimization of the transition time of a given pathway can be achieved by increasing the flux, J, decreasing the total metabolite concentration, σ, or both simultaneously. Searching for the enzyme profile that minimizes τ thus corresponds to seeking the solution that minimizes the metabolite concentrations and simultaneously maximizes the flux through the system.

It is noted as an aside that other authors have proposed similar optimization functions, which in fact are combinations of single objective functions. For instance, Cascante et al. (1996) proposed a *metabolic performance index* for the productivity of metabolic systems, which they defined as $\psi = J^2/\sigma$. The authors claim that ψ is a suitable criterion for evolutionary effectiveness in such a way that the best-adapted metabolic systems should exhibit high values of ψ. Hatzimanikatis, Floudas, and Bailey (1996a) proposed *flux selectivity* as a new function for characterizing a biochemical pathway. It is defined as the ratio of the desired flux over the sum of all possible fluxes, or, in other words, as a "flux fraction." The rationale for this definition was the maximization of a given flux in a network where several final products of interest occur simultaneously. It should also be mentioned that Conejeros and Vassiliadis (1998, 2000) have proposed methods for optimizing the accumulated biomass of a pathway, given a fixed final time. Their method is based on principles of control engineering and will not be discussed further.

The Indirect Multi-Objective Optimization Method (IMOOM)

The multi-objective analogue of the IOM approach is based on the same principles that govern the IOM approach. Specifically, the core of the optimization is again based on the linear steady-state equations of the corresponding S-system model. Thus, the steps of IMOOM are: translation of a model into S-system form, optimization in this form, and translation back into the original model.

In its general form, the formulation of an IMOOM problem may be written as

(1)   maximize $\{ f_1(\ln(\mathbf{X})) \}$
         maximize $\{ f_2(\ln(\mathbf{X})) \}$
         $\vdots$
         maximize $\{ f_k(\ln(\mathbf{X})) \}$
      subject to

(2)   steady-state equations, expressed in logarithms of variables
(3)   ln(dependent or independent variable) $\Leftrightarrow$ constant
(4)   ln(flux) $\Leftrightarrow$ constant
(5)   ln(Σ dependent or independent variables) $\leq$ constant

In comparison with a standard linear program, we now have several (k many) objective functions (1) that must be satisfied simultaneously. In this presentation, all objectives are maximizations, but minimizations are implicitly included if we allow the functions f_i to carry a negative sign. Note that these functions may depend on several or all variables, which is indicated by the vector notation $\mathbf{X}$.

The constraints (2)–(4) are the same as in the mono-objective optimization. We use the symbol $\Leftrightarrow$ to indicate the mathematical comparison of a quantity on the left side with a constant. This comparison could mean: greater or equal, less or equal, equal, or unconstrained. Constraint (5) accounts for limits on tolerable metabolic burden and cellular stress associated with enzyme overexpression and intermediate metabolite accumulation, as discussed in Chapter 6.

Accompanying the more complex mathematics of multi-objective optimization is the need for specific numerical algorithms. We use a software package that is based on an algorithm for linear multitask optimization (Jorge-Santiso 1992). The algorithm is a variant of Zeleny's multi-objective simplex algorithm, which allows the iterative calculation of all efficient extremes of a linear multitask program. This software is not the only choice. Other packages, such as ADBASE (Steuer 1995), which we also used successfully, and EFFACET (Isermann 1977a,b), are available and may be used for the same purposes.

For the multi-objective optimizations, the steady-state metabolite concentrations, enzyme activities, and flux constraints are the same as discussed previously. The stoichiometric constraint is omitted, because this slightly simplifies the system and is justified by the results described.

Mono-Objective Optimization

We begin with a scenario where we simplify the double objective of flux maximization and metabolite minimization to a single-objective task in which the transition time is minimized. This scenario allows us to compare results from IOM and IMOOM. For the ethanol example, we thus consider

$$\tau = \frac{\sigma}{V_4^-}, \tag{7.24}$$

where

$$\sigma = \sum_{i=1}^{5} X_i \tag{7.25}$$

and V_4^- is the ethanol flux V_{Eth} as it was given in Eq. (7.1).

Because the objective function is not in the form of a product of power-law functions, it is approximated by such a product. This is accomplished with methods of partial differentiation that were demonstrated several times (Chapter 2). Other constraints are the same as before. Furthermore, the methods are exactly as discussed in the section on IOM.

To analyze the results of this optimization, we compare them with a true multi-objective optimization, for which we define two simultaneous objective functions, namely, the maximization of V_{Eth} and the minimization of σ. A distinct feature of such a multi-objective optimization is its output: one typically obtains a *family* of admissible, efficient solutions, instead of the unique solution that is typical for mono-objective optimization (see Chapter 5).

Table 7.12. Multi-Objective Optimization (Two Tasks) of Ethanol Production in S. cerevisiae[a]

| | Optimized Metabolite, Enzyme, and Flux Profiles | | | | |
	1	2	3	4	5
Metabolite					
X_1	1.2	1.2	1.2	1.2	1.2
X_2	0.8	0.8	0.8	1.2	1.2
X_3	0.8	0.8	0.87	0.87	1.2
X_4	1.2	1.2	1.2	1.2	1.2
X_5	0.8	1.12	1.2	1.2	1.2
Enzyme					
X_6	25.2	33.91	36	39.8	39.8
X_7	23.3	31.1	332.9	33.15	33.21
X_8	28.77	44.28	48.23	35.96	36.02
X_9	38.06	50	50	50	41.28
X_{10}	29	41	43.68	43.82	43.17
X_{13}	50	50	50	50	50
Flux					
V_{Eth}	32.23	44.25	47.12	47.28	47.32
	Corresponding Responses Profiles				
σ	9.07	9.44	10.2	10.64	13.6
τ	$9.35 \cdot 10^{-3}$	$7.08 \cdot 10^{-3}$	$7.18 \cdot 10^{-3}$	$7.47 \cdot 10^{-3}$	$9.54 \cdot 10^{-3}$
ψ	$103.74 \cdot 10^3$	$188.09 \cdot 10^3$	$197.39 \cdot 10^3$	$190.37 \cdot 10^3$	$149.39 \cdot 10^3$

[a] Five optimal enzyme and metabolite profiles that result from simultaneous minimization of σ and maximization of the rate of ethanol production (V_{Eth}). Solution 2 coincides with the mono-objective minimization of the transition time $\tau = \sigma/V_{Eth}$. All entries show levels relative to baseline, that is, $X_i/(X_i)_{base}$ and $V_{Eth}/(V_{Eth})_{base}$. Enzymes not shown are kept constant at the basal values. Optimal values of performance are indicated by shaded boxes.

Carrying out the simultaneous minimization of σ and maximization of V_{Eth}, we obtain five different efficient solutions, which in some cases differ quite significantly. Some results are summarized in Table 7.12. Solutions 3, 4, and 5 show the highest V_{Eth}, but at the expense of the values for σ. These solutions turn out to be the worst if we consider the minimization of τ as the optimization criterion. At the other extreme is solution 1, which simultaneously has low values of V_{Eth} and σ, but again is not optimal with respect to τ. There is an obvious correlation between the flux and the concentration of intermediate metabolites, which illustrates the fact that minimization of σ and maximization of V_{Eth} are counteracting objectives. The solution showing the minimum value of τ is solution 2. This solution represent a compromise between V_{Eth} and σ and coincides exactly with the unique solution that we obtain from the mono-objective optimization, where the objective function is minimization of τ. If we choose metabolic performance, ψ, as the criterion of optimality, the best outcome is solution 3, which neither exhibits the best flux, nor the minimum value of τ, nor the minimum value of σ, but optimizes ψ.

Multi-Objective Optimization

Simultaneous objectives. To illustrate the suitability of the IMOOM approach further, let us explore a second, more complex scenario. We define six simultaneous objective functions. Specifically, we set out to minimize each of the metabolite concentrations, X_i, while optimizing ethanol production rate, V_{Eth}, at the same time. Thus, we may set up the problem in the following fashion.

The flux maximization has the same form as before, namely

$$\text{maximize } V_{Eth} = V_4^- = 0.0945 \cdot X_3^{0.05} X_4^{0.533} X_5^{-0.0822} X_{10} \tag{7.26}$$

or, equivalently in logarithms,

$$\text{maximize } 0.05 y_3 + 0.533 y_4 - 0.0822 y_5 + y_{10} \tag{7.27}$$

(Eqs. 7.8 and 7.9). Furthermore, we intend to minimize the intermediate metabolite concentrations and formulate as additional tasks:

$$\text{minimize } y_1, y_2, y_3, y_4, y_5, \tag{7.28}$$

which actually represents five different objectives.

Constraints. We allow limited manipulation of the enzymes responsible for ethanol production, as before. No stoichiometric constraint is imposed, because it was found unnecessary. The steady-state constraints ensure operation of the optimized solution under steady-state conditions.

Results. The optimal enzyme activity profiles are similar to those shown in Table 7.12. Table 7.13 shows the remaining results. There are ten efficient solutions. Most of the metabolite pools reach the maximum value compatible with the imposed 20% constraint, in spite of our intent to minimize these values. Solution 10 produces maximal flux, though not the maximal value of σ, which is attained by solution 5. Solution 9 yields the minimal values of σ and τ. This situation is typical for multi-objective optimization (see Chapter 5). Alternative solutions are optimal in some respect but not in another. There is no mathematical rationale for selecting one over the other solution, or even over a solution that is a compromise among the conflicting tasks, such as solution 7 or 8.

In general, studies of the natural evolution or technical optimization of metabolic systems with mono-objective approaches are limited by two facts. First, in most instances the exploration must be reduced to a small number of biological responses, lest experimenter, mathematics, and algorithms become overwhelmed. The "detour" of S-system formulation is helpful in allowing larger systems. Secondly, one must expect that organisms in vivo do not pursue one optimization goal in isolation (see Wilhelm and Brüggermann 2000), but that they weigh numerous goals against each other and ultimately exhibit a balanced distribution of desirable and suboptimal outcomes. In a biotechnological setting, this situation is often different, because the experimenter may be able to focus on just one objective, such as improved yield. The manipulated organisms would not necessarily be viable in the outside world, but this would be irrelevant if the organism grows in a controlled fermenter.

Table 7.13. Multi-Objective (Six Tasks) Optimization of Ethanol Production in S. cerevisiae[a]

	Optimal Profile									
	1	2	3	4	5	6	7	8	9	10
Metabolite										
X_1	1.2	1.2	1.2	1.2	1.2	1.2	1.2	1.2	1.2	1.2
X_2	0.8	1.2	0.8	1.2	1.2	1	0.8	1	0.8	1.04
X_3	1.2	1.2	1.2	1.2	1.2	1.2	0.84	1	0.86	1.05
X_4	1.2	1.2	1.2	1.2	1.2	1.2	1.2	1.2	1.2	1.2
X_5	0.8	0.8	0.8	0.8	0.9	0.98	0.87	0.98	0.96	1.11
Enzyme										
X_6	39.4	43.6	42.47	43.6	50	50	42.4	50	40.7	50
X_7	36.4	36.6	39.2	36.6	41.8	43.7	39.1	43.7	37.4	43.06
X_8	44.8	33.4	50	33.4	40	50	50	50	50	50
X_9	37.5	37.4	40.2	37.71	42.7	44.58	50	50	50	50
X_{10}	36.0	36.1	39.3	36.18	42.4	44.7	40	45.1	41.4	47.87
X_{11}	50	50	50	50	50	50	50	50	50	50
X_{12}	50	50	50	50	50	50	50	50	15.3	17.05
X_{13}	50	50	50	50	50	50	50	50	50	50
Response profiles										
V_{Eth}	40.86	41.0	44.27	41.0	47.43	49.8	44.3	49.8	45.5	52.45
σ	12.73	13.13	12.81	13.09	13.24	13.12	9.52	11.2	8.84	12.95
τ	$1.03 \cdot 10^{-2}$	$1.06 \cdot 10^{-2}$	$9.6 \cdot 10^{-2}$	$1.06 \cdot 10^{-2}$	$9.2 \cdot 10^{-2}$	$0.8 \cdot 10^{-2}$	$0.71 \cdot 10^{-2}$	$0.74 \cdot 10^{-2}$	$0.6 \cdot 10^{-2}$	$0.8 \cdot 10^{-2}$
ψ	$119.4 \cdot 10^3$	$116.4 \cdot 10^3$	$138.8 \cdot 10^3$	$116.4 \cdot 10^3$	$155.2 \cdot 10^3$	$187.4 \cdot 10^3$	$187.8 \cdot 10^3$	$202.6 \cdot 10^3$	$228.3 \cdot 10^3$	$197.4 \cdot 10^3$

[a] The ten optimal solutions emphasize either ethanol yield or reduced metabolic burden. No solution satisfies all tasks optimally. The decision among solutions is not a mathematical one, but must be based on the weight that is placed on each task by the experimenter. As before, results on metabolites and fluxes are presented as scaled values $(X_i)/(X_i)_{base}$, $V_{Eth}/(V_{Eth})_{base}$; the remaining quantities σ, τ, and ψ are presented unscaled.

If a multi-objective task can be reduced to a single objective, analysis is drastically simplified. In many cases, however, such a reduction is not really valid. If it is employed anyway, one may still obtain a viable solution, as we saw. However, the true multi-objective problem by its very nature may have many rather different solutions, and this diversity is lost during the reduction to a mono-objective problem.

Comparison with Observations on Simultaneous Overexpression of Glycolytic Enzymes

The model system and the multi-objective optimization methods can be directly applied to a recent experimental study by Smits et al. (2000). These authors were interested in enhancing the fermentation capacity in yeast by simultaneously overexpressing several glycolytic enzymes. In particular, the purpose of their studies was to unravel the quantitative relationships between internal metabolite levels and the glycolytic flux. Insights regarding these relationships yield a better understanding of the regulatory structure and ultimately promise targeted, effective manipulations of key glycolytic enzymes that could lead to significant advances in the biotechnology of ethanol production.

Smits and collaborators constructed recombinant *S. cerevisiae* strains with 1.1 to 3.4-fold increased activity levels of up to seven enzymes in the lower part of the glycolytic pathway, namely glyceraldehyde 3-phosphate dehydrogenase, phosphoglycerate mutase, phosphoglycerate mutase kinase, enolase, pyruvate kinase, pyruvate decarboxylase, and alcohol dehydrogenase. Smits et al. measured the intracellular concentrations of glucose 6-phosphate and fructose 1,6-bisphosphate and the glycolytic flux at different glucose concentrations in continuous and batch cultures. Additionally, they investigated the capacity of the recombinant strains to respond to increased cellular demands for ATP.

The aims of the work of Smits and co-workers are similar in purpose and philosophy to some of our optimization tasks, because they address the maximization of ethanol production rate while simultaneously minimizing the intermediate concentrations. It is interesting to compare some results. We use again Galazzo and Bailey's (1990, 1991) model of glycolysis with its S-system analogue (Eqs. 7.1–7.7), but amend it slightly to account better for the dynamics of ATP, as suggested by Schlosser, Riedy, and Bailey (1994). The details of these amendments are unimportant for the present discussion, and the interested reader is referred to Vera et al. (submitted) for specific information. The model serves as the basis for an indirect multi-objective optimization (IMOOM) that simulates the experimental design by Smits et al. (2000). The specific task is again to maximize ethanol production rate and simultaneously to minimize the intermediate concentrations, as described in Eqs. (7.26–7.28). The constraints also are the same as discussed previously.

The key result of interest is a comparison between the experimental magnitudes of metabolites and enzymes in the recombinant strain YHM7 (Smits et al. 2000) and those from two efficient solutions obtained with IMOOM (Table 7.14). Similar to the recombinant strain, the first efficient solution implies weak overexpression of the

Table 7.14. Predicted and Efficient Solutions Obtained from the Optimization of Ethanol-Producing Strains of *Saccharomyces cerevisiae*, According to Smits et al. (2000)[a]

	Recombinant YHM7 Strain		Efficient Solutions	
	Experimental	Predicted	1	2
	$(X_i)_{YHM7}/(X_i)_{CEN.PK.K45}$		$(X_i)_{optimum}/(X_i)_{CEN.PK.K45}$	
X_1 (Glucose)	≈ 1	0.97	1	1
X_2 (G6P)	≈ 1	1.13	1	0.97
X_3 (FDP)	≈ 1	0.55	0.54	0.54
X_4 (PEP)	≈ 1	0.17	0.54	1.68
X_5 (ATP)	≈ 1	1.25	1.1	0.93
σ_n	≈ 1	0.72	0.68	0.66
$X_9 (V_{GAPD})$	1.51	1.51	1.46	1.42
$X_{10} (V_{PK})$	2.52	2.52	1.25	0.86
ρ_n	≈ 0.76	1.07	1.02	0.98
θ_n	1.48	1.48	1.05	1.04
η_n	0.51	0.72	0.96	0.94

[a] Experimental values were obtained from continuous culture experiments with 20 g glucose/liter.

enzymes GAPD (X_{11}) and PK(X_{12}). This mild overexpression does not significantly change the ethanol rate production, as observed in the YHM7 strain. YHM7 shows a lower specific rate of glucose consumption and ethanol production, with a reduced specific growth rate that is presumably associated with the protein burden caused by the introduced alterations (Hauf, Zimmerman, and Müller 2000). The profile of the second efficient solution supports this observation: It shows that it is possible to improve the relevant outputs (intermediate concentrations, fluxes, and efficiency) through a reduction in the activities of the modified enzymes.

When the YHM7 enzyme profile is entered in the original model, we obtain the metabolite and enzyme profile shown in the third column of Table 7.14, labeled *Recombinant YHM7 strain, predicted*. The results demonstrate that the model generally predicts the experimental profile well. An exception is the predicted level of PEP (X_4), which is lower than observed. This discrepancy is interesting, because Smits et al. (2000) observed that PEP and some other metabolites were not influenced by the overexpression of the glycolytic enzymes in the lower part of the pathway.

The results can be interpreted in terms of other global characteristics. The first is the normalized total pathway metabolite concentration, σ_n, which is defined as the ratio of the total intermediate concentration at the optimum solution over the total intermediate concentration at the basal steady state; thus, $\sigma_n = (\Sigma X_i)_{optimum}/(\Sigma X_i)_{basal}$. An equivalent quantity characterizes the total enzyme concentration θ_n, which is defined as the ratio of all enzyme activities at the optimum solution over all enzyme activities at the basal steady state; $\theta_n = (\Sigma X_j)_{optimum}/(\Sigma X_j)_{basal}$. The ratio of the ethanol production rate at the optimum solution over the ethanol rate production at the basal steady state is represented as $\rho_n = (V_{Eth})_{optimum}/(V_{Eth})_{basal}$. Finally, we introduce the *biosynthetic effort efficiency* η_n, which is defined as the ratio ρ_n/θ_n.

Table 7.15. Dynamic Characterization of the Optimized and Basal Steady States in Terms of the Relaxation Time κ^a

	Relaxation Time, κ(s)		
	Basal	Efficient Solution 2	$\kappa_{optimized}/\kappa_{basal}$
X_1 (glucose int.)	0.91	0.45	0.49
ΣX_i	3.75	2.65	0.7
V_{Eth}	3.2	2.6	0.8

^{a} All times are given in seconds.

As Table 7.14 shows, the efficient solutions require lower concentrations for different combinations of intermediate metabolite concentrations than the observed solution, with σ_n values that are about a third lower. Other responses are also better in the efficient solutions. For instance, they have lower θ_n and larger η_n values than the YHM7 strain. Although ρ_n is lower in the observed solution, the superior values of θ_n and η_n in the efficient solutions suggest that these will be better fitted in a biotechnological setting.

The most significant goal for the recombinant strain was enhanced fermentation capacity. This capacity can be measured as the rate of carbon dioxide production. According to Smits et al. (2000), this rate was twice as high in the recombinant strain when the cellular demand for ATP was increased in response to of a glucose pulse, thus indicating a higher fermentation capacity. To test and compare the behavior of our optimized solutions, we can simulate responses to glucose pulses in the wildtype and the recombinant strain.

Table 7.15 shows the results of explorations of the dynamics of both glycolytic systems after a perturbation in the intracellular glucose concentration. The dynamics is characterized here with the relaxation time, κ, which is defined as the time needed to reach a metabolite concentration or flux value within 1% of the basal steady state, after the system had been perturbed. In this specific case, the perturbation consists of a small exogenous intracellular glucose bolus.

The recombinant, optimized strain (*efficient solution 2*) processes the glucose bolus much faster than the wildtype, requiring only about half the relaxation time for internal glucose, 70% for the total metabolite concentration, and 80% for the ethanol flux. One must remember that there are obvious delays between changes in glucose concentration in the medium and the corresponding changes in internal fluxes and intermediate concentrations, for which our model does not account. Nonetheless, even with this limitation, the IMOOM prediction qualitatively agrees well with the experimentally observed results.

DISCUSSION

What is the minimum set of enzymes that has to be modified in a given metabolic system, and by how much, to obtain a viable strain that produces the highest possible yield of a desired product? What is the most promising sequence of steps in the genetic engineering process toward this solution? How can we deal with multiple responses that should all be optimized but require conflicting action?

Valid answers to these questions would save enormous amounts of time and re-sources and provide us with a rational venue for enhancing the productivity of pro-cesses of biotechnological interest. Although we are still quite far away from solid answers, the repertoire of methods we have discussed here and in previous chapters constitutes a good starting point for further developments.

Indeed, others have begun to build upon S-system based optimization methods, albeit with different goals, including assessments of metabolic network structure (Hatzimanikatis, Floudas, and Bailey 1996a) and the characterization of parametric and experimental uncertainties during optimization (Petkov and Maranas 1997; Voit and Del Signore, 2001). It is interesting to note that Hatzimanikatis, Floudas, and Bailey (1996a) also identified optimal subsets of enzymes, but with a rather different method, namely mixed-integer linear optimization. In another study, Hatzimanikatis, Floudas, and Bailey (1996b) used methods of metabolic control analysis to obtain op-timal profiles. Although they used a different terminology, the underlying philosophy is similar to the S-system approach proposed here.

To achieve market significance with a novel biotechnological process, more than marginal improvements in performance are needed. Apparently, purely experimental approaches of random manipulation and selection have reached a plateau in many cases, and leaving this plateau in a quantum leap requires additional, effective strate-gies. The methods discussed here may become valuable contributors to such strategies in that they provide a screening tool and offer guidance before one engages in costly genetic manipulations.

In most pathways analyzed up to date, a large number of enzymes involved in the pathway must be modulated to achieve significant increases in production rate. This translates directly into significant experimental effort and a high degree of un-certainty. A reason for this situation is that these studies addressed metabolic path-ways of great industrial relevance, such as ethanol and citric acid production. These pathways are well known, and information about the kinetics and topology of the pathways is readily available. However, these pathways have also been subjected to experimental optimization for a long time, so that all "simple" solutions, involving just a few modulations, have already been tried out. Consequently, improvements predicted by mathematical optimization call for a larger number of specific alter-ations in enzymatic and other steps.

The situation is different if lesser-known or less relevant pathways are analyzed. For example, when we shifted the optimization task from ethanol production to increased production of glycerol or carbohydrates, even one or two alterations were predicted to have a noticeable effect. Similarly, a recent study of carnitine production (Alvarez-Vasquez et al. 2002; see also Chapter 8) shows that significant increases in yield are possible with reasonable effort, if this effort is specifically targeted. The authors report 60–70% increases in product, as a result of manipulating only two control variables. Given the medical importance of carnitine, such increases can become very valuable. The same study also indicates that the "detour" from an original model to an S-system and back (in other words, IOM or IMOOM) is worth the effort and leads to valid results. Thirdly, this study makes it clear that variables in the S-system model are not necessarily metabolites or enzymes. For instance, the authors define a dilution rate and a specific growth rate as independent variables, which are successfully subjected to optimization.

Very little has been done in terms of multi-objective optimizations of biochemical systems. This situation is not surprising because the inherent problems and challenges associated with such endeavor. The most prominent of these are caused by the nonlinear nature of the metabolic systems. They are difficult even in the case of only one objective function and become overwhelming if several tasks must be optimized simultaneously. The indirect optimization methods discussed in this chapter offer help. They translate the core of the nonlinear problem to a much more convenient linear problem, thereby avoiding many of the difficulties encountered in the original formulation of the optimization task. The translation requires approximation of the kinetic functions in the original model with products of power-law functions. This approximation by its nature incurs inaccuracies in representation, which are very difficult to assess in general. The comparisons between different models and their approximations in this chapter, combined with a rich literature addressing this very topic, indicate that the approximation error is often rather small, thereby suggesting that the indirect optimization methods, using the S-system form as an intermediary, is a very promising strategy.

REFERENCES

Alvarez-Vasquez, F., M. Cánovas, J.L. Iborra, and N.V. Torres: Modelling and optimization of continuous L-(-)-carnitine production by high-density *Escherichia coli* cultures, 2002.

Archer, D.B., D.A. MacKenzie, and D.J. Jeenes: Genetic engineering: Yeasts and filamentous fungi. In: C. Ratledge and B. Kristiansen (Eds.), *Basic Biotechnology* (Chapter 5). Cambridge University Press, Cambridge, U.K., 2001.

Bothast, R.J, N.N. Nichols, and B.S. Dien: Fermentations with new recombinant organisms. *Biotechnol. Prog.* 15(5), 867–75, 1999.

Brown, C.E., J.M. Taylor, and L.M. Chan: The effect of pH on the interaction of substrates and effector to yeast and rabbit muscle pyruvate kinase. *Biochim. Biophys. Acta* 829(3), 342–7, 1985.

Cascante, M., R. Curto, and A. Sorribas: Comparative characterization of the fermentation pathway of *Saccharomyces cerevisiae* using biochemical systems theory and metabolic control analysis: Steady-state analysis. *Math. Biosci.* 130, 51–69, 1995.

Cascante, M., M. Lloréns, E. Meléndez-Hevia, J. Puigjaner, F. Montero, and E. Martí: E. The metabolic productivity of the cell factory. *J. Theor. Biol.* 182, 317–25, 1996.

Conejeros R., and V.S. Vassiliadis: Analysis and optimization of biochemical process reaction pathways. 1. Pathway sensitivities and identification of limiting steps. *Ind. Eng. Chem. Res.* 37, 4699–708, 1998.

Conejeros R., and V.S. Vassiliadis: Dynamic biochemical reaction process analysis and pathway modification predictions. *Biotechnol. Bioeng.* 68(3), 285–97, 2000.

Curto, R., A. Sorribas, and M. Cascante: Comparative characterization of the fermentation pathway of Saccharomyces cerevisiae using biochemical systems theory and metabolic control analysis: Model definition and nomenclature. *Math. Biosci.* 130, 25–50, 1995.

Curto, R., E.O. Voit, A. Sorribas, and M. Cascante: Validation and steady-state analysis of a power-law model of purine metabolism. *Biochem. J.* 324, 761–75, 1997.

Curto, R., E.O. Voit, A. Sorribas, and M. Cascante: Mathematical models of purine metabolism in man. *Math. Biosci.* 151, 1–49, 1998a.

Curto, R., E.O. Voit, A. Sorribas, and M. Cascante: Analysis of abnormalities in purine metabolism leading to gout and to neurological dysfunctions in man. *Biochem. J.* 329, 477–87, 1998b.

Davies, S.E.C., and K.M. Brindle: Effects of overexpression of phosphofructokinase on glycolysis in the yeast *Saccharomyces cerevisiae. Biochemistry* 331, 4729–35, 1992.

de Atauri, P., R. Curto, J. Puigjaner, A. Cornish-Bowden, and M. Cascante: Advantages and disadvantages of aggregating fluxes into synthetic and degradative fluxes when modeling metabolic pathways. *Eur. J. Biochem.* 265, 671–9, 1999.

Easterby, J.S.: A generalized theory of the transition time for sequential enzyme reactions. *Biochem. J.* 199, 155–61, 1981.

Enfors, S.-O.: Baker's yeast. In: C. Ratledge and B. Kristiansen (Eds.), *Basic Biotechnology* (Chapter 17). Cambridge University Press, Cambridge, U.K., 2001.

Ferreira, A.E.N.: PLAS©. http://correio.cc.fc.ul.pt/~aenf/plas.html, 2000.

Galazzo, J.L., and J.E. Bailey: In vivo nuclear magnetic resonance analysis of immobilization effects on glucose metabolism of yeast *S. cerevisiae. Biotechnol. Bioeng.* 33, 1283–9, 1989.

Galazzo, J.L., and J.E. Bailey: Fermentation pathway kinetics and metabolic flux control in suspended and immobilized *S. cerevisiae. Enzyme Microbiol. Technol.* 12, 162–72, 1990.

Galazzo, J.L., and J.E. Bailey: Errata. *Enzyme Microbiol. Technol.* 13, 363–71, 1991.

Gong, C.S., N.J. Cao, J. Du, and G.T. Tsao: Ethanol production from renewable resources. *Adv. Biochem. Eng. Biotechnol.* 65, 207–41, 1999.

Hatzimanikatis, V., C., Floudas, and J.E. Bailey: Optimization of regulatory architectures in metabolic reaction networks. *Biotechnol. Bioeng.* 52(4), 485–500, 1996a.

Hatzimanikatis, V., C., Floudas, and J.E. Bailey: Analysis and design of metabolic reaction networks via mixed-integer linear optimization. *AIChE J.* 42(5), 1277–92, 1996b.

Hauf, J., F.K. Zimmerman, and S. Müller: Simultaneous genomic overexpression of seven glycolytic enzymes in the yeast *Saccharomyces cerevisiae. Enzyme Microbiol. Technol.* 26, 688–98, 2000.

Heinish, J: Isolation and characterization of the two structural genes encoding for phosphofructokinase in yeast. *Mol. Gen. Genet.* 202, 75–80, 1986.

Heinrich, R., and E. Hoffmann: Kinetic parameters of enzymatic reactions in states of maximal activity. An evolutionary approach. *J. Theor. Biol.* 151, 249–83, 1991.

Heinrich, R., and S. Schuster: *The Regulation of Cellular Systems.* Chapman and Hall, New York, 1996.

Heinrich, R., S.M. Rapoport, and T.A. Rapoport: Metabolic regulation and mathematical models. *Prog. Biophys. Mol. Biol.* 32, 1–82, 1977.

Heinrich, R., E. Hoffmann, and H.-G. Holzhütter: Calculation of kinetic parameters of a reversible enzymatic reaction in states of maximal activity. *Biomed. Biochim. Acta* 49, 891–902, 1990.

Heinrich, R., S. Schuster, and H.-G. Holzhütter: Mathematical analysis of enzymatic reaction systems using optimization principles. *Eur. J. Biochem.* 201, 1–21, 1991.

Hess, B., and T. Plesser: Temporal and spatial order in biochemical systems. *Ann. N.Y. Acad. Sci.* 316, 203–13, 1978.

Hynne, F., S. Dane, and P.G. Sørensen: Full-scale model of glycolysis in *Saccharomyces cerevisiae. Biophys. Chem.* 94(1–2), 121–63, 2001.

Isermann, H.: *EFFACET Operating Manual.* University of Frankfurt, Germany, 1977a.

Isermann, H.: The enumeration of the set of all efficient solutions for a linear multiple objective program. *Operations Res. Quart.* 28(3), 711–25, 1977b.

Jeffries, T.W., and N.Q. Shi: Genetic engineering for improved xylose fermentation by yeasts. *Adv. Biochem. Eng.-Biotechnol.* 65, 117–61, 1999.

Jeong, H., B. Tombor, R. Albert, Z.N. Oltvai, and A.-L. Barabási: The large-scale organization of metabolic networks. *Nature* 407, 651–4, 2000 (supplementary material to be found at http://www.nd.edu/~networks/cell/supply.htm).

Jorge-Santiso, J.: Aspectos metodológicos y computacionales de la optimización multiobjetivo. El caso lineal. Master's thesis, Universidad de La Laguna, 1992.

Leicester, H.M.: *Development of Biochemical Concepts from Ancient to Modern Times.* Harvard University Press, Cambridge, MA, 1974.

Levenberg, K.: A method for the solution of certain non-linear problems in least squares. *Quart. Appl. Math.* 2, 164–8, 1944.

Marquardt, D.W.: An algorithm for least-squares estimation of nonlinear parameters. *J. SIAM* 11(2), 431–41, 1963.

Michalewicz, Z.: *Genetic Algorithms + Data Structure = Evolution Programs.* Springer, Berlin, 1992.

Michalewicz, Z., and N. Attia: Evolutionary optimization of constrained problems. In: *Proceedings of the 3rd Annual Conference on Evolutionary Programming* (pp. 150–65). World Scientific, Singapore, 1994.

Moré, J., G. Burton, and H. Kenneth: *User guide for MINIPACK-1.* Argonne National Labs Report ANL-80-74. Argonne, IL, 1980.

Niederberger, P., R. Prasad, G. Miozzari, and H. Kacser: A strategy for increasing an in vivo flux by genetic manipulations. The tryptophan system of yeast. *Biochem. J.* 287, 473–9, 1992.

Petkov, S.B., and C.D. Maranas: Quantitative assessment of uncertainty in the optimization of metabolic pathways. *Biotechnol. Bioeng.* 56(2), 145–61, 1997.

Pettersson, G.: Evolutionary optimization of the catalytic efficiency of enzymes. *Eur. J. Biochem.* 206, 289–95, 1992.

Pretorius, I.S.: Tailoring wine yeast for the new millennium: Novel approaches to the ancient art of winemaking. *Yeast* 16(8), 675–729, 2000.

Rothman, L., and E. Cabib: Allosteric properties of yeast glycogen synthetase. I. General kinetic study. *Biochemistry* 6(7), 2098–112, 1967.

Savageau, M.A.: *Biochemical Systems Analysis. A Study of Function and Design in Molecular Biology.* Addison-Wesley, Reading, MA, 1976.

Savinell, J.M., and B.Ø. Palsson: Optimal selection of metabolic fluxes for *in vivo* measurement. II. Application to *Escherichia coli* and *hybridoma* cell metabolism. *J. Theor. Biol.* 155(2), 215–42, 1992.

Schaaff, I., Heinish, J., and F.K. Zimmermann: Overproduction of glycolytic enzymes in yeast. *Yeast* 5, 285–90, 1989.

Schlosser, P.M., G. Riedy, and J.E. Bailey: Ethanol production in baker's yeast: Application of experimental perturbation techniques for model development and resultant changes in flux control analysis. *Biotechnol. Prog.* 10, 141–54, 1994.

Schuster, S., and R. Heinrich: Minimization of intermediates concentrations as a suggested optimality principle for biochemical networks. I. Theoretical analysis. *J. Math. Biol.* 29, 425–42, 1991.

Schuster, S., R. Schuster, and R. Heinrich: Minimization of intermediates concentrations as a suggested optimality principle for biochemical networks. II. Time hierarchy, enzymatic rate laws, and erytrocyte metabolism. *J. Math. Biol.* 29, 443–55, 1991.

Shanks, J.V., and J.E. Bailey: Estimation of intracellular sugar phosphate concentrations in Saccharomyces cerevisiae using ^{31}P magnetic resonance spectroscopy. *Biotechnol. Bioeng.* 32, 1138–52, 1988.

Shiraishi, F., and M.A. Savageau: The tricarboxylic acid cycle in *Dictyostelium discoideum.*

I. Formulation of alternative kinetic representations. *J. Biol. Chem.* 267, 22912–18, 1992a.

Shiraishi, F., and M.A. Savageau: The tricarboxylic acid cycle in *Dictyostelium discoideum*. II. Evaluation of model consistency and robustness. *J. Biol. Chem.* 267, 22919–25, 1992b.

Shiraishi, F., and M.A. Savageau: The tricarboxylic acid cycle in *Dictyostelium discoideum*. III. Analysis of steady state and dynamic behaviour. *J. Biol. Chem.* 267, 22926–33, 1992c.

Shiraishi, F., and M.A. Savageau: The tricarboxylic acid cycle in *Dictyostelium discoideum*. IV. Resolution of discrepancies between alterntative methods of analysis. *J. Biol. Chem.* 267, 22934–43, 1992d.

Smits, H.P., J. Hauf, S. Müller, T.J. Hobley, F.K. Zimmermann, B. Hahn-Hägerdal, J. Nielsen, and L. Olsson: Simultaneous overexpression of enzymes of the lower part of glycolysis can enhance the fermentative capacity of Saccharomyces cerevisiae. *Yeast* 16, 1325–34, 2000.

Sorribas, A., R. Curto, and M. Cascante: Comparative characterization of the fermentation pathway of *Saccharomyces cerevisiae* using biochemical systems theory and metabolic control analysis: Model validation and dynamic behavior. *Math. Biosci.* 130, 71–84, 1995.

Stephanoupoulos, G., and T.W. Simpson: Flux amplification in complex metabolic networks. *Chem. Eng. Sci.* 52(15), 2607–27, 1997.

Steuer, R.: ADBASE. Multiple Objective Linear Programming Package. Operating Manual. Faculty of Management Science, University of Georgia, Athens, GA, 1995.

Su, S., and P.J. Russel: Adenylate kinase from bakers' yeast. 3. Equilibria: Equilibrium exchange and mechanism. *J. Biol. Chem.* 243, 3826–33, 1968.

Thomas, S., and D. Fell: The role of multiple enzyme activation in metabolic flux control. *Adv. Enzyme Reg.* 38, 65–85, 1998.

Torres, N.V.: Application of the transition time of metabolic systems as a criterion for optimization of metabolic processes. *Biotechnol. Bioeng.* 44, 291–6, 1994.

Torres, N.V., E.O. Voit, C. Glez-Alcón, and F. Rodriguez: An indirect optimization method for biochemical systems: Description of method and application to the maximization of the rate of ethanol, glycerol, and carbohydrate production in *Saccharomyces cerevisiae*. *Biotechnol. Bioeng.* 55(5), 758–72, 1997.

Vera, J., P. de Atauri, M. Cascante, and N.V. Torres: Multi-criteria optimization of biochemical systems by linear programming. Application to the ethanol production by *Saccharomyces cerevisiae*, submitted.

Voit, E.O.: *Computational Analysis of Biochemical Systems. A Practical Guide for Biochemists and Molecular Biologists*. Cambridge University Press, Cambridge, U.K, 2000.

Voit, E.O., and M. Del Signore: Assessment of effects of experimental imprecision on optimized biochemical systems. *Biotechnol. Bioeng.* 74(5), 443–8, 2001.

Voit, E.O., and A.E.N. Ferreira: Buffering in models of integrated biochemical systems. *J. Theor. Biol.* 191, 429–38, 1998.

Voit, E.O., and M.A. Savageau: Analytical solutions to a generalized growth equation. *J. Math. Anal. Appl.* 103(2), 380–6, 1984.

Voit, E.O., and M.A. Savageau: Accuracy of alternative representations for integrated biochemical systems. *Biochemistry* 26, 6869–80, 1987.

Wilhelm, T., and R. Brüggermann: Goal functions for the developments of natural systems. *Ecol. Model.* 132, 231–46, 2000.

Wilkinson, K.D., and I.A. Rose: Isotope trapping studies of yeast hexokinase during steady state catalysis. A combined rapid quench and isotope trapping technique. *J. Biol. Chem.* 254(24), 12567–72, 1979.

Yang, S.T., and W.C. Deal, Jr.: Metabolic control and structure of glycolytic enzymes. VI. Competitive inhibition of yeast glyceraldehyde 3-phosphate dehydrogenase by cyclic adenosine monophosphate, adenosine triphosphate, and other adenine-containing compounds. *Biochemistry* 8(7), 2806–13, 1969.

CHAPTER EIGHT

Conclusions

The two pathways we discussed in the previous chapters, namely citric acid accumulation in the mold *A. niger* and ethanol production in yeast, are good and at the same time not-so-good examples. They are good for the illustration of the proposed methods, because they are well known and relevant. Many kinetic features of the enzymatic and transport steps are readily available in the literature, and the glycolytic and citric acid pathways themselves, as well as diverging side branches, are well known in the biochemical community. Furthermore, both pathways are of undoubted industrial relevance, so that every predicted improvement in yield could potentially translate into biotechnological strategies of interest.

The examples are not ideal for a very closely related reason. Because of the strong industrial interest in ethanol and citric acid, uncounted scientists have been charged over the years with finding ways to increase their yield. As a consequence, it seems that "simple" means of improvement, consisting of alterations in one or two enzymes or transport steps, have already been found by experimental methods, such as random mutagenesis and selection. Consistent with the infeasibility of simple solutions, the predictions made from our mathematical optimizations require targeted and specific changes in several steps simultaneously. This translates directly into considerable experimental effort, which so far has precluded implementation in the laboratory, against which we could compare our results.

The conjecture that singular changes in well-studied systems are often ineffectual is supported by experimental evidence that has been surprising and disappointing at times. For instance, efforts to enhance the rate of ethanol production in yeast through the overexpression of each of the glycolytic enzymes unexpectedly failed; even simultaneous increases in the activities of pairs of enzymes, such as pyruvate kinase and phosphofructokinase or pyruvate decarboxylase and alcohol dehydrogenase, did not increase the rate of ethanol production (Schaaff, Heinisch, and Zimmermann 1989). In fact, simultaneous overexpression of various glycolytic operons in *Zymomonas mobilis*, where 50% of the cytoplasmic enzymes are glycolytic enzymes (An et al. 1991), caused a counterintuitive *decrease* in glycolytic flux (Snoep et al. 1995).

Although prominent, glycolysis is not the only example. Experiments overexpressing citrate synthase in *Aspergillus niger* failed to amplify citric acid accumulation (Ruijter et al. 2000). The cloning of all genes involved in the biosynthesis of penicillin and their overexpression in an industrial strain did not improve penicillin yield (Bailey 1999). Lysine production did not improve when the alleged rate-limiting step of the pathway was amplified (Stephanopoulos 2001). An *E. coli* knockout, lacking all pyruvate kinase activity, exhibited essentially the same central carbon flux ratios and the same macroscopic characteristics in terms of growth rate, biomass yield, and glucose uptake rate as the parent strain (Bailey 1999).

In the following, we briefly review two pathway analyses and optimizations that deal with the opposite type of example. They address lesser-known metabolic systems, which require some assumptions that compensate for missing data, but promise significant improvements in product with very reasonable effort. Indeed, the prescribed specific alterations in independent variables of the first system were so straightforward that experimentalists implemented them in the lab and confirmed the results predicted by the S-system model and its optimization. We also discuss these two cases, because they contain modeling strategies and aspects that were not included in the former examples. Following these two smaller case studies, we discuss some future trends we envision in the area of pathway analysis and optimization and indicate what role the new tools of molecular biology and genomics may play in their implementation.

L-CARNITINE PRODUCTION IN HIGH-DENSITY CULTURES OF *ESCHERICHIA COLI*

L-carnitine is a trimethylated amino acid (Figure 8.1) that plays the role of an obligatory cofactor in the transformation of free long-chain fatty acids into acylcarnitines and their transport into the mitochondrial matrix. It was discovered in 1905, but its important role in human metabolism was not established until 1955. Carnitine deficiency was identified as a metabolic syndrome in 1972, and for about two decades, carnitine has been used as a dietary supplement. Carnitine is important, because mitochondrial fatty acid oxidation is the primary energy source in heart and skeletal muscle. Well-nourished people seldom suffer from carnitine deficiency, but a number of diseases and syndromes benefit from carnitine supplementation. They include childhood cardiomyopathy, coronary vascular disease, muscular myopathies, hypoglycemia, anorexia, chronic fatigue, male infertility caused by idiopathic asthenospermia, scleroderma, diphtheria, and Rett syndrome. Dietary carnitine supplemention also seems to be beneficial for patients on dialysis and individuals with HIV disease. In the latter case, carnitine has been proposed as a component of nutritional

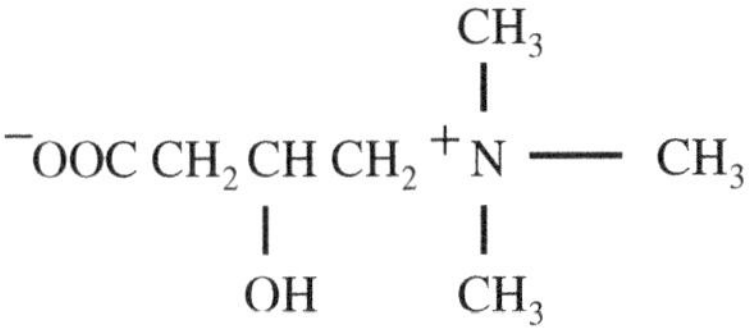

$$^-OOC\ CH_2\ CH\ CH_2\ {}^+N \underset{\underset{CH_3}{|}}{\overset{\overset{CH_3}{|}}{-}} CH_3$$
$$\underset{OH}{|}$$

Figure 8.1. Chemical structure of carnitine.

treatment, because it ameliorates mitochondrial toxicity, which is a common side effect of currently used highly active antiviral therapies (Patrick 2000). Kelly (1998) and Winter and Buist (2000) recently reviewed some of the therapeutic effects of L-carnitine.

Although it is possible to synthesize carnitine with methods of analytical chemistry, the result is a racemic mixture from which the L-carnitine isomer has to be isolated in a second step. This step is very expensive; purifying a commodity and separating mixtures often incur between 50–90% of the total costs of a chemical process (Duncan and Reimer 1998). An alternative that is rapidly gaining in importance is the microbial production of L-carnitine from a relatively cheap substrate like crotonobetaine. A crucial advantage of this biotechnological process is that the result is not compromised by the presence of D-carnitine. Among the most efficient strains for this purpose is *Escherichia coli* O44 K74, a strain that produces isomeric L-carnitine in worthwhile quantities and is suitable for large-scale industrial production (Obón et al. 1999).

To understand L-carnitine production better – and ultimately to manipulate it for improved yield – Cánovas et al. (2002) formulated the process for high-density cultures of *E. coli* with a relatively simple, unstructured mathematical model. Based on assumptions common in biochemical engineering, the authors formulated the model in terms of rational functions, following mostly the tradition of generalized Michaelis–Menten rate functions. They tested the model and found it adequate in quantitatively describing carnitine metabolism under different experimental conditions.

Alvarez-Vasquez et al. (forthcoming) made it their goal to optimize L-carnitine production, based on the model proposed by Cánovas et al. (2002) and using methods of BST and S-system optimization. Preliminary analysis of the model indicated that it was unnecessary to account for the metabolic transformation of crotonobetaine into γ-butyrobetaine, which is catalyzed by crotonobetaine reductase (Roth et al. 1994), because the flux through this step and the γ-butyrobetaine concentration itself turned out to be negligible. This fact prompted Alvarez-Vasquez and his collaborators to simplify the original model to one with only four dependent variables, namely glycerol, G, crotonobetaine, Cr, carnitine, C, and the biomass in the culture, X. The describing differential equations read

$$\dot{G} = [Q \cdot G_{in}] - \left[Q \cdot G + \frac{\mu_{\max,g} \cdot G \cdot \left(K_{i1} \cdot 2.3^{-Go/K_{ig}} + K_{i2} \right) \cdot X}{\left(K_{go,1} + K_{go,2} + G \right) \cdot \left(Yx_{go,1} + Yx_{go,2} \cdot e^{-Go/K_{yg}} \right)} \right]$$

$$\dot{Cr} = \left[\frac{X \cdot (K_{A1} + K_{A3}) \cdot C \cdot V_2}{K_2 + C} + Q \cdot Cr_{in} \right]$$
$$\qquad - \left[\frac{X \cdot (K_{A1} + K_{A3}) \cdot Cr \cdot V_1}{K_1 + Cr} + Q \cdot Cr \right]$$

$$\dot{C} = \left[\frac{X \cdot (K_{A1} + K_{A3}) \cdot Cr \cdot V_1}{K_1 + Cr} \right] - \left[\frac{X \cdot (K_{A1} + K_{A3}) \cdot C \cdot V_2}{K_2 + C} + Q \cdot C \right]$$

$$\dot{X} = \left[\frac{\mu_{\max,g} \cdot G \cdot \left(K_{i1} \cdot 2.3^{-Go/K_{ig}} + K_{i2} \right) \cdot X}{K_{go,1} + K_{go,2} + G} \right] - [u_e \cdot X].$$

The details of the model are not so important here, and the only reason we explicitly show the equations is to give an impression of their complexity. It is obvious that the model contains several nonlinearities that complicate analysis and optimization. For instance, a numerical algorithm is needed for computing the steady state, and because the steady state is required as a constraint for optimization tasks under batch conditions, the numerical search for steady-state values would have to be executed for every step of the optimization routine.

Alvarez-Vasquez et al. (forthcoming) circumvented this problem by reformulating the model as an S-system. They directly followed the methods and guidelines described in Chapter 2 and illustrated in Chapters 3, 6, and 7. The result consisted of the same four dependent variables and of five independent variables, namely the input glycerol and crotonobetaine concentrations, along with the flow rate of the fermentation system, the maximum specific growth rate, and the maximal activity of carnitine dehydratase. We note this explicitly to emphasize that independent variables in S-system models are not necessarily metabolite concentrations or enzyme activities, but may represent essentially any factors that affect some of the processes in the system. Alvarez-Vasquez et al. used standard BST methodology to assure stability, robustness, and reasonable dynamic behavior. Once the authors were convinced of the quality of the S-system model, they optimized the model with the methods described in this book.

Several of the results of the optimization are particularly interesting.

1. The kinetic model of Cánovas et al. (2002) and the S-system model yielded very similar optimal profiles for the independent variables.
2. A single increase in the dilution rate up to the permitted maximum of 50% translated into a 50% increase in the rate of carnitine production. This is strikingly different from our former examples of citric acid and ethanol production, where a singular alteration did not result in any noticeable improvements in product flux. It is noted that the logarithmic gain of the carnitine production flux with respect to dilution rate is 1. Even though the logarithmic gain is a quantity defined for infinitesimally small changes, it here predicts the effects of large alterations well.
3. An increase in input crotonobetaine concentration, in addition to the 50% increase in dilution rate, predicted a total amplification of carnitine flux of about 70%.
4. The next variable to be altered, the maximal activity of carnitine dehydratase, only increased carnitine production by a few additional percent.
5. Interestingly, the variable with the highest logarithmic flux gain of carnitine production, namely the input concentration of glycerol, turned out to be almost irrelevant for optimization purposes. This shows that logarithmic gains are valuable, but not always sufficient, tools for prediction.

After the optimization analysis had produced these theoretical results, the experimental group of Iborra tested some of the predictions (Alvarez-Vasquez et al., forthcoming). This was relatively easy to do, because the two most important variables to be altered were the dilution rate and the input crotonobetaine concentration, which were both readily manipulated. As Alvarez-Vasquez et al. report, the theoretical and experimental findings were very close. For instance, just changing the dilution rate was predicted to increase carnitine flux by 50%, and the experiments showed a 54% increase. Changing both, dilution rate and input crotonobetaine concentration were predicted from the mathematical optimization to improve carnitine flux by 68%, and the experiments resulted in a 74% increase. These results are very encouraging, because they lend quantitative support to the original kinetic model, its reformulation as an S-system, the method of optimization with the S-system model, and the translation back to the kinetic model.

TRYPTOPHAN BIOSYNTHESIS IN *ESCHERICHIA COLI*

Tryptophan is an essential amino acid. It is produced for medical purposes, such as the treatment of depression (Nelson 2000) and chronic fatigue syndrome (Werbach 2000). It would also be a valuable feed additive, could it be produced at a reasonable price. Like other amino acids, industrial tryptophan is mostly synthesized chemically, which is rather expensive and, as in the previous case of carnitine, has the significant disadvantage of a racemic mixture of D- and L-isomers. By contrast, microbiological production yields exclusively the biologically active L-isomer and is therefore much more attractive. Large-scale bacterial production of tryptophan has been initiated some while ago, but yields are still low in comparison to the microbial production of other amino acids (Jetton and Sinskey 1995; Eggeling, Pfefferle, and Sahm 2001).

The development of a rational strategy for improving tryptophan yield is a typical task of metabolic engineering. It requires a deeper understanding of the biosynthetic and catabolic pathways governing tryptophan metabolism in vivo, as well as a valid mathematical model describing these pathways within the wider web of microbial metabolism. Rudimentary models of this type have appeared in the literature since the 1960s (Goodwin 1965; Griffith 1968a,b; Bliss, Painter, and Marr 1982; Sinha 1988; Sen and Liu 1990). More recently, Xiu, Zeng, and Deckwer (1997) proposed a dynamical model that accounts for both feedback inhibition of the biosynthetic enzymes and repression of the *trp* operon by tryptophan. The model also explicitly considers the growth rate and the demand for tryptophan during protein synthesis. It has the following form

$$\frac{dP}{dt} = K_p \cdot E \frac{K_I^2}{K_I^2 + P^2} - (K_3 + \mu) \cdot P - n \cdot k_2[P_n \cdot R]$$

$$- \left(P_{pro}^m + \beta \cdot \mu\right) \cdot \mu \cdot C \frac{P}{P + K_s}$$

$$\frac{dM}{dt} = K_m \cdot D \cdot [O] - (K_1 + \mu) \cdot M$$

$$\frac{dE}{dt} = K_e \cdot M - (K_2 + \mu) \cdot E.$$

The three dependent variables are the intracellular tryptophan concentration, P, the mRNA concentration, M, and an enzyme, E, that serves as a simplified representative for the activity of the synthetic pathway of tryptophan. Several parameters are relevant. D is the gene dosage; $[O]$ represents free operator; K_p, K_m, and K_e are rate constants for the formation of the species denoted by subscripts; K_1, K_2, and K_3 are the rate constants for the degradation of mRNA, enzyme, and tryptophan; K_I is the concentration of tryptophan leading to 50% inhibition; K_s is a saturation constant; C the molar percentage of tryptophan in cellular protein; P_{pro}^m is the maximum protein concentration; β expresses the influence of growth rate on the protein content of cells, and μ is the specific growth rate of cells. The four terms in the first equation model production of tryptophan, degradation and dilution, demand of tryptophan bound to the apo-repressor, and consumption for protein synthesis.

Xiu et al. (1997) used this model to investigate the effects of the growth rate on the synthesis of tryptophan and the stability of the system under different parameter settings. They observed that in a metabolically stable setting, the level of repressor and the strength of feedback inhibition crucially affected the target variable, namely tryptophan concentration. Xiu and collaborators also concluded that effective tryptophan synthesis would require the cellular growth rate to be as low as possible. Under this condition, the tryptophan concentration is high enough to maintain metabolic stability, which is required for improved production during continual operation. However, a high demand for tryptophan reduces the intracellular tryptophan concentration, thereby risking instability. Thus, any manipulation or maximization of tryptophan synthesis must satisfy two conflicting tasks. On one hand, the efflux of tryptophan must be increased. On the other hand, the tryptophan concentration must remain high for metabolic stability.

Avoiding the complications caused by the nonlinear structure of Xiu's model, Marín-Sanguino and Torres (2000) used for their analysis an extension of the indirect optimization method, which was introduced in Chapter 7. The extension consisted of a stepwise optimization, in which each cycle consisted of an indirect optimization. Specifically, Xiu's model was first formulated as an S-system model, and the optimization constraints were set rather tightly, thus allowing only modest improvements. The S-system results were then translated back into Xiu's model and tested for stability and robustness. In the next step, Xiu's optimized model was used to parameterize the next S-system model, which was subsequently optimized.

Conceptually, this iterative procedure resembles a dynamic, piecewise approximation of Xiu's model with a series of S-system models. As opposed to a static piecewise approximation, however, Xiu's model was parameterized for every optimization cycle with values for the independent variables that had been optimized in the previous cycle with the corresponding S-system model. Marín-Sanguino and Torres (2000) found this iterative procedure useful, because a former one-step indirect optimization had led to differences between the S-system and the original model that the authors considered too large.

Marín-Sanguino and Torres (2000) successfully confirmed stability, robustness, and dynamic properties of the model, and this encouraged them to proceed with the optimization. They set the following conditions. The objective function was straightforwardly defined as maximization of tryptophan production. Constraints included, as in earlier examples (Chapters 6 and 7), steady-state equations, boundaries on the growth rate μ, boundaries on the strain level of the *trp* operon repressor, boundaries on the inhibition constant K_I of the representative enzyme E, and boundaries on tryptophan consumption and secretion. Some parameters were fixed at their nominal values, because they were considered uncontrollable. Examples included the rate of mRNA degradation K_1, and the saturation constant of tryptophan, K_s. As in earlier examples, constraints were also imposed on the dependent variables. Choosing the same default as we proposed in earlier chapters, the authors allowed the dependent variables to vary within a 20% range of their nominal values. With these objectives and constraints, the optimization task consisted of a regular linear program.

The optimization yielded a number of interesting results, including the following.

1. The optimum configuration of parameter values yields a new stable steady state where the concentrations of tryptophan, mRNA, and enzymes are within physiologically feasible ranges.
2. All parameters to be modified are amenable to manipulation by standard techniques of microbiology, biochemistry, or recombinant DNA technology.
3. Improvements in tryptophan yield are possible in a stepwise fashion.
4. The optimal solution requires increases in the maximum rate of tryptophan synthesis and the inhibition constant of tryptophan up to the maximally allowed levels. The solution furthermore calls for decreases in growth rate and the activity level of the *trp* operon.
5. Alterations in other parameters do not improve tryptophan yield.
6. Doubling the transport through the cytoplasmic membrane alone is predicted to lead to more than twice the baseline tryptophan production rate.
7. Altering four parameters (maximum rate of tryptophan synthesis, inhibition constant of tryptophan, growth rate, and *trp* operon) promises a four-fold increase in product.
8. If the growth rate is considered uncontrollable and thus fixed at its observed baseline level, alterations in the remaining three parameters are still expected to amplify the baseline rate more than twice.

As in the carnitine example, the tryptophan case demonstrates that independent variables, which are subjected to optimization, may include not only enzymes and substrates, but also other factors, such as growth rates and kinetic constants. The example again lends support to our conjecture that lesser-known pathways may promise improvements in yield that require alterations in only one, two, or a few processes.

TRENDS IN PATHWAY ANALYSIS AND THE ROLE OF GENE TECHNOLOGY AND GENOMICS

Future metabolic engineering will enormously benefit from modern methods of gene technology and genomics. Recombinant DNA manipulation and the transfection and transformation of cells promise sweeping improvements in the biotechnological production of complex organic molecules (Archer, MacKenzie, and Jeenes 2001; Harwood and Wipat 2001). These will be complemented by metabolic flux and network analyses aimed at identifying optimal combinations and magnitudes of genetic alterations. This merging of biotechnological techniques with methods of biomathematical systems analysis is at the very core of metabolic engineering (Nielsen 1998; Stephanopoulos 1998; Koffas et al. 1999). Bailey (1999) asserted that "the metabolic engineering goal of identifying genes that confer a particular phenotype is conceptually and methodologically congruent with central issues in drug discovery and functional genomics."

Experiments that are routinely carried out in modern biology labs are able to equip cells with additional copies of some of their native genes or with genes that are entirely new to them (Koffas et al. 1999; Ratledge and Kristiansen 2001). These experiments are not limited to the academic lab: 12.8 million hectares of genetically modified crops in 1997 attest to the immense potential of these techniques of genetic engineering (Smith 2001).

Clearly, the expression of introduced genes has the potential to affect the global metabolic flow pattern in an organism and holds promise for optimal redistributions of fluxes, in which the biosynthetic production of a desired metabolite is greatly enhanced. Examples like *E. coli* manufacturing indigo, golden rice with carotene genes from daffodils, and *Streptomyces* producing polyketides as starting backbones for the construction of antibiotics are just the beginning of a biotechnological and metabolic engineering revolution in which "superior cells" will be created specifically for pharmaceutical, medical, agricultural, or environmental remediation purposes (Lowe 2001a; Stephanopoulos and Kelleher 2001; Vandevivere and Verstraete 2001). An illustration of the speed of development is the biotechnological production of enzymes with microbes, which has increased at an annual rate of 12% over the past decade (Lowe 2001b). Some strategies for yield improvement through gene manipulation are easily imagined, but there will also be many surprises. As one amazing example, Zaslavskaia et al. (2001) succeeded in converting an obligate photoautotrophic organism into a light-independent organism through the introduction of a single human gene coding for a glucose transporter. This alteration constitutes a fundamental change in the metabolic make-up of the organism and opens entirely new possibilities, such as the greatly simplified use of algae for fermentation processes.

Complementing the benefits from molecular biology and genetic engineering, genomics will add another dimension to future metabolic engineering. For instance, it is reasonable to expect that genome-wide characterizations of the responses of microorganisms and cell lines to specific stimuli will be of prominent importance. These types of experiments show which genes are turned on or off, which are up- or downregulated, and which are unaffected in a well-defined situation. The current gold rush for known or unknown genes that strongly respond to some stimulus of interest is only the beginning of the exploitation of genome-wide data. The translation of altered gene expression patterns into changes in fluxes or enzymatic reactions is the next logical step, because it will shed light on the global metabolic functioning of an organism.

Insights from the translation of genome responses into biochemical and physiological responses will be of benefit at two levels of metabolic organization. The first is the stoichiometry and distribution of fluxes, which is probably best addressed mathematically with *metabolic flux analysis* (MFA). The second is the full kinetic and regulatory structure of integrated metabolic networks, which are the focus of *biochemical systems theory* (BST) and *metabolic control analysis* (MCA).

Metabolic Flux Analysis

Modern methods of biology, systems analysis, and engineering have made it possible to assemble genome-scale metabolic networks for some microbial organisms (Covert et al. 2001). These networks show all major pathways and are mathematically characterized by matrix equations that reflect the global stoichiometry of the organism. The experimental determination of intracellular fluxes, combined with the mathematical analysis of the distribution of these fluxes, is collectively called *metabolic flux analysis* (MFA; Aiba and Matsuoka 1979; Koffas et al. 1999). If physicochemical constraints and metabolic demands are considered in addition to the stoichiometry of fluxes, one also speaks of *flux-based analysis* (FBA; Edwards and Palsson 1998). Considerable metabolic engineering effort will be directed toward large-scale analyses and manipulations of fluxes.

A crucial issue of MFA is the identification of *principal branch points* or *nodes* of the networks, at which the incoming metabolic flux is split and subsequently proceeds through separate pathways toward different products. According to Stephanopoulos et al. (1998, p. 427), "one focus of metabolic engineering research should be on altering nodal split ratios as the main mechanism of improving product yields." Of course, global metabolic networks have very many branch points, but experience has shown that only a few of them are typically principal in a sense that changes in their flux split ratio would have noticeable effects on the improvement of a desired product. For instance, a network model of growth and penicillin production in *Penicillium chrysogenum* was found to have only four principal nodes (van Gulik et al. 2000). Once the important branch points are identified, they may be classified as *flexible, weakly rigid*, or *strongly rigid*, a categorization that is accomplished by means of perturbation experiments characterizing the control at these branch points (Stephanopoulos, Aristidou, and Nielsen 1998). As indicated by the terminology, flexible branch points may be manipulated by the experimenter, whereas the flux split ratios at weakly or strongly rigid branch points can only be altered with difficulty or not at all, respectively. Rigidity is usually due to strong regulation of the enzymes that control the diverging branches.

Once the network of interest is established and the nodes are classified, matrix algebra allows the calculation of a flux distribution map from measured (or theoretically obtained) accumulation rates of metabolites. The calculation is usually carried out for networks under constant physiological conditions over extended periods of time. However, one can also, to some degree, derive intracellular flux maps during transient phases, as long as the intracellular metabolites are in steady state and the rates of extracellular metabolites can be measured online (Takiguchi, Shimizu, and Shioya 1997). The flux estimates can be validated experimentally through measurements of isotope enrichment in key metabolites of the network (Stephanopoulos, Aristidou, and Nielsen 1998).

The characterization of the flux pattern in a metabolic network permits different types of analysis. Of greatest importance in our context is the optimization of flux distribution. As discussed in Chapter 5, this optimization is feasible, because the stoichiometric matrix of the network usually allows a multidimensional solution space, within which all flux distributions are admissible. Furthermore, the optimization task consists of a linear program, whose objective function is the maximization of some metabolite concentration or flux and whose constraints include boundaries on metabolites, fluxes, and enzyme activities that the system has to satisfy. The optimization is not necessarily restricted to intracellular metabolites. Xie and Wang (1996) have shown that the same type of flux analysis can be used for the design of improved cell culture media.

At present, most global flux analyses address microorganisms. However, the biotech and pharmaceutical industries are increasingly adding mammalian cell cultures to their repertoire (for a recent introduction, see Vriezen, van Dijken, and Häggström 2001). Mammalian cells pose special challenges, in that they are much more fragile than microbes. Under normal batch culture conditions, they grow only to low densities, because they accumulate inhibitory byproducts of glucose and glutamine metabolism, such as lactate and ammonia. Nonetheless, they show great promise that makes attempts to overcome the technical challenges worthwhile. In contrast to prokaryotes, which biotech companies initially used to produce relatively simple nonglycosylated proteins like insulin and human growth factor, mammalian cell cultures are able to manufacture and secrete complex pharmaceutically interesting proteins, such as glycosylated, complex glycoproteins. Biopharmaceuticals presently produced by mammalian cells include erythropoetin, Factor VIII, tissue plasminogen activator, and granulocyte colony-stimulating factor (Stephanopoulos, Aristidou, and Nielsen 1998).

Genomics will play an increasingly important role for metabolic flux analysis, because it characterizes organismic functioning at a global level. In principle, genomic experiments can quantify the simultaneous responses of all genes of an organism to a stimulus or, even better, provide time series data of such responses. These global responses can be superimposed on the metabolic flux map, thereby revealing what types of redistribution patterns of fluxes are actually used by the organism.

Genome Information and Biochemical Systems Theory

By definition, MFA leads to a linear system, whose variables are flux rates. Even in flux-based analysis (FBA), which accounts for additional physiological and other constraints, the underlying mathematical structure remains to be that of a constrained linear system. Indeed, a major attraction of MFA and FBA is that no enzymatic or other kinetic information, which is the cause of nonlinearities, is needed to carry out the analysis (Edwards and Palsson 1998). Although the linearity of flux systems is appealing, it has its drawbacks. Stephanopoulos, Aristidou, and Nielsen (1998) warn that effector concentrations strongly affect nodal rigidity, which is consequential because the identification and classification of nodes and their flexibility or rigidity is one of the crucial steps of MFA.

As we have discussed throughout the book, BST takes a different approach that fully incorporates enzymatic and regulatory information. The purpose of this brief section is thus merely to indicate how genomic information might be useful in biochemical systems analyses. Three immediate areas of potential benefit are presently evident. The first is an interpretation of genomic data in terms of enzyme activities. The second is the use of genomic stimulus–response data for static parameter estimations, and the third may be the use of genomic time series data for parameterization or validation of dynamic model responses.

Metabolic interpretation of genomic data. A recent preliminary investigation indicates how genomic and metabolic information can complement each other (Voit and Radivoyevitch 2000). In this study, we connected gene expression data of glycolysis in *Saccharomyces cerevisiae* after heat shock with a metabolic S-system model from the literature (Curto, Sorribas, and Cascante 1995, see also Chapter 7). The analysis was based on heat-shock related spotted array data, which were provided among expression profiles of over 2,000 genes under different experimental conditions (http://rana.Stanford.EDU/clustering/Figure2. txt; see also Eisen et al. 1998). To illustrate our ideas as simply as feasible, we focused exclusively on the degrees of expression in glycolytic enzymes ten minutes after heat shock and simplistically interpreted them as corresponding changes in enzyme activities.

The data show an activity profile that is everything but intuitive. The initial steps of glucose transport and the glucokinase and hexokinase catalyzed conversion of glucose into glucose 6-phosphate are increased ten- to twenty-fold. By contrast, the subsequent glucose 6-phosphate isomerase and phosphofructokinase steps are essentially unchanged. Later processes of glycolysis are slightly increased in activity (about two- to three-fold). Branches channeling material from glucose 6-phosphate toward glycogen and trehalose and toward the pentose cycle show four-fold amplified activities. It has been noted frequently that the accuracy of gene expression data is not necessarily very high. Even if this is true in the present case, the observed differences in magnitudes of expression are quite striking.

The typical fashion of analyzing array data is clustering, where those genes are considered to be in the same class that show a similar expression pattern. Looking at the limited case of glycolysis after heat shock, as just described, one immediately sees that clustering provides no explanation. To explain the observed expression pattern, we employed the S-system model of glycolysis proposed by Curto, Sorribas, and Cascante (1995) and analyzed it in two ways. First, we studied the effects of changing enzyme levels in accordance with the observed degree of gene expression, and secondly, we studied hypothetical expression profiles and their advantages and drawbacks. The results were quite interesting.

First, enzyme activities that were altered in accord with the observed heat shock responses led to a well-coordinated, stable metabolic state. In this state, the production of metabolites needed for a coordinated heat shock response, namely trehalose, ATP, and NADPH, was noticeably increased, but none of the intermediate metabolites accumulated, with the exception of glucose 6-phosphate. This is important, because unneeded intermediates can cause metabolic burden that impedes growth and viability of the organism (Bentley et al. 1990; Snoep et al. 1995; Smits et al. 2000).

The second part of our analysis dealt with the question of whether the cell could have accomplished the same results with less expression effort. We performed numerous simulations with the S-system model, altering enzyme activities to different degrees, either individually or in combination. These simulations showed that all hypothetical profiles tested were less effective and more costly because they required stronger upregulation or had physiological disadvantages. For instance, when we increased only the initial hexose transporter step of the pathway, several intermediate metabolites assumed unreasonably high values, thereby causing unnecessary metabolic burden.

This particular study clearly is preliminary, but it indicates, nevertheless, how different types of information can be integrated with methods of BST. In the given situation, information was available about gene expression and about the enzymatic and kinetic properties of the pathway. This redundancy allowed us to compare predictions made on one level with observations on the other. In future analyses, only one type of data may be available. If these data are sufficiently reliable, and if a model structure is known, the same type of analysis may be able to make testable predictions, either on metabolic consequences of a particular expression profile or, in the reverse direction, on expression patterns needed to evoke an observed metabolic response.

A missing link in our preliminary system is the feedback exerted by proteins or other metabolites on the expression of genes. It is known that these signaling pathways are complex, robust, and very efficient, often involving dozens of proteins and/or lipids (Bray 1995; Weng, Bhalla, and Iyenger 1999; Auge et al. 2000; Abrieu, Doree, and Fisher 2001; Hannun, Luberto, and Argraves 2001; Kyriakis and Avruch 2001). Once these pathways are understood well enough to permit quantitative analysis (for a prototype model, see, e.g., Bhalla and Iyenger 1999), signaling molecules might become an attractive target for manipulation and, thus, for metabolic engineering. Signaling pathways are so intriguing for this purpose because they have, by design, high gains, which implies that small changes have strong effects. To overexpress some gene or upregulate some process artificially, it would be sufficient to supply minute amounts of the appropriate signaling molecule. One could imagine that such a manipulation might be precise and elegant and, possibly, extremely specific.

Static parameter estimation. Genome and proteome data may become important for the estimation of parameter values in metabolic models (Voit, in press). Two types of experiments are needed for a "static" estimation. One consists of traditional measurements of metabolite concentrations before and after a stimulus and the other of an assessment of corresponding changes in gene expression profile or, better, protein profile. It is useful to scale the values of' the dependent and independent variables following the stimulus with respect to the baseline values before the stimulus, $\tilde{X}_j / X_j (j = 1, \ldots, n + m)$, which corresponds to $\tilde{y}_j - y_j$ in logarithmic coordinates.

If the metabolic system assumes a new steady state after the stimulus, it may be evaluated analytically from the equation

$$\vec{y}_D = \mathbf{M} \cdot \vec{b} + \mathbf{L} \cdot \vec{y}_I$$

that was derived in Chapter 2 (Eq. 2.26). The vectors $\vec{y}_D$ and $\vec{y}_I$ contain the (logarithms of the) dependent and independent variables, respectively. The elements of $\mathbf{M}$ are the sensitivities with respect to changes in parameter values, and the elements of $\mathbf{L}$ are the logarithmic gains. Because the stimulus has no effect on the control structure of the system, the kinetic orders, the rate constants, and consequently the term $\mathbf{M} \cdot \vec{b}$ and the matrix $\mathbf{L}$ are unaltered. Thus, subtracting the two steady-state equations, before and after the stimulus, from each other, one obtains

$$\tilde{\vec{y}}_D - \vec{y}_D = \mathbf{L} \cdot (\tilde{\vec{y}}_I - \vec{y}_I).$$

This equation alone does not permit the complete determination of parameter values, but comprises a set of constraints that the elements of $\mathbf{L}$, and thus the parameters g_{ij} and h_{ij}, have to satisfy.

If N or more independent sets of experiments are executed and one codes the differences as $\vec{w}_p = \tilde{\vec{y}}_D - \vec{y}_D$ and $\vec{v}_p = \tilde{\vec{y}}_I - \vec{y}_I$ for each experiment p, the vectors can be collected in matrices, yielding

$$(\vec{w}_1, \vec{w}_2, \ldots) = \mathbf{L} \cdot (\vec{v}_1, \vec{v}_2, \ldots),$$

which, under ideal conditions allows the complete characterization of $\mathbf{L}$, either with algebraic or statistical methods (Lanczos 1964). Obviously, these types of estimations will have to be accompanied by analyses of errors propagating from the original data.

Much has been written about the importance of the logarithmic gain matrix $\mathbf{L}$. Because its elements consist of algebraic combinations of kinetic orders and are independent of the rate constants, $\mathbf{L}$ alone is not sufficient to identify a system uniquely (Sorribas and Cascante 1994). However, every independent logarithmic gain measurement constitutes a constraint on the admissible parameter set of the system. If additional information about the system is available, as is usually the case, the constraints eventually lead to a system of equations that is compatible only with a few sets – or even a unique set – of kinetic order parameters.

Parameter estimation from genomic time series data. Each microarray experiment yields DNA or RNA data on the expression of large sets of genes at a given point in time. If several such experiments are carried out in short sequence over a period of time, following some stimulus, the result is a collection of time courses, each one describing the expression dynamics of a particular gene. Similarly, it is to be expected that very soon protein chips will produce the analogous results at the metabolite and enzyme level. These time courses contain very valuable information, and if they can be produced with sufficient reliability and accuracy, parameter estimation will have a new, unprecedented data resource.

For instance, assume for simplicity that the system contains four dependent and five independent variables and that measurements of the dependent variables are available for P time points t_p ($p = 1, \ldots, P$). The independent variables do not change during the experiment and retain the same value for all t_p. Smoothing methods of various degrees of sophistication can be employed to estimate from these data the instantaneous changes of each dependent variable at each time point. Each instantaneous change corresponds to the derivative of the variable at the given time point and thus to the left hand-side of one of the differential equations in the model.

As an illustration, consider the S-system equation of the second variable in the system with four dependent and five independent variables. It has the form

$$\dot{X}_2 = \alpha_2 \prod_{j=1}^{4+5} X_j^{g_{2j}} - \beta_2 \prod_{j=1}^{4+5} X_j^{h_{2j}}.$$

Given measurements and slopes at time points t_p ($p = 1, \ldots, P$), one has available P algebraic equations that pertain to X_2, and, similarly, P equations each for the other three dependent variables. The equations pertaining to X_2 read

Estimated slope of $X_2(t_1) \approx \dot{X}_2(t_1) = \alpha_2 X_1^{g_{21}}(t_1) X_2^{g_{22}}(t_1) \ldots X_4^{g_{24}}(t_1) X_5^{g_{25}} \ldots X_9^{g_{29}}$

$$- \beta_2 X_1^{h_{21}}(t_1) X_2^{h_{22}}(t_1) \ldots X_4^{h_{24}}(t_1) X_5^{h_{25}} \ldots X_9^{h_{29}}$$

Estimated slope of $X_2(t_2) \approx \dot{X}_2(t_2) = \alpha_2 X_1^{g_{21}}(t_2) X_2^{g_{22}}(t_2) \ldots X_4^{g_{24}}(t_2) X_5^{g_{25}} \ldots X_9^{g_{29}}$

$$- \beta_2 X_1^{h_{21}}(t_2) X_2^{h_{22}}(t_2) \ldots X_4^{h_{24}}(t_2) X_5^{h_{25}} \ldots X_9^{h_{29}}$$

$$\vdots$$

Estimated slope of $X_2(t_P) \approx \dot{X}_2(t_P) = \alpha_2 X_1^{g_{21}}(t_P) X_2^{g_{22}}(t_P) \ldots X_4^{g_{24}}(t_P) X_5^{g_{25}} \ldots X_9^{g_{29}}$

$$- \beta_2 X_1^{h_{21}}(t_P) X_2^{h_{22}}(t_P) \ldots X_4^{h_{24}}(t_P) X_5^{h_{25}} \ldots X_9^{h_{29}}$$

(Voit 2000, Chapter 5). For four dependent variables and P time points, one thus has $4 \cdot P$ equations, from which the system parameters are to be determined, for instance, with nonlinear regression.

THE FUTURE HAS BEGUN

Out of many roots, the interdisciplinary field of metabolic engineering has emerged with full force. Merely a decade old (Bailey 1991), metabolic engineering has gained such momentum and shows such potential that the biotech, pharmaceutical, and agricultural industries are hiring experts – and promising novices – with backgrounds in molecular biology, genetic engineering, bioinformatics, and computational biology at a staggering rate. Many major funding agencies around the world are now offering support to multidisciplinary research groups that combine biology with computational analysis and reductionist science with systems thinking. New programs, departments, centers, and entire institutes are being formed with the goal of integrating biological data into systems and transforming observation into insight and understanding.

Metabolic engineering focuses on the applied aspects of these modern efforts of systems biology. It greatly benefits from the advances in molecular biology and genomics and embraces mathematical approaches of representing and manipulating complex systems. It uses new insights from these endeavors and, at the same time, provides them with a purpose by promising practical application and the betterment of pressing issues in medicine, pharmacy, agriculture, many branches of industry, and the environment.

Maybe the most exciting aspect of metabolic engineering is that it truly draws from disciplines that only a short while ago were seen as too different to interact effectively. Under the auspices of metabolic engineering, experts from the established disciplines of molecular biology, genetics, chemical engineering, control theory, and systems analysis are merging their knowledge in a novel fashion that makes unique use of the advances in each parent discipline. Genetic engineering allows the introduction of genes in host organisms, genomics provides the complete gene expression status of an organisms at a given time point, bioinformatics helps organize and explain genomic data, systems analysis integrates local information into global pictures of functional networks, control theory assesses the regulation of these networks, mathematics and computer science predict optimal metabolic arrangements, and molecular biology implements promising alterations in suitable microorganisms and cell lines. The potential of such a multipronged, multidisciplinary approach cannot be overestimated. The scope and novelty of ideas emerging from collaboration across the diverse scientific subspecialties engaged in metabolic engineering are the harbinger of a paradigm shift, which Kuhn (1962) proposed as the driving force for scientific maturation and progress. The rapid development and spread of the concepts of metabolic engineering will shape biotechnology and medicine, and in doing so, fundamentally change society of the twenty-first century and its expectations of science.

REFERENCES

Abrieu, A., M. Doree, and D. Fisher: The interplay between cyclin-B-Cdc2 kinase (MPF) and MAP kinase during maturation of oocytes. *J. Cell Sci.* 114(2), 257–67, 2001.

Aiba, S., and M. Matsuoka: Identification of metabolic model: Citrate production from glucose by *Candida lipolytica. Biotechnol. Bioeng.* 21, 1373–86, 1979.

Alvarez-Vasquez, F., M. Cánovas, J.L. Iborra, and N.V. Torres: Modelling and optimization of continuous L-(-)-carnitine production by high-density *Escherichia coli* cultures. *Biotechnol. Bioeng.* (in press).

An, H., R.P. Scopes, M. Rodríguez, K.-F. Kesvav, and L.O. Ingram: Gel electrophoretic analysis of *Zymomonas mobilis* glycolytic and fermentative enzymes: Identification of an alcohol dehydrogenase II as a stress protein. *J. Bacteriol.* 173, 5975–82, 1991.

Archer, D.B., D.A. MacKenzie, and D.J. Jeenes: Genetic engineering: Yeasts and filamentous fungi. In: C. Ratledge and B. Kristiansen (Eds.), *Basic Biotechnology* (Chapter 5). Cambridge University Press, Cambridge, U.K., 2001.

Auge, N., A. Negre-Salvayre, R. Salvayre, and T. Levade: Sphingomyelin metabolites in vascular cell signaling and atherogenesis. *Prog. Lipid Res.* 39(3), 207–29, 2000.

Bailey, J.: Toward a science of metabolic engineering. *Science* 252, 1668–74, 1991.

Bailey, J.E.: Lessons from metabolic engineering for functional genomics and drug discovery. *Nature Biotechnol.* 17, 616–18, 1999.

Bentley, W.E., N. Mirjalili, D.C. Andersen, R.H. Davis, and D.S. Kompala: Plasmid-encoded protein: The principal factor in the "metabolic burden" associated with recombinant bacteria. *Biotechnol. Bioeng.* 35, 668–81, 1990.

Bhalla, U.S., and R. Iyenger: Emergent properties of networks of biological signaling pathways. *Science* 283, 381–7, 1999.

Bliss, R.D., P.R. Painter, and A.G. Marr: Role of feedback inhibition in stabilizing the classical operon. *J. Theor. Biol.* 97, 177–93, 1982.

Bray, D.: Protein molecules as computational elements in living cells. *Nature* 376, 307–12, 1995.

Cánovas, M., J.R. Maiquez, J.M. Obón, and J.L. Iborra: Modelling of the biotransformation of crotonobetaine into L(-)-carnitine by *Escherichia coli* strains. *Biotechnol. Bioeng.* 77, 764–775, 2002.

Covert, M.W., C.H. Schilling, I. Famili, J.S. Edwards, I.I. Goryanin, E. Selkov, and B.Ø. Palsson: Metabolic modeling of microbial strains *in silico*. *Trends Biochem. Sci.* 26, 179–86, 2001.

Curto, R., A. Sorribas, and M. Cascante: Comparative characterization of the fermentation pathway of *Saccharomyces cerevisiae* using biochemical systems theory and metabolic control analysis. Model definition and nomenclature. *Math. Biosci.* 130, 25–50, 1995.

Duncan, T.M., and J.A. Reimer: *Chemical Engineering Design and Analysis: An Introduction.* Cambridge University Press, Cambridge, U.K., 1998.

Edwards, J., and B.Ø. Palsson: How will bioinformatics influence metabolic engineering? *Biotechnol. Bioeng.* 58, 162–9, 1998.

Eggeling, L., W. Pfefferle, and H. Sahm: Amino acids. In: C. Ratledge and B. Kristiansen (Eds.), *Basic Biotechnology* (Chapter 13). Cambridge University Press, Cambridge, U.K., 2001.

Eisen, M.B., P.T. Spellman, P.O. Brown, and D. Botstein: Cluster analysis and display of genome-wide expression patterns. *Proc. Natl. Acad. Sci. USA* 95, 14863–8, 1998.

Goodwin, B.C.: Oscillatory behavior in enzymatic control processes. *Adv. Enzyme Reg.* 3, 425–38, 1965.

Griffith, J.S.: Mathematics of cellular control processes. I. Negative feedback to one gene. *J. Theor. Biol.* 2, 202–8, 1968a.

Griffith, J.S.: Mathematics of cellular control processes. II. Positive feedback to one gene. *J. Theor. Biol.* 2, 209–16, 1968b.

Hannun, Y.A., C. Luberto, and K.M. Argraves: Enzymes of sphingolipid metabolism: From modular to integrative signaling. *Biochemistry* 40(16), 4893–903, 2001.

Harwood, C.R., and A. Wipat: Genome management and analysis: Prokaryotes. In: C. Ratledge and B. Kristiansen (Eds.), *Basic Biotechnology* (Chapter 4). Cambridge University Press, Cambridge, U.K., 2001.

Jetton, M.S.M., and A.J. Sinskey: Recent advances in physiology and genetics of amino acid-producing bacteria. *Crit. Rev. Biotechnol.* 15(1), 73–103, 1995.

Kelly, G.S.: L-Carnitine: Therapeutic applications of a conditionally-essential amino acid. *Alt. Med. Rev.* 3(5), 345–60, 1998.

Koffas, M.A., C. Roberge, K. Lee, and G. Stephanopoulos: Metabolic engineering. *Annu. Rev. Biomed. Eng.* 1, 535–57, 1999.

Kuhn, T.S.: *The Structure of Scientific Revolutions.* The University of Chicago Press, Chicago, IL, 1962.

Kyriakis, J.M., and J. Avruch: Mammalian mitogen-activated protein kinase signal transduction pathways activated by stress and inflammation. *Physiol. Rev.* 81(2), 807–69, 2001.

Lanczos, C.: *Applied Analysis.* Prentice Hall, Englewood Cliffs, NJ, 1964.

Lowe, D.: Antibiotics. In: C. Ratledge and B. Kristiansen (Eds.), *Basic Biotechnology* (Chapter 16). Cambridge University Press, Cambridge, U.K., 2001a.

Lowe, D.: Production of enzymes. In: C. Ratledge and B. Kristiansen (Eds.), *Basic Biotechnology* (Chapter 18). Cambridge University Press, Cambridge, U.K., 2001b.

Marín-Sanguino, A., and N.V. Torres: Optimization of tryptophan production in bacteria. Design of a strategy for genetic manipulation of the tryptophan operon for tryptophan flux maximization. *Biotechnol. Prog.* 16(2), 133–45, 2000.

Nelson, J.C.: Augmentation strategies in depression 2000. *J. Clin. Psychiat.* 61(Suppl. 2), 13–19, 2000.

Nielsen, J.: Metabolic engineering: Techniques for analysis of targets for genetic manipulations. *Biotechnol. Bioeng.* 58, 125–32, 1998.

Obón, J.M., J.R. Maiquez, M. Cánovas, H.P. Kleber, and J.L. Iborra: High density *E. coli* cultures for continuous L(-)-carnitine production. *Appl. Microbiol. Biotechnol.* 51, 760–4, 1999.

Patrick, L.: Nutrients and HIV: Part three – N-acetylcysteine, alpha-lipoic acid, L-glutamine, and L-carnitine. *Alt. Med. Rev.* 5(4), 290–305, 2000.

Ratledge, C., and B. Kristiansen (Eds.): *Basic Biotechnology*. Cambridge University Press, Cambridge, U.K., 2001.

Roth, S., K. Jung, R.K. Hommel, and H.P. Kleber: Crotonobetaine reductase from *Escherichia coli*. A new inducible enzyme of aerobic metabolism of L-(-)-carnitine. *Antoine van Leuwenhoek J. Microbiol. Serol.* 65, 63–9, 1994.

Ruijter, C.J.G., H. Panneman, X. Ding-Bang, and J. Visser: Properties of *Aspergillus niger* citrate synthase and effects of citA overexpression on citric acid production. *FEMS Microbiol. Lett.* 184, 35–40, 2000.

Schaaff I., J. Heinisch, and F.K. Zimmermann: Overproduction of glycolytic enzymes in yeast. *Yeast* 5(4), 285–90, 1989.

Sen, A.K., and W.-M. Liu: Dynamic analysis of genetic control and regulation of aminoacid synthesis: The tryptophan operon in *Escherichia coli*. *Biotechnol. Bioeng.* 35, 185–94, 1990.

Sinha, S.: Theoretical study of tryptophan operon: Application in microbial technology. *Biotechnol. Bioeng.* 31, 117–24, 1988.

Smith, J.E.: Public perception of biotechnology. In: C. Ratledge and B. Kristiansen (Eds.), *Basic Biotechnology* (Chapter 1). Cambridge University Press, Cambridge, U.K., 2001.

Smits, H.P., J. Hauf, S. Müller, T.J. Hobley, F.K. Zimmermann, B. Hahn-Hägerdal, J. Nielsen, and L. Olsson: Simultaneous overexpression of enzymes of the lower part of glycolysis can enhance the fermentative capacity of *Saccharomyces cerevisiae*. *Yeast* 16, 1325–34, 2000.

Snoep, J.L., L.P. Yomano, H.V. Westerhoff, and L.O. Ingram: Protein burden in *Zymomonas mobilis*: Negative flux and growth control due to overproduction of glycolytic enzymes. *Microbiology* 141, 2329–37, 1995.

Sorribas, A., and M. Cascante: Structure identifiability in metabolic pathways: Parameter estimation in models based on the power-law formalism. *Biochem. J.* 298, 303–11, 1994.

Stephanopoulos, G.: Metabolic enginnering. *Biotechnol. Bioeng.* 58, 119–20, 1998.

Stephanopoulos, G.: A platform of flux and gene expression measurements for metabolic engineering and drug discovery (Presentation). *Biological Information Processing and Systems Workshop*, Clemson University, Clemson, SC, January 19–20, 2001.

Stephanopoulos, G., and J. Kelleher: How to make a superior cell. *Science* 292, 2024–5, 2001.

Stephanopoulos, G.N., A.A. Aristidou, and J. Nielsen: *Metabolic Engineering. Principles and Methodologies*. Academic Press, San Diego, CA, 1998.

Takiguchi, N., H. Shimizu, and S. Shioya: An online physiological-state recognition system for the lysine fermentation process – based on a metabolic reaction model. *Biotechnol. Bioeng.* 55, 170–81, 1997.

Vandevivere, P., and W. Verstraete: Environmental applications. In: C. Ratledge and

B. Kristiansen (Eds.), *Basic Biotechnology* (Chapter 24). Cambridge University Press, Cambridge, U.K., 2001.

van Gulik, W.M., W.T.A.M. de Laat, J.L. Vinke, and J.J. Heijnen: Application of metabolic flux analysis for the identification of metabolic bottlenecks in the biosynthesis of penicillin-G. *Biotechnol. Bioeng.* 68(6), 602–18, 2000.

Voit, E.O.: *Computational Analysis of Biochemical Systems. A Practical Guide for Biochemists and Molecular Biologists* (xii + 532 pp.). Cambridge University Press, Cambridge, U.K., 2000.

Voit, E.O.: Models-of-data and models-of-processes in the post-genomic era. *Math. Biosci.*, in press.

Voit, E.O., and T. Radivoyevitch: Biochemical systems analysis of genome-wide expression data. *Bioinformatics* 16(11), 1023–37, 2000.

Vriezen, N., J.P. van Dijken, and L. Häggström: In: C. Ratledge and B. Kristiansen (Eds.), *Basic Biotechnology* (Chapter 21). Cambridge University Press, Cambridge, U.K., 2001.

Weng, G., U.S. Bhalla, and R. Iyenger: Complexity in biochemical signaling systems. *Science* 284, 92–6, 1999.

Werbach, M.R.: Nutritional strategies for treating chronic fatigue syndrome. *Alt. Med. Rev.* 5(2), 93–108, 2000.

Winter, S.C., and N.R. Buist: Cardiomyopathy in childhood, mitochondrial dysfunction, and the role of L-carnitine. *Am. Heart J.* 139(2, Pt 3), S63–9, 2000.

Xie, L., and D.I.C Wang: Material balance studies on animal cell metabolism using a stoichiometrically based reaction network. *Biotechnol. Bioeng.* 52, 579–90, 1996.

Xiu, Z.-L., A.-P. Zeng, and W.-D. Deckwer: Model analysis concerning the effects of growth rate and intracellular tryptophan level on the stability and dynamics of tryptophan biosynthesis in bacteria. *J. Biotechnol.* 58, 125–40, 1997.

Zaslavskaia, L.A., J.C. Lippmeier, C. Shih, D. Ehrhardt, A.R. Grossman, and K.E. Apt: Trophic conversion of an obligate photoautotrophic organism through metabolic engineering. *Science* 292, 2073–5, 2001.

Author Index

Subject Index